# Situation Awareness
# Analysis and Measurement

# Situation Awareness Analysis and Measurement

*Edited by*

## Mica R. Endsley
## Daniel J. Garland

*SA Technologies, Inc.*

## CRC Press
Taylor & Francis Group
Boca Raton   London   New York

CRC Press is an imprint of the
Taylor & Francis Group, an **informa** business

First published 2000 by Lawrence Erlbaum Associates, Inc.

Published 2008 by CRC Press
Taylor & Francis Group
6000 Broken Sound Parkway NW, Suite 300
Boca Raton, FL 33487-2742

© 2000 by Taylor & Francis Group, LLC
CRC Press is an imprint of Taylor & Francis Group, an Informa business

No claim to original U.S. Government works

ISBN 13: 978-0-8058-2134-5 (pbk)

**Visit the Taylor & Francis Web site at**
**http://www.taylorandfrancis.com**

**and the CRC Press Web site at**
**http://www.crcpress.com**

Cover design by Kathryn Houghtaling Lacey

**Library of Congress Cataloging-in-Publication Data**

Situation awareness analysis and measurement
edited by Mica R. Endsley, Daniel J. Garland.
p. cm.
Includes bibliographical references and index.
ISBN 0-8058-2133-3 (cloth : alk. paper)
ISBN 0-8058-2134-1 (pbk. : alk. paper)
1. Human engineering. 2. Awareness. 3. Aeronautics—Human factors. 4. Automation—Human factors.
I. Endsley, Mica R. II. Garland, Daniel J.
TA166 .S57 2000
620.8'2—dc21                               99-057237
                                            CIP

# Situation Awareness Analysis and Measurement

*Edited by*

## Mica R. Endsley
## Daniel J. Garland

*SA Technologies, Inc.*

CRC Press
Taylor & Francis Group
Boca Raton  London  New York

CRC Press is an imprint of the
Taylor & Francis Group, an **informa** business

First published 2000 by Lawrence Erlbaum Associates, Inc.

Published 2008 by CRC Press
Taylor & Francis Group
6000 Broken Sound Parkway NW, Suite 300
Boca Raton, FL 33487-2742

© 2000 by Taylor & Francis Group, LLC
CRC Press is an imprint of Taylor & Francis Group, an Informa business

No claim to original U.S. Government works

ISBN 13: 978-0-8058-2134-5 (pbk)

**Visit the Taylor & Francis Web site at
http://www.taylorandfrancis.com**

**and the CRC Press Web site at
http://www.crcpress.com**

Cover design by Kathryn Houghtaling Lacey

**Library of Congress Cataloging-in-Publication Data**

Situation awareness analysis and measurement
edited by Mica R. Endsley, Daniel J. Garland.
p. cm.
Includes bibliographical references and index.
ISBN 0-8058-2133-3 (cloth : alk. paper)
ISBN 0-8058-2134-1 (pbk. : alk. paper)
1. Human engineering. 2. Awareness. 3. Aeronautics—Human factors. 4. Automation—Human factors.
I. Endsley, Mica R. II. Garland, Daniel J.
TA166 .S57 2000
620.8'2—dc21                                99-057237
                                                  CIP

*When you can measure what you are speaking about,*
*and express it in numbers, you know something about it;*
*but when you cannot measure it,*
*when you cannot express it in numbers,*
*your knowledge is of a meager and unsatisfactory kind*

—*William Thomson, Lord Kelvin (1824–1907)*

# Contents

## Part III   Special Topics in Situation Awareness

# Preface

The evolution of *situation awareness* (SA) as a major area of study within human factors research has followed a consistent and steady upward trend since the 1980s. As designers in the aerospace industry struggled to provide pilots with this elusive commodity, a need surfaced to understand how pilots gathered, sorted, and processed great stores of information in their dynamic environments. Pilots could explain their need for SA and its importance to them, but little else was known about how to create cockpit designs that supported SA. Many new technologies were nonetheless being proposed for enhancing SA, thus the need also existed for evaluation of the effects of these technologies on SA in order to address this need properly in the system design process.

From this beginning, the early seeds of SA as an area of study were formed in the late 1980s. The foundations of a theory of how people acquired and maintained SA have developed and along with it, several methods for measuring SA in system design evaluation. The 1990s have seen the expansion of this early work to include many other domains and research objectives. From its beginnings in the cockpit realm, more recent work has expanded to include air traffic control, medicine, control rooms, ground transportation, maintenance, space, and education. Research objectives have also expanded from one of system design and evaluation to focuses on training, selection, and more basic research on the cognitive processes themselves. These extensions have shown us both how ubiquitous SA is and how elusive it can be.

Significant efforts have been put forth to develop an understanding of SA. Progress in this area largely depends on the development of a solid body of conceptual and methodological knowledge on how to measure SA. The systematic assessment of SA is a necessary precursor to testing developing theories of SA, exploring factors related to individual differences in

SA, the evaluation of system designs and new training techniques that purport to improve SA, and a myriad of related issues.

This book focuses on the issue of how to measure SA. In Part I, chapter 1, by Mica Endsley, and 2, by Richard Pew, provide a good backdrop on the construct itself; what we currently know about what it is and isn't, and how people get it and maintain it. These chapters lay the foundation for the rest of the book, which examines different approaches for measuring SA. As the field is really in its early stages of study, much of the existing theory and information is speculative and based on related work in the psychological sciences. Providing solid recommendations on which system designs or training programs will improve SA really depends on arming researchers with the tools needed to conduct research and to test new concepts. That is the objective of this book.

Part II contains nine chapters that each address different approaches for measuring SA. The selection of an approach is dependent on many factors including the research objectives, the tools that are available, and the constraints of the testing environment. Most researchers will be well served to have several approaches in their toolkits as each has distinct advantages and disadvantages and all may provide unique insights into the SA construct. Understanding the approaches that are available can best be understood through the schematic in Fig. P.1.

SA is a highly critical aspect of human decision making and consistent with work on naturalistic decision making. In chapter 3, Gary Klein discusses the work he and his colleagues have carried out using the critical incident technique to explore decision making by experts in naturalistic settings. Many aspects of SA were unearthed in these studies. He examines the benefits of exploring SA and decision making through naturally occurring field research that yields rich data that may not be available in more closely controlled laboratory studies of the phenomena. Consistent with this work, in chapter 4 Mark Rodgers, Richard Mogford, and Barry Strauch explore research that has worked backward from databases on incidents and accidents to uncover information on SA. They draw on the work that has been done in aircraft accident investigations and air traffic control error investigations to show the information that can be gleaned from such analyses. They also provide insights and recommendations for those in other domains who seek to better understand human error through such analyses.

Chapters 5, 6, and 7 discuss SA measures that have been developed to directly assess SA during actual or simulated operations. Chapter 5, by Debra Jones, provides an overview of subjective techniques for the self-rating of SA by operators performing a task. A frequently used approach, the validity and sensitivity of these metrics are discussed. Chapter 6, by Herbert Bell and Don Lyon, compliments this effort by exploring the issues surrounding subjective rating of operators by outside observers.

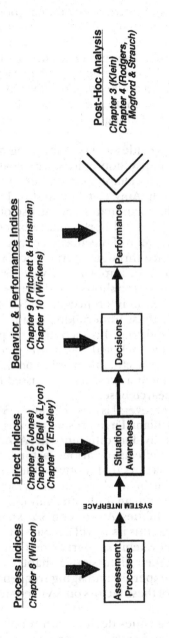

FIG. P.1 Roadmap of SA measurement approaches.

They have many useful insights on ways of obtaining good ratings and methods for side-stepping the potential pitfalls in this approach. Another frequently used approach, in chapter 7 by Mica Endsley, provides a thorough review of the use of simulation freezes and SA queries to obtain a direct and objective measure of SA. The development of queries for different domains is discussed along with recommendations for administering the technique. Numerous studies pertinent to the validity, reliability, and sensitivity of this measure are reviewed along with examples of the rich data provided by the technique.

In chapter 8, Glenn Wilson provides a very interesting and insightful analysis of the possibility of using psychophysiological measures to study the processes that underlie SA. This chapter stimulates a very new direction for SA measurement research. Amy Pritchett and R. John Hansman follow in chapter 9 with a careful examination of the use of carefully designed studies that provide an expected performance-based response to measure SA. Their work has been some of the most successful in measuring SA by inference from performance through operator actions. In chapter 10, Christopher Wickens delivers an important caveat for such efforts, however. His empirically based research shows that different performance measures can lead to entirely different inferences about SA. This work carries with it important lessons for those undertaking to infer SA from performance. Michael Vidulich rounds out this section in chapter 11 with a meta-analysis of studies that have included SA measurement. This new comparative analysis provides very useful information on the relative sensitivity of the different measurement approaches discussed here and provides a good addition to the research base.

Part III includes an overview of several issues relevant to SA research. In chapter 12, Leo Gugerty and William Tirre present a series of studies designed to examine the locus of individual difference in SA. Although much research on SA only looks at the effects of various factors on groups, this work provides insights into the factors that motivate the large individual differences that are observed in SA abilities in many settings. Cheryl Bolstad and Thomas Hess follow with an in-depth analysis in chapter 13 of the effects of aging on these SA abilities. As work on SA expands out of the cockpit into many other domains, this issue will become an even more important factor to consider. Wayne Shebilske, Barry Goettl, and Daniel Garland compliment this work with their discussion in chapter 14 of expertise and SA. As the development of expertise and aging frequently go hand in hand, any consideration of one of these factors on SA needs to keep both issues in mind.

The last two chapters explore issues dealing with training to improve SA. Based on an analysis of SA and crew resource management training, Carolyn Prince and Eduardo Salas, chapter 15, develop a set of guidelines for training SA in aircrews. Finally, in chapter 16, Mica Endsley and

Michelle Robertson examine the approaches and methods for training SA in individuals. They also review the development and evaluation of a program for training SA in teams of maintenance workers. Although work on methods for training to improve SA is in its early stages, these two chapters provide good information on future directions for this work.

This book provides researchers, designers, and students with state-of-the-art information on how to address SA measurement in their work. Designing systems that will always keep operators at a high level of SA may be a goal that is never quite reachable, particularly given the level of complexity of today's systems. The measurement approaches discussed here, however, provide use with tools for insuring that our future systems approach this important goal as closely as possible.

—*Mica R. Endsley*
—*Daniel J. Garland*

# Contributors

**Herbert H. Bell** is a branch chief and senior research psychologist with the Air Force Research Laboratory, Warfighter Training Research Division in Mesa, Arizona. His research interests focus on applied research involving the acquisition and maintenance of combat mission skills. He is currently investigating the relation between specific training experiences and mission competency. Dr. Bell received his PhD in experimental psychology from Vanderbilt University.

**Cheryl A. Bolstad** is a doctoral candidate in psychology at North Carolina State University and research associate with SA Technologies in Marietta, Georgia. She received her master's in Psychology from Florida Atlantic University and a bachelor's degree in computer applications in psychology from the University of Colorado. Before returning to school for her doctorate, Ms. Bolstad was employed by Monterey Technologies and Northrop Corporation where she conducted research in the area of human computer interaction and situation awareness. In 1994, she began her doctoral program to further her interest in the field of aging. Since that time she has coauthored several papers related to the field of aging and cognition.

**Mica R. Endsley** is president of SA Technologies in Marietta, Georgia where she specializes in situation awareness issues in advanced aviation systems. Prior to forming SA Technologies, she was a visiting associate professor at MIT in the Department of Aeronautics and Astronautics and associate professor of industrial engineering at Texas Tech University. She was an engineering specialist for the Northrop Corporation, serving as principal investigator of a research and development program focused on the areas of situation awareness, mental workload, expert systems, and interface design for the next generation of fighter cockpits. She received a PhD in industrial and systems engineering from the University of Southern California. Dr. Endsley has been conducting research on situation awareness, decision making and automation in aircraft, air traffic control, and aviation maintenance for the past 15 years. She is the author of more than 100 scientific articles and reports on numerous subjects including the implementation of technological change, the impact of automation, the design of expert system interfaces, new methods for knowledge elicitation for artificial intelligence system development, pilot decision making, and situation awareness.

**Daniel J. Garland** is a senior research associate with SA Technologies. His research has focused on a number of areas addressing the effective interaction and integration of humans with aviation system technologies. He has conducted and managed research on information management and presentation, advanced display techniques, integrated displays, selection and training, situation awareness, workplace and workspace evaluation and design, and distributed decision-making and associated cognitive factors in air traffic control, air traffic management, airport security, and aviation support systems. Dr. Garland has presented his work at several national and international professional meetings, is author or coauthor of more than 80 scientific articles and reports addressing human factors and information systems issues, and has edited six books. Dr. Garland is principal editor (with J. A. Wise & V. D. Hopkin) of the 1999 *Handbook of Aviation Human Factors* published by Lawrence Erlbaum Associates, Inc. Prior to joining SA Technologies in 1999, Dr. Garland led the development of the Department of Human Factors and Systems at Embry-Riddle Aeronautical University (ERAU), serving as its first department chair. He also led the development of ERAU's bachelor of science degree program in Applied Experimental Psychology/Human Factors and the master of science degree program in Human Factors and Systems. In addition, he has served as a visiting researcher at the FAA William J. Hughes Technical Center. His work focused on the working memory processes related to air traffic control tactical operations. He has also served as a visiting researcher at Williams AFB, Arizona. His work focused on the decision processes and strategies that may be used in designing training technology and methods to improve decision processing in dynamic decision environments. Dr. Garland received a PhD in experimental psychology from the University of Georgia in 1989.

**Barry P. Goettl** received a bachelor's degree in psychology from the University of Dayton in 1981, and a PhD degree in engineering psychology from the University of Illinois at Urbana-Champaign in 1987. From 1987 until 1992 he was an assistant professor at Clemson University Department of Psychology. He is currently a research psychologist for the Air Force

Research Laboratory (AFRL/HEAS) at Lackland AFB. His research interests include training and complex skill acquisition, computer-based training, attention and workload, visual displays, tracking, and motor control.

**Leo J. Gugerty** is an assistant professor in the Psychology Department at Clemson University. He received his PhD in experimental psychology from the University of Michigan in 1989. His research focuses on individual differences in the performance of complex, real-time tasks.

**R. John Hansman** is currently a professor in the Department of Aeronautics and Astronautics at MIT, where he is head of the Humans and Automation Division. He is director of the MIT Aeronautical Systems Laboratory and of the MIT International Center for Air Transportation. His research focuses on a broad range of flight safety topics ranging from aviation weather hazards, such as icing and windshear, to pilot vehicle interface issues. Professor Hansman has served on numerous advisory and technical committees including the Congressional Aeronautical Advisory Committee, the AIAA Atmospheric Environment Technical Committee, and the FAA Research and Development Subcommittee on the National Airspace System. Dr. Hansman also has over 4,600 hours of flight experience in airplanes, helicopters, and sailplanes.

**Thomas M. Hess** is a professor of psychology at North Carolina State University whose interest is in understanding both the positive and negative influences of aging on cognitive functioning. He has edited three books and published numerous research articles addressing this topic. He is also a Fellow in the American Psychological Association, American Psychological Society, and Gerontological Society of America and is currently on the editorial boards of *Journal of Gerontology: Psychological Sciences and Aging, Neuropsychology*, and *Cognition*.

**Debra G. Jones** is currently a research associate with SA Technologies of Marietta, Georgia. She received her PhD and master's in industrial engineering from Texas Tech University where she was an Air Force Office of Scientific Research Fellow. She received a bachelor's in psychology from the University of Alabama. Her research interests focus on human decision making and situation awareness in aviation systems.

**Gary Klein** is chairman and chief scientist of Klein Associates. He has performed research on naturalistic decision making in a wide variety of task domains and settings, including fire fighting, aviation, command and control, market research, and software troubleshooting. Based on these and related projects, he has developed significant new models of proficient decision making. His research interests include the study of individual and team decision making under conditions of stress, time pressure, and uncertainty. Dr. Klein has furthered the development and application of a decision-centered approach to system design and training programs and has edited two books on naturalistic decision making. Currently, he is extending the naturalistic decision making framework to cover planning and problem solving. He received his PhD in experimental psychology from the University of Pittsburgh in 1969.

**Don R. Lyon** is a senior research psychologist with Raytheon Training and Services under contract with the Air Force Research Laboratory, Warfighter Training Research Division, Mesa, Arizona. His interests include basic and applied research in human cognitive processes, including situation awareness, visual attention, spatial working memory, and performance measurement. He is also conducting research on issues involved in learning to control uninhabited air vehicles remotely from a ground location. Dr. Lyon has a PhD in experimental psychology from the University of Oregon.

**Ricard H. Mogford** received his bachelor's degree in psychology from York University in 1974 and his master's degree in psychology from Sonoma State University in 1978. He worked for 10 years in vocational rehabilitation and clinical neuropsychology. Dr. Mogford obtained his PhD in experimental psychology and human factors from Carleton University in 1990. He worked for a Transport Canada contractor on the Canadian Automated Air Traffic System and was employed for six years as a contractor at the FAA William J. Hughes Technical Center in Atlantic City, New Jersey. There he was involved in a variety of air traffic control human factors projects. He was employed by the FAA as an engineering research psychologist for the Human Factors Branch at the Technical Center and is currently at NASA–Ames Research Center. Dr. Mogford's research interests focus on human-machine interaction in air traffic control systems. He has extensive experience managing large-scale simulation studies.

**Richard W. Pew** is a principal scientist at BBN Technologies. He holds a bachelor's in electrical engineering from Cornell University (1956), a master's degree in psychology from Harvard University (1960), and a PhD in psychology with a specialization in engineering psychology from the University of Michigan (1963). He has 35 years of experience in human factors, human performance, and experimental psychology as they relate to systems design and development. Throughout his career he has been involved in the development and utilization of human performance models and in the conduct of experimental and field studies of human performance in applied settings. He spent 11 years on the faculty of the psychology department at Michigan where he was involved in human performance teaching research and consulting before moving to BBN in 1974. He was the first chairman of the National Research Council Committee on Human Factors, and has been president of the Human Factors Society and president of Division 21 of the American Psychological Association, the division concerned with engineering psychology. He has also been chairman of the Biosciences Panel for the Air Force Scientific Advisory Board. Dr. Pew has more than 60 book chapters, articles, and technical reports to his credit.

**Carolyn Prince** is a researcher with the University of Central Florida. She was formerly with the Science and Technology Division at the Naval Air Warfare Center Training Systems Division in Orlando, Florida. She holds a PhD degree in industrial/organizational psychology from the University of South Florida. Her research interests are focused on aviation team training, aircrew coordination training, team situation awareness, aeronautical decision making, and team performance measurement.

**Amy R. Pritchett** is an assistant professor in the School of Industrial and Systems Engineering and a joint assistant professor in the School of Aerospace Engineering at Georgia Tech. Her research interests include flight deck design, air traffic control, and flight simulation. She received her PhD, master's, and bachelor's degrees from the Department of Aeronautics and Astronautics at the Massachusetts Institute of Technology and is a licensed pilot.

**Michelle M. Robertson** is a researcher for the Liberty Mutual Research Center for Safety and Health in Hopkinton, Massachusetts. She is conducing applied research projects in ergonomics and management, specifically focusing on teamwork, human factors training in aviation maintenance, training evaluation, office worker productivity, and organizational factors. Prior to joining Liberty Mutual, she was a senior research manager in the Product Research Group at Herman Miller, Inc. Dr. Robertson also spent 12 years on the faculty in Ergonomics and Management at the Institute of Safety and System Management at the University of Southern California. She has published over 70 research papers and has presented her work nationally and internationally at aviation, ergonomics, and management conferences and symposiums, including a book chapter entitled "Maintenance Resource Management" for the FAA Office of Aviation Medicine Human Factors Guide for Aviation Maintenance. Dr. Robertson holds a PhD in instructional technology, a master's in systems management, and a bachelor's in human factors. She is a board certified professional ergonomist.

**Mark D. Rodgers** received his PhD in experimental psychology from the University of Louisville in 1991. Upon graduation he took a position as an engineering research psychologist at the FAA's Civil Aeromedical Institute (CAMI). While at CAMI, Dr. Rodgers was responsible for directing the development and initiating the national deployment of the Systematic Air Traffic Operations Research Initiative (SATORI) for use as an air traffic training and incident investigation tool. Dr. Rodgers worked at CAMI four years prior to joining the FAA Office of System Architecture and Investment Analysis in Washington, D.C. In his current position Dr. Rodgers is chief scientific and technical advisor for human factors in the Office of Aviation Research. His research interests include the assessment of taskload, human performance, system performance and modeling, situation awareness, and issues related to complexity in air traffic control and their application to the evaluation of ATC technology.

**Eduardo Salas** was a senior research psychologist and head of the Training Technology Development Branch of the Naval Air Warfare Center Training Systems Division. In 1984, he received his PhD in industrial and organizational psychology from Old Dominion University. Dr. Salas has coauthored over 100 journal articles and book chapters and has coedited five books. He is on the editorial boards of *Human Factors, Personnel Psychology, Military Psychology, Interamerican Journal of Psychology,* and *Training Research Journal.* His research interests include team training and performance, training effectiveness, tactical decision making under stress, team decision making, performance measurement, and learning strategies for teams. Dr. Salas is a Fellow of the American Psychological Association and a recipient of the Meritorious Civil Service Award from the Department of the Navy. He is currently on the faculty at the University of Central Florida.

**Wayne L. Shebilske** is chair of the Department of Psychology at Wright State University. Previously he was a visiting professor at Embry-Riddle University and professor of psychology at Texas A&M University. He received his PhD in psychology from the University of Wisconsin in 1974.

**Barry Strauch** is the chief of the Human Performance Division of the National Transportation Safety Board's Office of Aviation Safety. He joined the Safety Board as a human performance investigator in 1983 and became an investigator-in-charge of major aviation accident investigations in 1986. In 1990 he became the deputy chief of the Major Investigations Division before assuming his present position in 1993. Before joining the Safety Board, Dr. Strauch was on the faculty of the University of Louisville where he taught psychology and at Embry-Riddle Aeronautical University where he taught psychology and conducted human factors research in aviation. He earned a bachelor's in psychology from New York University, and master's and PhD degrees in educational psychology from Pennsylvania State University.

**William C. Tirre** is a senior research psychologist at the Information Training Branch, Warfighter Training Division, Human Effectiveness Directorate of the Air Force Research Laboratory, Brooks Air Force Base, San Antonio, Texas where he has worked since 1982. He received his PhD in educational psychology from the University of Illinois at Urbana-Champaign in 1983, and his master's in general-experimental psychology from Illinois State University in 1977.

**Michael A. Vidulich** received his PhD in experimental psychology from the University of Illinois in 1983. He worked for four years in the human performance and workload group at NASA Ames Research Center. Since 1987, he has been at Wright-Patterson Air Force Base as an Air Force research psychologist. His primary areas of work are developing metrics to predict and assess mental workload and situation awareness, and the application of those metrics to the evaluation of Air Force Systems.

**Christopher D. Wickens** is currently a professor of experimental psychology, head of the Aviation Research Laboratory, and associate director of the Institute of Aviation at the University of Illinois at Urbana-Champaign. He also holds an appointment in the department of Mechanical and Industrial Engineering at the Beckman Institute of Science and Technology. He received his bachelor's degree from Harvard College in 1967 and his PhD from the University of Michigan. His research interests involve the application of the principles of human attention, perception, and cognition to modeling operator performance in complex environments, and to designing displays to support that performance. Particular interest has focused on aviation, air traffic control, and data visualizations. Dr. Wickens is a member and Fellow of the Human Factors Society. He received the Society's Jerome H. Ely Award in 1981 for the best article in the Human Factors Journal and the Paul M. Fitts Award in 1985 for outstanding contributions to the education and training of human factors specialists by the Human Factors Society. He was elected to the Society of Experimental Psychologists. He was also elected Fellow of the American Psychological Association, and in 1993, he received the Franklin Taylor Award for Outstanding Contributions to Engineering Psychology from Division 21 of that association. He served as a distinguished visiting professor at the Department of Behavioral Sciences and Leadership, U. S. Air Force Academy in 1983–1984 and 1991–1992. From 1995 to 1998, he chaired a National Research Council Panel to examine the human factors of automation of the air traffic control system.

**Glenn F. Wilson** is a senior research psychologist at the United States Air Force Research Laboratory's Human Effectiveness Directorate in the Crew System Interface division. His research is aimed at understanding how operators function in complex environments using psychophysiological measures.

# Part I

# INTRODUCTION AND OVERVIEW

# Chapter 1

# Theoretical Underpinnings of Situation Awareness: A Critical Review

Mica R. Endsley
*SA Technologies, Inc.*

The enhancement of operator *situation awareness* (SA) has become a major design goal for those developing operator interfaces, automation concepts, and training programs in a wide variety of fields, including aircraft, air traffic control (ATC), power plants, and advanced manufacturing systems. This dramatic growth in interest in SA, beginning in the mid-1980s and accelerating through the 1990s, was spurred on by many factors, chief among them the challenges of a new class of technology.

One can easily see that SA has always been needed in order for people to perform tasks effectively. Prehistoric man undoubtedly needed to be aware of many cues in his environment in order to successfully hunt and keep from being hunted. For many years, having good SA was largely a matter of training and experience—learning the important cues to watch for and what they meant.

With the advent of the machine age, our emphasis shifted to creating a new class of tools to help people perform tasks, largely those physical in nature. The computer age and now the information age have followed rapidly on the heels of basic mechanization. The tools provided are no longer simple; they are amazingly complex, focused on not just physical tasks, but elaborate perceptual and cognitive tasks as well. The pilot of today's aircraft, the air traffic controller, the power plant operator, the anesthesiologist: all must perceive and comprehend a dazzling array of data that is often changing very rapidly. I have taken to calling this challenge the information gap (Fig. 1.1).

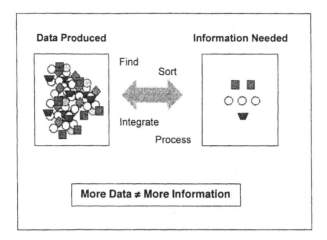

FIG. 1.1.   The information gap.

Today's systems are capable of producing a huge amount of data, both on the status of their own components and on the status of the external environment. Due to achievements in various types of datalink and internet technologies, systems can also provide data on almost anything anywhere in the world. The problem with today's systems is not a lack of information, but finding what is needed when it is needed.

Unfortunately, in the face of this torrent of data, many operators may be even less informed than ever before. This is because there is a huge gap between the tons of data produced and disseminated and the operator's ability to find the necessary bits and process them together with the other bits to arrive at the actual information required for their decisions. This information must be integrated and interpreted correctly as well, frequently a tricky task. This problem is real and ongoing whether the job is in the cockpit or behind a desk. It is becoming widely recognized that more data does not equal more information. Issues of automation and "intelligent systems" have frequently only exacerbated the problem, rather than aided it (Endsley & Kiris, 1995; Sarter & Woods, 1995).

The criteria for what we are seeking from system designs have correspondingly changed. In addition to designing systems that provide the operator with the needed information and capabilities, we must also ensure that it is provided in a way that is usable cognitively as well as physically. We want to know how well the system design supports the operator's ability to get the needed information under dynamic operational constraints (i.e., How well does it bridge the information gap?). This design objective and measure of merit has been termed SA.

## WHAT IS SA?

Most simply put, SA is knowing what is going on around you. Inherent in this definition is a notion of what is important. SA is most frequently defined in operational terms. Although someone not engaged in a task or objective might have awareness (e.g., sitting under a tree idly enjoying nature), this type of individual has been largely outside the scope of human factors design efforts. Rather, we have been concerned mostly with people who need SA for specific reasons. For a given operator, therefore, SA is defined in terms of the goals and decision tasks for that job. The pilot does not need to know everything (e.g., the copilot's shoe size and spouse's name), but does need to know a great deal of information related to the goal of safely flying the aircraft. Surgeons have just as great a need for SA; however, the things they need to know about will be quite different and dependent on a different set of goals and decision tasks.

Although the "elements" of SA vary widely between domains, the nature of SA and the mechanisms used for achieving SA can be described generically. (A determination of the elements of SA for different domains is discussed in more detail in chap. 8.) It is the goal of this chapter to provide a foundation for understanding the construct that is SA. This foundation is important both for creating systems that support SA and for creating tools that effectively measure SA.

Many definitions of SA have been developed. Some are very closely tied to the aircraft domain and some are more general. (See Dominguez, 1994, or Fracker, 1988, for a review.) That is, many are tied to the specifics of one domain, aircraft piloting, from whence the term originated. SA is now being studied in a variety of domains, however; education, driving, train dispatching, maintenance, and weather forecasting are but a few of the newer areas in which SA has been receiving attention.

A general definition of SA that has been found to be applicable across a wide variety of domains describes SA as "the perception of the elements in the environment within a volume of time and space, the comprehension of their meaning and the projection of their status in the near future" (Endsley, 1988, p. 97). Shown in Fig. 1.2, this definition helps to establish what "knowing what is going on" entails.

*Level 1 SA: Perception.* Perception of cues (Level 1 SA) is fundamental. Without a basic perception of important information, the odds of forming an incorrect picture of the situation increase dramatically. Jones and Endsley (1996) found that 76% of SA errors in pilots could be traced to problems in the perception of needed information (due to either failures or shortcomings in the system or problems with cognitive processes).

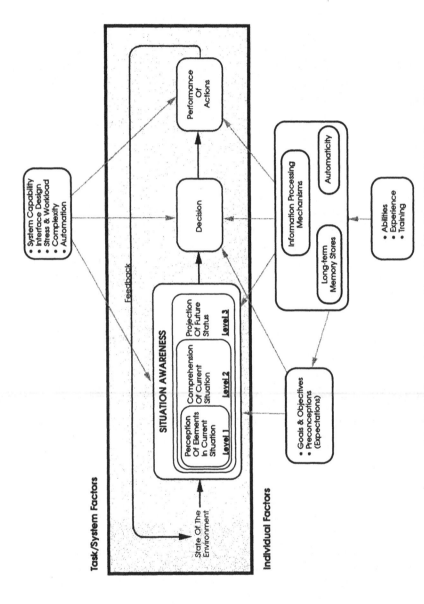

FIG. 1.2. Model of SA in dynamic decision making (from Endsley, 1995b). *Note.* From "Toward a theory of situation awareness in dynamic systems" by M. R. Endsley. In *Human Factors, 37*(1), 32–64, 1995. Copyright 1995 by the Human Factors and Ergonomics Society. Reprinted with permission.

*Level 2 SA: Comprehension.*　SA as a construct goes beyond mere perception, however. It also encompasses how people combine, interpret, store, and retain information. Thus, it includes more than perceiving or attending to information; it includes the integration of multiple pieces of information and a determination of their relevance to the person's goals (Level 2 SA). This is analogous to having a high level of reading comprehension as compared to just reading words. Twenty percent of SA errors were found to involve problems with Level 2 SA (Jones & Endsley, 1996).

Flach (1995) pointed out that "the construct of situation awareness demands that the problem of meaning be tackled head-on. Meaning must be considered both in the sense of subjective interpretation (awareness) and in the sense of objective significance or importance (situation)" (p. 3). A person with Level 2 SA has been able to derive operationally relevant meaning and significance from the Level 1 SA data perceived. As Flach pointed out, this aspect of SA sets it apart from earlier psychological research and places it squarely in the realm of ecological realism.

*Level 3 SA: Projection.*　At the highest level of SA, the ability to forecast future situation events and dynamics (Level 3 SA) marks operators who have the highest level of understanding of the situation. This ability to project from current events and dynamics to anticipate future events (and their implications) allows for timely decision making. In almost every field I have studied (aircraft, ATC, power plant operations, maintenance, medicine), I have found that experienced operators rely heavily on future projections. It is the mark of a skilled expert.

*Temporal Aspects of SA.*　Time, both the perception of time and the temporal dynamics associated with events, plays an important role in the formulation of SA. First, time itself has appeared as an important component of SA in many domains (Endsley, 1993b, 1994; Endsley, Farley, Jones, Midkiff, & Hansman, 1998; Endsley & Robertson, 1996; Endsley & Rodgers, 1994). A critical part of SA is often understanding how much time is available until some event occurs or some action must be taken. The phrase "within a volume of space and time" contained in the definition of SA derives from the fact that operators constrain the parts of the world (or situation) that are of interest to them based not only on space (how far away some element is), but also on how soon that element will have an impact on the operator's goals and tasks. Time is a strong part of Level 2 SA (comprehension) and Level 3 SA (projection of future events).

The dynamic nature of real-world situations is a third important temporal aspect of SA. The rate at which information changes is that part of SA regarding the current situation that also allows for projection of future situations (Endsley, 1988, 1995c). A situation's dynamic nature dictates that as the situation is always changing, so the person's SA must constantly change or be rendered outdated and inaccurate. In highly dynamic environments, this forces the human operator to adapt many cognitive strategies for maintaining SA. Adams, Tenney, and Pew (1995) emphasized the importance of the dynamics of both situations and cognitive processes in their model of SA. Sarter and Woods (1991) also discussed the importance of the temporal aspects of the situation for SA.

**SA and Decision Making**

The Endsley model, shown in Fig. 1.2, shows SA as a stage separate from decision making and performance. SA is depicted as the operator's internal model of the state of the environment. Based on that representation, operators can decide what to do about the situation and carry out any necessary actions. SA therefore is represented as the main precursor to decision making; however, many other factors also come into play in turning good SA into successful performance.

SA is clearly indicated as a separate stage in this model rather than as a single combined process. This is for several reasons. First, it is entirely possible to have perfect SA, yet make an incorrect decision. For example, battle commanders may understand where the enemy is and the enemy's capabilities, yet select a poor, or inappropriate, strategy for launching an attack. They may have inadequate strategies or tactics guiding their decision processes. They may be limited in decision choices due to organizational or technical constraints. They may lack the experience or training to have good, well developed plans of action for the situation. Individual personality factors (such as impulsiveness, indecisiveness, or riskiness) may also make some individuals prone to poor decisions. A recent study of human error in aircraft accidents found that 26.6% involved situations where there was poor decision making even though the aircrew appeared to have adequate SA to make a correct decision (Endsley, 1995b). Conversely, it is also possible to make good decisions even with poor SA even if only by luck.

This characterization is not meant to dispute the important role of SA in the decision-making process or the integral link between SA and decision making in many instances, particularly where experienced decision makers are involved. Klein's work in the area of recognition-primed decision making shows strong evidence of a direct link between situation recognition-classification and associated action selection (Klein, 1989; Klein, Calderwood, & Clinton-Cirocco, 1986). Where such learned linkages exist, they undoubtedly may be activated frequently in the decision process. Adams et al. (1995) and Smith and Hancock (1994) also discussed the integral relation between SA and decision making. Decisions are formed by

SA and SA is formed by decisions. These are certainly views with which I agree. I nevertheless feel it is important to recognize that SA and decision making need not be coupled as one process and in practice frequently are not.

The human operator makes a conscious choice in the decision to implement the linked recognition-primed decision action plan or to devise a new one. This behavior can be seen in combat tasks, for instance, where people often wish to be unpredictable. In the many instances where no action plan is readily linked to the recognized situation, a separate decision as to what to do must take place. Just as easily as SA and decision making can be linked in practice, they can also be unlinked. Although this distinction may be purely theoretical, it is made here for the purpose of clarity of discussion. SA is not decision making and decision making is not SA. This distinction has implications for the measurement of SA.

Furthermore, the link between human decision making and overall performance is indirect in many environments. A desired action may be poorly executed due to physical error, other workload, inadequate training, or system problems. The system's capabilities may limit overall performance. In some environments, such as the tactical aircraft domain, the action of external agents (e.g., enemy aircraft) may also create poor performance outcomes from essentially good decisions and vice versa. Therefore, for theoretical purposes, SA, decision making, and performance can be seen as distinct stages that can each affect the other in a circular ongoing cycle, yet which can be decoupled through various other factors.

**Who Needs SA?**

The earliest discussions of SA undoubtedly derive from the pilot community going back as far as World War I (Press, 1986). Spick (1988) provided a fascinating fighter pilot's view of the importance of SA in combat aircraft. He traced SA to one of Clausewitz's four main "frictions" of war.

Although an emphasis on SA in systems design and early scientific work on SA and its measurement originated in the fighter aircraft domain (Arbak, Schwartz, & Kuperman, 1987; Endsley, 1987, 1988; Fracker, 1988; IC3, 1986; Marshak, Kuperman, Ramsey, & Wilson, 1987; Taylor, 1990), the use and application of the term has rapidly spread to other domains. The most prominent driver of this trend has been technology. The growth and complexity of electronic systems and automation has driven designers to seek new methodological frameworks and tools for dealing effectively with these changes in domains from driving to space operations to medicine.

It should be clearly noted, however, that technological systems do not provide SA in and of themselves. It takes a human operator to perceive information to make it useful. Nor do they create the need for SA. No matter how a person does their job, the need for SA has always existed even in what would not normally be considered heavily cognitive or technological fields. In sports, such as football or hockey, for instance, the importance of SA in selecting and running plays is readily apparent. The highlighted emphasis on SA in current system design has occurred because we can now do more to help provide good SA through decision aids and system interfaces, and we are far more able to actually hinder SA through these same efforts if, as designers, we fail to adequately address the SA needs of the operator.

**How Do We Get SA?**

In its simplest terms, SA is derived from all of our various sources of information as shown in Fig. 1.3 (Endsley, 1995c, 1997). Cues may be received through visual, aural, tactile, olfactory or taste receptors. Some cues may be overt (e.g., a system alarm) and some so subtle that they are registered only subconsciously (e.g., the slight change in the hum of an engine). As we move toward the use of remote operators in many domains (e.g., unmanned air vehicles, remote maintenance), a major challenge will be providing sufficient information through a remote interface to compensate for the cues once perceived directly.

It is important to note that system designers tend to focus on the information provided through the system and its operator interface, but this is

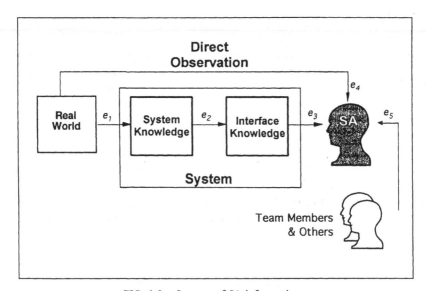

FIG. 1.3.   Sources of SA information.

not the only source of SA. In many domains, operators may be able to directly view and hear information from the environment itself (e.g., pilots, drivers, machine operators, ATC tower controllers), although in some cases they may not (e.g., an enroute air traffic controller who only has remotely transmitted information). It is important, therefore, that analyses of the SA provided by system designs also take into account the information that operators also derive from other means. That is, what is the value added (or subtracted) by a given system after taking into account information one already gets via other means and the potential cost in terms of interference with that information? For instance, the U.S. Army is currently considering providing helmet mounted displays to foot soldiers. Any analysis of this technology needs to consider not only the SA provided by the new system, but it must also consider whether the system may be interfering with SA that the soldier is able to obtain without the new technology (e.g., information on the local environment).

In addition to directly perceived information, as shown in Fig. 1.3, the system's sensors (or its datalink and internet components) collect some subset of all available information from the system's environment and internal system parameters. Of all the data the system possesses, some portion is displayed to the operator via its user interface. Of this information, the operator perceives and interprets some portion, resulting in SA.

It is critical to note that this is not a passive process of receiving displayed information, but one in which the operator may be very actively involved. For instance, the operator in many systems can control which information is displayed (e.g., through menu selection) and which information is attended to. The operator may also be able to control which information the system collects by sending out commands to get certain information from linked systems or by setting the direction and coverage of the sensors. People are therefore very active participants in the situation assessment process, with SA guiding the process and the process resulting in the SA.

The role of others in the process of developing SA has also received attention. Verbal and nonverbal communication with others (including radio communication, hand signals, and "wing-tipping" by other pilots) has historically been found to be an important source of SA information. Even in situations with restricted visual cues, air traffic controllers report that they get a great deal of information just from the voice qualities of the pilot's radio communications and can deduce such information as the pilot's experience level, stress, familiarity with English instructions, level of understanding of clearances, and need for assistance.

Operators also rely on each other for confirming their own SA. Jentsch, Barnett and Bowers (1997) and Gibson, Orasanu, Villeda, and Nygren (1997), for instance, found that it was much rarer for a single pilot to lose SA than for the whole crew to lose SA when accidents and incidents occurred.

Although these results undoubtedly reflect the fact that the second crew-member with SA would most likely prevent the situation from becoming an accident in the first place, it also speaks of the effectiveness of the process. Thus, SA is derived from a combination of the environment, the system's displays, and other people as integrated and interpreted by the individual.

## THEORIES OF SA

Simply describing the many sources of SA information does little to convey the intricate complexities of how people pick and choose information, weave it together, and interpret it in an ongoing and ever-changing fashion as both situations and operator goal states change. Despite the claims by some operators that SA is derived through instinct (Spick, 1988) or extra-sensory perception (Klein, 1995), I believe that we can shed some light on the cognitive processes that come into play when acquiring and maintaining SA.

Several researchers have put forth theoretical formulations for depicting the role of numerous cognitive processes and constructs on SA (Adams et al., 1995; Endsley, 1988, 1995c; Fracker, 1988; Smith & Hancock, 1995; Taylor, 1990; Taylor & Selcon, 1994; Tenney, Adams, Pew, Huggins, & Rogers, 1992). There are many commonalties in these efforts pointing to essential mechanisms that are important for SA and that have a direct bearing on the appropriateness of proposed measures of SA. The key points are discussed here, however, the reader is directed to these references for more details on each model.

In previous work (Endsley, 1988, 1990b, 1995c), I proposed a framework model based on an information-processing theory (Wickens, 1992). The model is used as an umbrella for discussing other theories and more recent research relevant to SA theory. The cognitive mechanisms that are important for the development of SA are shown in Fig. 1.4.

### Working Memory and Attention

Several factors will influence the accuracy and completeness of the SA that individual operators derive from their environment. First, humans are limited by working memory and attention. The way attention is employed in a complex environment with multiple competing cues is essential in determining which aspects of the situation will be processed to form SA. Once taken in, information must be integrated with other information, compared to goal states, and projected into the future—all heavily demanding on working memory.

Much can be said about the importance of attention on SA. How people direct their attention has a fundamental impact on which portions of the

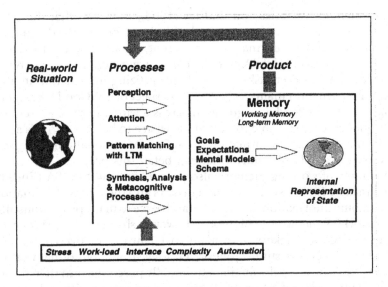

FIG. 1.4. Mechanisms and processes involved in SA.

environment are incorporated into their SA. In this light, both the perceptual salience of environmental cues (i.e., the degree to which they draw attention) and the individual's meaningful direction of attention are important. Numerous factors will dictate how people direct their attention in acquiring information, including learned scan patterns and information sampling strategies, goals, expectations, and other information already processed.

Several recent studies have confirmed the role of attention in SA. In Endsley and Smith (1996), we found that fighter pilots' attention to targets on a tactical situation display was directly related to the importance of those targets in their tactical tasks. Another study (Endsley & Rodgers, 1998) showed that air traffic controllers tend to reduce attention to less important information as their taskload (numbers of aircraft) increases. Gugerty (1998) found that drivers paid more attention to cars in front of or close to them than to those behind or farther away. This reflects the distribution of attention based on the operational importance of the vehicles. Clearly, the experienced operators in these three tasks deployed their attention in ways that are consistent with operational goals.

Attention to information is prioritized based on how important that information is perceived to be. It should also be pointed out, however, that even experienced operators can err in this process and neglect to attend to certain information over other information. Adams et al. (1995) described the challenges aircrews face in dynamically juggling many competing tasks and pieces of information and the role of attention in managing these challenges. In Jones and Endsley (1996), we found that the single most frequent causal factor associated with SA errors involved situations where all the needed information was present, but was not attended to by the operator (35% of total SA errors). This was most often associated with distraction due to other tasks. Correctly prioritizing information in a dynamic environment remains a challenging aspect of SA. Good SA requires enough awareness of what is going on across a wide range of SA requirements (global SA) to be able to determine where to best focus one's attention for more detailed information (local SA). Attentional narrowing is a well-known example of failures that can befall this process.

The limits of working memory also pose a constraint on SA (Endsley, 1988; Fracker, 1988). Novice decision makers and those in novel situations must combine information, interpret it, and strive to make projections all in limited working memory. (This model, however, shows that in practice, experienced decision makers have many mechanisms for overcoming this bottleneck.) An earlier study (Jones & Endsley, 1996) found that working memory losses (where information was initially perceived and then forgotten) explained 8.4% of SA errors (frequently associated with task distractions and interruptions). Many other failures in ability to form Levels 2 and 3 SA may also have a working memory limitation come into play.

Durso and Gronlund (in press) reported four strategies actively used by operators to reduce the working memory load associated with SA, including information prioritization, chunking, "gistification" of information (such as encoding only relative values of information if possible), and restructuring the environment to provide external memory cues. Despite these workarounds, Gugerty and Tirre (1997) found strong evidence of the importance of working memory in discriminating between people with higher and lower levels of SA. As Adams et al. (1995) pointed out, even experienced operators may be faced with so much information that attention and working memory constraints can be an issue.

### Long-Term Memory and Working Memory Connection

To view SA as a function of either working memory or long-term memory would probably be erroneous. In Endsley (1990, 1995a), for instance, I showed that experienced pilots could report on relevant SA information for 5 to 6 minutes following freezes in an aircraft simulation without the memory decay that would be expected from information stored in working memory. This was hypothesized to support a model of cognition that shows working memory to be an activated subset of long-term memory (Cowan, 1988), as shown in Fig. 1.5. In this model, information is proceeding directly from sensory memory to long-term memory which is necessary for pattern-recognition and coding. Those portions of the environment that are salient remain in working memory as a highlighted subset of

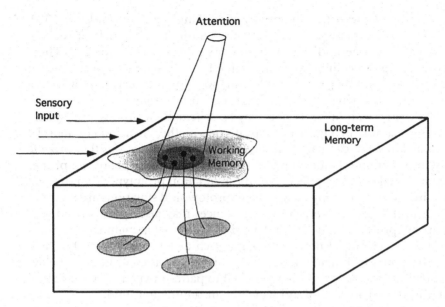

FIG. 1.5.  Working memory as an activated subset of long-term memory.

long-term memory through either localized attention or automatic activation. In this way, information from the environment may be processed and stored in terms of the activated mental model or schema (i.e., it provides the current situation values of these more abstract models). Thus activated, these schemata provide a rich source of data for bringing to bear on the situation including mechanisms for processing the data (i.e., forming Level 2 and Level 3 SA) and default values for filling in missing information (e.g., a Cessna is probably going a certain speed).

Durso and Gronlund (in press) reached a similar conclusion; they drew on a model by Ericsson and Kintch (1995) in which pointers in working memory point to information stored in long-term memory. Adams et al. (1995) also discussed pointers from working memory to long-term memory. Sarter and Woods (1991) emphasized the importance of information that can be activated from long-term memory to support limited working memory. In this sense, SA is a unique product of acquired external information, working memory processes, and the internal long-term memory stores activated and brought to bear on the formation of the internal representation.

## Long-Term Memory, Mental Models, and SA

Although long-term memory stores may take many forms, the idea of mental models has received much support. Long-term memory stores in the form of mental models or schemata are hypothesized to play a major role in dealing with the limitations of working memory (Endsley, 1988, 1995c). With experience, operators develop internal models of the systems they operate and the environments in which they operate. These models help direct limited attention in efficient ways, provide a means of integrating information without loading working memory, and provide a mechanism for generating projection of future system states.

Associated with these models may be schemata of prototypical system states as shown in Fig. 1.6. Critical cues in the environment may be matched to such schemata to indicate prototypical situations that provide instant situation classification and comprehension. Scripts of the proper actions to take may be attached to these situation prototypes thereby simplifying decision making. Schemata of prototypical situations are incorporated in this process and in many instances may also be associated with scripts to produce single-step retrieval of actions from memory, thus providing for very rapid decision making such as has been noted by Klein (1989). The use of mental models in achieving SA is considered to be dependent on the ability of the individual to pattern match between critical cues in the environment and elements in the mental model.

In this sense, SA, or the situation model, is the current state of the mental model. For example, one can have a mental model of a car engine in general, but the situation model (SA) is the state it is believed to be in now

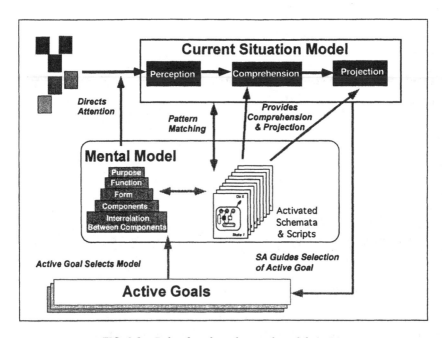

FIG. 1.6. Role of goals and mental models in SA.

(e.g., it has a given RPM, a given level of fuel, or is at a given temperature). This situational model captures not only the person's representation of the various parameters of the system, but also a representation of how they relate in terms of system form and function to create a meaningful synthesis—a gestalt comprehension of the system state. So, in addition to containing the current value of engine temperature and its rate of change, the situation model also includes an understanding of the impact of that state on the system and on projected events (e.g., it will overheat).

The concept of a mental model is useful in that it provides a mechanism for guiding attention to relevant aspects of the situation, integrating information perceived to form an understanding of its meaning, and projecting future states of the system based on its current state and an understanding of its dynamics. Without such a device, integration and projection would be particularly onerous tasks, yet experts appear to be able to perform them with ease. Using the information developed by Holland, Holyoak, Nisbett, and Thagood (1986), I provided a more detailed description of how mental models and schemata may be developed and formed through experience (Endsley, 1995c).

Other researchers also have posited a strong relationship between SA and mental models. Sarter and Woods (1991) stated that adequate mental models are a prerequisite for achieving SA. Mogford (1997) discussed the relation between mental models and SA. He stated that a mental model is the underlying knowledge that is the basis for SA. "Information in the mental model influences and structures the data held in SA and directs attention" (p. 333). According to Mogford, as information moves from the mental model into SA it increases in awareness, although its retention may decrease.

The use of mental models is not all positive for SA, however. Fracker (1988) stated that while schemata may be very useful for facilitating situation assessment by providing a reduction in working memory demands, they can also lead to significant problems with biasing in the selection and interpretation of information that may create errors in SA. Supporting this position, in previous research, we found that approximately 7% of SA errors were traced to a poor or insufficient mental model (Jones & Endsley, 1996). Additionally, we found that 6.5% of SA errors involved the use of the incorrect mental model to process information, thus arriving at an incorrect understanding of the situation. Another 4.6% of SA errors involved overreliance on default values in the mental model. Together these three problems with the use of mental models accounted for approximately 18% of SA errors (most of the 20.3% Level 2 SA error cases).

Jones (1997) further investigated the impact of mental models on the development of Level 2 SA. Specifically, she examined a type of Level 2 SA problem that has been labeled a representational error: incoming cues and information are misinterpreted based on an incorrect mental model. She manipulated the mental model of air traffic controllers by first introducing certain erroneous information. She then provided cues that should have indicated that the earlier information could not be correct. These cues were either "bizarre" (could not be possible with the mental model), "irrelevant" (could be possible with the mental model, although unlikely and more attuned with an alternate model), "unexpected" (a cue which was not expected to occur with the mental model), or "absence of expected" (in which a cues anticipated by the mental model did not happen). Interestingly, only 35% of the cues overall resulted in the controller detecting the error in the mental model. The "absence of expected" cues were the most successful for overcoming the representational error (56%) and demonstrated the real depth of information provided by mental models. Overall, however, the study primarily demonstrated the resilience of a person's initial assessment of a situation in guiding subsequent interpretations of later information. Although this strategy can be effective (Mosier & Chidester, 1991), it has also been related to errors in process control (Carmino, Idee, Larchier Boulanger, & Morlat, 1988) and medicine (Klein, 1993). Anchoring and confirmation bias undoubtedly come into play.

Building on the relation between SA and mental models, Endsley, English, and Sundararajan (1997) demonstrated that measures of SA (as momentary snapshots of the mental model) could be reconstructed to form a computer model of the mental model of expert fighter pilots. Although it is difficult to say that any computer model is identical to a human's internal model, the resultant computer model performed with an accuracy rate equivalent to the experts thereby demonstrating the utility of exploiting such a relationship. The problems involved in building good computer and analytic models of mental models have been well aired (Wilson & Rutherford, 1989). Overall, mental models and schemata are believed to be central mechanisms for overcoming limited working memory constraints to form SA under very challenging conditions.

### Pattern Matching and Other Cognitive Processes

Pattern matching goes hand in hand with schemata and mental models in facilitating the development of SA. There is considerable evidence that experienced decision makers use pattern matching to recognize perceived information as a particular exemplar of a known class of situations (Dreyfus, 1981; Hinsley, Hayes, & Simon, 1977; Kaempf, Wolf, & Miller, 1993; Klein, 1989; Klein et al., 1986). This pattern-matching process can occur almost instantly (Hinsley et al., 1977; Kaempf et al., 1993; Secrist & Hartman, 1993). Hartman and Secrist (1991) attributed near-threshold processing of environmental cues as an important basis for pattern matching.

Federico (1995) found that situations tended to be classified based on both surface features and deeper structural features. Amongst the naval operators in his study, surface features tended to include contact information, critical cues and similar platforms, patterns, and sequences. Deep structures included the importance attributed to similar contexts and other background information. Federico found that experts attended more to the issue of context than did novices. He did not find evidence for differences between novices and experts in the number of schemata or the accessibility to them.

In Endsley and Bolstad (1994), we found evidence of the importance of pattern-matching skills in distinguishing between fighter pilots with higher and lower levels of SA. It should be noted, however, that SA is not solely dependent on pattern matching. Cohen, Freeman, and Wolf (1996) presented a good discussion of the role of certain metacognitive skills in augmenting pattern matching and in forming the basis for developing situation assessment in cases where pattern-matching fails. Kaempf, Klein, Thordsen, and Wolf (1996) found that pattern matching to situation prototypes accounted for 87% of the decisions by tactical commanders with another 12% employing a story building process (most likely made possible through mental models). Commanders also tended to evaluate courses of action through mental simulation 18% of the time thus actively developing Level 3 SA. A combination of pattern matching, conscious analysis, story building, mental simulation, and metacognitive processes all may be used by operators at various times to form SA.

### Goals

Goals are central to the development of SA. Essentially, human information processing in operating complex systems is seen as alternating between data-driven (bottom–up) and goal-driven (top–down) processing. This process is viewed as critical in the formation of SA (Endsley, 1988, 1995c). In goal-driven processing, attention is directed across the environment in accordance with active goals. The operator actively seeks information needed for goal attainment and the goals simultaneously act as a filter in interpreting the information that is perceived. In data-driven processing, perceived environmental cues may indicate new goals that need to be active. Dynamic switching between these two processing modes is important for successful performance in many environments.

In defining SA as "adaptive, externally directed consciousness," Smith and Hancock (1995) viewed SA as purposeful behavior that is directed toward achieving a goal in a specific task environment. They furthermore pointed out that SA is therefore dependent on a normative definition of task performance and goals that are appropriate in the specific environment.

Within any environment, operators typically will have multiple goals that may shift in importance. At any one time, only a subset of these goals may be actively pursued. In the model shown in Fig. 1.6, the individual's goals serve the following critical functions (Endsley, 1995c):

1. The active goals direct the selection of the mental model. For example, the goal of "diagnose the warning light" would activate the mental model associated with that particular system and diagnostic behavior.

2. That goal and its associated mental model are used to direct attention in selecting information from the environment. They serve to direct scan patterns and information acquisition activities. For this reason, the selection of the correct goal(s) is extremely critical for achieving SA. If the individual is pursuing the wrong goal (or a less important goal), critical information may not be attended to or may be missed.

3. Goals and their associated mental models are used to interpret and integrate the information to achieve comprehension. The goal determines the "so what" of the information. For example, a flight level of 10,000 feet can only be interpreted in light of the goal of the desired altitude (e.g., the clearance altitude). The associated mental model is also critical in that the selection of the wrong mental model may lead to the misinterpretation of the information.

This is essentially a top–down, goal-driven process in which goals actively guide information selection and processing. Simultaneously, a bottom–up process occurs when information perceived is processed to form SA and a given situation assessment triggers the selection of a new goal. So, although a pilot may be busily engaged in the goal of navigation, the chiming of an emergency engine light triggers the activation of a new goal. This then directs selection of a new mental model and attention focus in the environment.

The alternating between bottom–up and top–down processing is one of the most important mechanisms underlying SA. Failures in this process can be particularly damaging to SA, including failures to process information needed to assure proper goal selection and the resultant inattention to needed information (e.g., attentional narrowing) or misinterpretation of perceived information (Level 2 SA errors).

Adams et al. (1995) pointed to a similar cycle of SA directing perception and perception leading to SA based on Neisser's (1976) model of the perceptual cycle. They did not, however, specifically address the role of goals and goal juggling in this process. A recognition of goals is key to understanding SA. Without understanding operator goals, the information in the environment has no meaning. In the words of Flach (1996), the construct of SA essentially allows researchers and designers to address the issue of meaning, something that has been heretofore all too frequently lacking in our science.

## Expectations

Additionally, preconceptions or expectations influence the formation of SA. People may have certain expectations about what they should see or hear in a particular environment. This may be due to mental models, prior experiences, instructions, or other communications. These expectations will influence how attention is deployed and the actual perception of information absorbed. That is, there is a tendency for people to see what they expect to see (or hear).

Expectations may be formulated based on the active mental model and prior expectations. They also may be developed through instructions or other communications. Pilots, for example, frequently develop strong expectations based on the preflight briefing. These expectations can provide a very efficient mechanism for filtering the data in the environment to a subset that is expected to be useful. Adams et al. (1995) stated that "anticipation reflects the focal schema's preparedness for certain information" (p. 89).

Adelman, Bresnick, Black, Marvin, and Sak (1996) reported that contextual information, provided by early information, acts to trigger different information processing strategies, thus influencing which information is attended to and how conflicting information is explained. This description is consistent with the model presented here and with Jones (1997), who found that even highly experienced air traffic controllers developed quite elaborate stories to explain away conflicting information that was inconsistent with the mental model created by earlier information. In an earlier study (Jones & Endsley, 1996), we found that in approximately half of the cases where incorrect mental models had been the source of a SA error, information was mismatched to fit the false expectation created by the incorrect mental model and thereby misinterpreted.

In Taylor, Endsley, and Henderson (1996), we demonstrated that expectations can form a two-edged sword. Teams who were given a well-developed set of expectations based on a premission video briefing tended to fall into traps easily when events and situations did not unfold as expected. They also did not develop the same type of group skills for active problem solving as did the other teams who were not given the video briefing. This demonstrated that although expectations can be a useful mechanism for SA, they also can contain certain dangers.

## Automaticity

Finally, automaticity is another mechanism developed with experience that can influence SA. With experience, the pattern-recognition and action-selection sequence can become so highly routine that a level of automaticity is developed (Logan, 1988). This mechanism provides good performance with a very low level of attention demand in certain well-understood environments. In this sense, automaticity can positively affect SA by reducing demands on limited attention resources, particularly for demanding physical tasks. SA can be negatively impacted by the automaticity of cognitive processes, however, due to a reduction in responsiveness to novel stimuli. Attention may not be given to information that is outside the routine sequence. Thus, SA may suffer when that information is important.

An issue of discussion has centered on whether people acting with automaticity really need to have a high level of conscious SA. A good example is the person driving over a familiar route without really thinking about it: turning and stopping where appropriate yet unable to consciously remember the past few minutes of the trip. As opposed to the view of performance presented here, where environmental cues are interpreted to form SA, produce decisions, and then actions (Fig. 1.7), automaticity would be seen to short circuit that process. A more direct stimulus–response pairing would be elicited and conscious awareness and decision making would thus be very low. Although the advantage of automaticity is the low level of attention required by the task, it also has significant hazards for SA and performance. Specifically, such direct stimulus–response pairing does not allow for integration of new information. In addition, people under automaticity tend to be nonreceptive to novel events and errors tend to occur when there must be a change in the learned pattern.

Although people may indeed perform many tasks with a fair degree of automaticity, it can be argued that in the highly cognitive, complex sys-

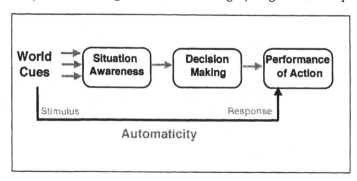

FIG. 1.7.  Automaticity in cognitive processes.

tems that typically interest us, the need for SA remains. In most real-world domains, people must perform based on more than just stimulus–response. They need to combine information and anticipate events that are beyond their experience. They must be proactive, not just reactive. They must use goal-driven, not just data-driven, processing. These aspects of successful performance require SA and are at odds with the automaticity of cognitive behaviors.

## Processes Versus Product: Situation Assessment and Situation Awareness

Figure 1.4 differentiates between cognitive processes (situation assessment) involved in achieving SA and the resultant state of knowledge about the situation (SA). It is widely recognized that each affects the other in an ongoing cycle. In my earlier model, the alternation of data-driven and goal-driven processes is seen as a central mechanism in this process. The goals, expectations, mental model, and current situation model all act to affect the processes employed (see Fig. 1.4). The processes determine the resulting SA. Adams et al. (1995) stressed the importance of the interrelationship between one's state of knowledge, or SA, and the processes used to achieve that knowledge. Framed in terms of Neisser's (1976) model of perception and cognition, they pointed out that one's current knowledge influences the process of acquiring and interpreting new knowledge in an ongoing cycle. This agrees with Sarter and Woods' (1991) statement that SA is "the accessibility of a comprehensive and coherent situation representation which is continuously being updated in accordance with the results of recurrent situation assessments" (p. 52). Smith and Hancock (1994) further supported this proposition by stating that "SA is up-to-the minute comprehension of task relevant information that enables appropriate decision making under stress. As cognition-in-action (Lave, 1988), SA fashions behavior in anticipation of the task-specific consequences of alternative actions" (p. 59). For purposes of clarity, however, situation assessment, as an active process of seeking information from the environment, is defined separately from SA the result of that process.

## SA MEASUREMENT

### Why Measure SA?

Some people have feared that SA will be a concept limited in utility because we cannot measure it. That perhaps SA is important to operators, but its lack of measurability makes it ultimately circular from a scientific standpoint. I believe the opposite is true. SA is a beneficial concept precisely because we can measure it. The direct measurement of SA provides great insight into how operators piece together the vast array of available information to form a coherent operational picture. The measurement of SA provides a useful index for evaluating system design and training techniques and for better understanding human cognition. As an intervening variable between stimulus and response, the measurement of SA provides far greater diagnosticity and sensitivity than is typically available from performance measures.

*Design Evaluation.* One of the chief reasons for measuring SA has been for the purpose of evaluating new system and interface designs. To determine the degree to which new technologies or design concepts actually improve or degrade operator SA, it is necessary to systematically evaluate them based on a measure of SA, thus providing a determination of which ideas have merit and which have unforeseen negative consequences. Explicit measurement of SA during design testing determines the degree to which design objectives have been met.

For many systems, operator SA and workload can be directly measured during design testing as well as operator performance measures. High-level performance measures (as collected during the limited conditions of simulation testing) are often not sufficiently granular or diagnostic of differences in system designs. Thus, although one system design concept may be superior to another in providing the operator with needed information in a format that is easier to assimilate with operator needs, the benefits of this may go unnoticed during the limited conditions of simulation testing or due to extra effort on the part of operators to compensate for a design concept's deficiencies.

If SA is measured directly, it should be possible to select design concepts that promote SA, and thus increase the probability that operators will make effective decisions and avoid poor ones. Problems with SA, frequently brought on by data overload, nonintegrated data, automation, complex systems that are poorly understood, excess attention demands, and many other factors, can be detected early in the design process and corrective changes made to improve the design.

*Evaluation of Training Techniques.* In addition to evaluating design concepts, a measure of SA may also be useful for evaluating the impact of training techniques on SA. Although relatively little work has been conducted to date on the impact of different training approaches on operator SA, theoretically, the same principles as those for evaluating design concepts apply.

*Investigating the SA Construct.* The measurement of SA is essential for conducting studies to empirically examine factors that may effect SA, such as individual abilities and skills, or the effectiveness of different processes and strategies for acquiring SA, and for investigating the nature of the SA construct itself. The theories discussed here can be expanded and developed further only through such efforts.

### Requirements for SA Measures

To adequately address these goals, however, the veracity of available SA measures must be established. Ultimately, validity and reliability must be provided for any SA measurement technique that is used. It is essential to show that a metric actually measures the construct it claims to measure and is not a reflection of other processes, provides the required insight in the form of sensitivity and diagnosticity, and does not substantially alter the construct in the process thereby providing biased data and altered behavior. In addition, it can be useful to establish the existence of a relation between the measure and other constructs as would be predicted by theory. To this end, the implications of SA information for SA measurement approaches are discussed.

### Implications of SA Theory for Measurement of SA

Several implications can be drawn from the theoretical viewpoints of SA that have significant implications for developing and selecting measures of SA (Endsley, 1996).

*Processes Versus States.* First, SA as defined here is a *state* of knowledge about a dynamic environment. This is different than the *processes* used to achieve that knowledge. Different individuals may use different processes (information acquisition methods) to arrive at the same state of knowledge, or may arrive at different states of knowledge based on the same processes due to differences in the comprehension and projection of acquired information or the use of different mental models or schemata. Measures that tap into situation assessment processes, therefore, may provide information of interest in understanding how people acquire information; however, they will only provide partial and indirect information regarding a person's level of SA. Although there may be experimental need for both types of measures, great care should be taken in attempting to infer one from the other.

*SA, Decision Making, and Performance Disconnect.* Second, just as there may be a disconnect between the processes used and the resultant SA, there may also be a disconnect between SA and the decisions made. With high levels of expertise in well-understood environments, there may be a direct SA-decision link, whereby understanding what the situation is leads directly to selection of an appropriate action from memory. This is not always the case, however. Individuals can still make poor decisions with good SA.

The relation between SA and performance, therefore, can be viewed as a probabilistic link. Good SA should increase the probability of good decisions and good performance, but does not guarantee it. Conversely, poor SA increases the probability of poor performance, however, in many cases does not create a serious error. For instance, being disoriented in an aircraft is more likely to lead to an accident when flying at low altitude than when flying at high altitude. Lack of SA about one's opponent in a fighter aircraft may not be a problem if the opponent also lacks SA. In relation to SA measurement, these issues indicate that behavior and performance measures are only indirect indices of operator SA.

*Attention.*    The way in which a person deploys his or her attention in acquiring and processing information has a fundamental impact on SA. Particularly in complex environments where multiple sources of information compete for attention, which information people attend to has a substantial influence on their SA. Therefore, design changes that influence attention distribution either intentionally or inadvertently, can have a big impact on SA. Similarly, measurement techniques that artificially influence attention distribution should be avoided, as they may well change the construct that is being measured in the process.

*Memory.*    Direct measures of SA tap into a person's knowledge of the state of the dynamic environment. This information may be resident in working memory for a short period of time or in long-term memory to some degree and under certain circumstances. A significant issue for measures that attempt to tap into memory is to what degree people can report on the mental processes that make this information accessible.

Automaticity may influence memory recall. With automaticity, there is very little awareness of the processes used. A careful review of the literature regarding automatic processing, however, reveals that although people may not be able to accurately report the processes used in decision making, they are usually aware of the situation itself at the time (Endsley, 1995c). A low level of attention may make this information difficult to obtain from memory after the fact, however.

Time also affects the ability of people to report information from memory. With time there is a rapid decay of information in working memory, thus only long-term memory access may be available. Nisbett and Wilson (1977) demonstrated that recall of mental processes after the fact tends to be overgeneralized, oversummarized, and overrationalized, and thus may not be an accurate view of the actual SA possessed in a dynamic sense.

Real-time, immediate access to information from memory can also be difficult as this process may influence ongoing performance, decision processes, and SA itself. Direct access to a person's memory stores can be problematic, therefore, and indicates that careful strategies for obtaining this information must be employed.

*Workload.* The relation between SA and workload has also been theorized to be important. Taylor (1990) included consideration of resource supply and demand as central to SA. Adams et al. (1995) also discussed task management problems involving prioritizing, updating task status, and servicing tasks in a queue as central to SA. Earlier, however, I showed that for a large range of the spectrum, SA and workload can vary independently, diverging based on numerous factors (Endsley, 1993a). Only when workload demands exceed maximum human capacity is SA necessarily at risk. SA problems may also occur under low workload (due to vigilance problems) or when workload is in some moderate region. SA and workload, although interrelated in certain circumstances, are essentially independent constructs in many ways.

As people can make trade-offs between the level of effort extended and how much they feel they need to know, it is important that both SA and workload be measured independently in the evaluation of a design concept. A particular design may improve or diminish SA, yet workload may remain stable. That is, operators may be putting forth the same amount of effort and getting more (or fewer) rewards in terms of the SA achieved. With other designs, it may be that operators are able to maintain the same level of SA, yet may have to work much harder. In order to get a complete understanding of the effects of a particular design concept, therefore, both SA and workload must be measured during design testing.

## How Much SA Is Enough?

This is perhaps the hardest question to answer. As SA and performance are only linked probabilistically, there is really no set threshold of SA that can guarantee a given level of performance. That is, it may be possible through luck to perform well on occasion with only a very low level of SA. Conversely, a person with even a high level of SA will not always perform well. It is probably better to think of SA in relative terms. As the level of SA possessed by an individual increases, the probability of making good decisions and performing well also increases. Therefore, in evaluating new design concepts or training programs it will be necessary to make relative comparisons. We are left with the question of whether a given design increases operator SA as compared to a previous design, or does it provide better SA than that possessed by rivals (e.g., enemy forces in a military context).

In making such assessments, it is also fair to say that there is no such thing as too much SA. More SA is always better. Because SA is operationally defined as that which one really needs to know, gaining more information and understanding these elements is always better and increases the likelihood of effective decision making and performance. (Although a person can have too much extraneous information that interferes with accessing and processing needed information, extraneous information, by definition, is not part of SA.)

In making this statement, however, it should be carefully noted that one should ensure that SA on one aspect of the situation is not gained at the cost of another, equally important aspect. If ideal SA is perfect knowledge on all relevant aspects of the situation, then this establishes a yardstick against which to measure. People rarely will reach this level; however, it allows us to determine the degree to which different system designs allow the operator to approach the ideal level of SA. This may vary for different elements of the situation, as SA has generally been found to be a multidimensional construct and not a unitary one (Endsley, 1990b; Endsley & Rodgers, 1998; Endsley, Selcon, Hardiman, & Croft, 1998). Therefore, as designers, we constantly seek to raise operator SA across a wide range of SA requirements.

## CONCLUSION

Numerous approaches to measuring SA have been proposed and are reviewed in subsequent chapters. Ultimately, each class of measures may have certain advantages and disadvantages in terms of the degree to which the measure provides an index of SA. Additionally, the objectives of the researcher and the constraints of the testing situation will also have a considerable impact on the appropriateness of a given measure of SA. Certain classes of measures may be highly suitable for qualitative investigations of SA processes, for instance, yet be inadequate for design testing, and vice versa. Regardless, it is vital that the veracity of any measure used for measuring SA be established so that informed research and design testing can take place.

## REFERENCES

Adams, M. J., Tenney, Y. J., & Pew, R. W. (1995). Situation awareness and the cognitive management of complex systems. *Human Factors, 37*(1), 85–104.

Adelman, L., Bresnick, T., Black, P. K., Marvin, F. F., & Sak, S. G. (1996). Research with Patriot air defense officers: Examining information order effects. *Human Factors, 38*(2), 250–261.

Arbak, C. J., Schwartz, N., & Kuperman, G. (1987). Evaluating the panoramic cockpit controls and displays system. In *Proceedings of the Fourth International Symposium on Aviation Psychology* (Vol. 1, pp. 30–36). Columbus: The Ohio Sate University.

Carmino, A., Idee, E., Larchier Boulanger, J., & Morlat, G. (1988). Representational errors: Why some may be termed diabolical. In L. P. Goodstein, H. B. Anderson, & S. E. Olsen (Eds.), *Task, errors and mental models* (pp. 240–250). London: Taylor & Francis.

Cohen, M. S., Freeman, J. T., & Wolf, S. (1996). Metarecognition in time-stressed decision making: Recognizing, critiquing, and correcting. *Human Factors, 38*(2), 206–219.

Cowan, N. (1988). Evolving conceptions of memory storage, selective attention, and their mutual constraints within the human information processing system. *Psychological Bulletin, 104*(2), 163–191.

Dominguez, C. (1994). Can SA be defined? In M. Vidulich, C. Dominguez, E. Vogel, & G. McMillan (Eds.), *Situation awareness: Papers and annotated bibliography* (AL/CF-TR-1994–0085; pp. 5–15). Wright-Patterson AFB, OH: Armstrong Laboratory.

Dreyfus, S. E. (1981). *Formal models vs. human situational understanding: Inherent limitations on the modeling of business expertise* (ORC 81-3). Berkeley: Operations Research Center, University of California.

Durso, F. T., & Gronlund, S. D. (in press). Situation awareness. In F. T. Durso, R. Nickerson, R. Schvaneveldt, S. Dumais, M. Chi, & S. Lindsay (Eds.), *Handbook of applied cognition*. New York: Wiley.

Endsley, M. R. (1987). *SAGAT: A methodology for the measurement of situation awareness* (NOR DOC 87–83). Hawthorne, CA: Northrop.

Endsley, M. R. (1988). Design and evaluation for situation awareness enhancement. In *Proceedings of the Human Factors Society 32nd Annual Meeting* (Vol. 1, pp. 97–101). Santa Monica, CA: Human Factors Society.

Endsley, M. R. (1990a, March). *Objective evaluation of situation awareness for dynamic decision makers in teleoperations.* Paper presented at the Engineering Foundation Conference on Human–Machine Interfaces for Teleoperators and Virtual Environments, Santa Barbara, CA.

Endsley, M. R. (1990b). *Situation awareness in dynamic human decision making: Theory and measurement.* Unpublished doctoral dissertation, University of Southern California, Los Angeles.

Endsley, M. R. (1993a). Situation awareness and workload: Flip sides of the same coin. In R. S. Jensen & D. Neumeister (Eds.), *Proceedings of the Seventh International Symposium on Aviation Psychology* (Vol. 2, pp. 906–911). Columbus: Department of Aviation, The Ohio State University.

Endsley, M. R. (1993b). A survey of situation awareness requirements in air-to-air combat fighters. *International Journal of Aviation Psychology, 3*(2), 157–168.

Endsley, M. R. (1994). *Situation awareness in FAA Airway Facilities Maintenance Control Centers (MCC): Final Report.* Lubbock: Texas Tech University.

Endsley, M. R. (1995a). Measurement of situation awareness in dynamic systems. *Human Factors, 37*(1), 65–84.

Endsley, M. R. (1995b). A taxonomy of situation awareness errors. In R. Fuller, N. Johnston, & N. McDonald (Eds.), *Human factors in aviation operations* (pp. 287–292). Aldershot, England: Avebury Aviation, Ashgate Publishing Ltd.

Endsley, M. R. (1995c). Toward a theory of situation awareness in dynamic systems. *Human Factors, 37*(1), 32–64.

Endsley, M. R. (1996). Situation awareness measurement in test and evaluation. In T. G. O'Brien & S. G. Charlton (Eds.), *Handbook of human factors testing and evaluation* (pp. 159–180). Mahwah, NJ: Lawrence Erlbaum Associates.

Endsley, M. R. (1997, April). *Communication and situation awareness in the aviation system.* Paper presented at the Aviation Communication: A Multi-Cultural Forum, Prescott, AZ.

Endsley, M. R., & Bolstad, C. A. (1994). Individual differences in pilot situation awareness. *International Journal of Aviation Psychology*, *4*(3), 241–264.

Endsley, M. R., English, T. M., & Sundararajan, M. (1997). The modeling of expertise: The use of situation models for knowledge engineering. *International Journal of Cognitive Ergonomics*, *1*(2), 119–136.

Endsley, M. R., Farley, T. C., Jones, W. M., Midkiff, A. H., & Hansman, R. J. (1998). *Situation awareness information requirements for commercial airline pilots* (ICAT–98–1). Cambridge: Massachusetts Institute of Technology International Center for Air Transportation.

Endsley, M. R., & Kiris, E. O. (1995). The out-of-the-loop performance problem and level of control in automation. *Human Factors*, *37*(2), 381–394.

Endsley, M. R., & Robertson, M. M. (1996). Team situation awareness in aviation maintenance. In *Proceedings of the 40th Annual Meeting of the Human Factors and Ergonomics Society* (Vol. 2, pp. 1077–1081). Santa Monica, CA: Human Factors & Ergonomics Society.

Endsley, M. R., & Rodgers, M. D. (1994). Situation awareness information requirements for en route air traffic control. In *Proceedings of the Human Factors and Ergonomics Society 38th Annual Meeting* (Vol. 1, pp. 71–75). Santa Monica, CA: Human Factors & Ergonomics Society.

Endsley, M. R., & Rodgers, M. D. (1998). Distribution of attention, situation awareness, and workload in a passive air traffic control task: Implications for operational errors and automation. *Air Traffic Control Quarterly*, *6*(1), 21–44.

Endsley, M. R., Selcon, S. J., Hardiman, T. D., & Croft, D. G. (1998). A comparative evaluation of SAGAT and SART for evaluations of situation awareness. In *Proceedings of the Human Factors and Ergonomics Society 42nd Annual Meeting* (Vol. 1, pp. 82–86). Santa Monica, CA: Human Factors & Ergonomics Society.

Endsley, M. R., & Smith, R. P. (1996). Attention distribution and decision making in tactical air combat. *Human Factors*, *38*(2), 232–249.

Ericsson, K. A., & Kintsch, W. (1995). Long term working memory. *Psychological review*, *102*, 211–245.

Federico, P. A. (1995). Expert and novice recognition of similar situations. *Human Factors*, *37*(1), 105–122.

Flach, J. M. (1995). Situation awareness: Proceed with caution. *Human Factors*, *37*(1), 149–157.

Flach, J. M. (1996). Situation awareness: In search of meaning. *Cseriac Gateway*, *VI*(6), 1–4.

Fracker, M. L. (1988). A theory of situation assessment: Implications for measuring situation awareness. In *Proceedings of the Human Factors Society 32nd Annual Meeting* (Vol. 1, pp. 102–106). Santa Monica, CA: Human Factors Society.

Gibson, J., Orasanu, J., Villeda, E., & Nygren, T. E. (1997). Loss of situation awareness: Causes and consequences. In R. S. Jensen & L. A. Rakovan (Eds.), *Proceedings of the Eighth International Symposium on Aviation Psychology* (Vol. 2, pp. 1417–1421). Columbus: The Ohio State University.

Gugerty, L. (1998). Evidence from a partial report task for forgetting in dynamic spatial memory. *Human Factors*, *40*(3), 498–508.

Gugerty, L., & Tirre, W. (1997). Situation awareness: A validation study and investigation of individual differences. In *Proceedings of the Human Factors and Ergonomics Society 40th Annual Meeting* (Vol. 1, pp. 564–568). Santa Monica, CA: Human Factors & Ergonomics Society.

Hartman, B. O., & Secrist, G. E. (1991). Situational awareness is more than exceptional vision. *Aviation, Space and Environmental Medicine*, *62*, 1084–9.

Hinsley, D., Hayes, J. R., & Simon, H. A. (1977). From words to equations. In P. Carpenter & M. Just (Eds.), *Cognitive processes in comprehension*. Hillsdale, NJ: Lawrence Erlbaum Associates.

Holland, J. H., Holyoak, K. F., Nisbett, R. E., & Thagard, P. R. (1986). *Induction: Processes of inference, learning and discovery*. Cambridge, MA: MIT Press.

IC3 (1986). *Intraflight Command, Control and Communications Symposium Final Report*. Unpublished manuscript.

Jentsch, F., Barnett, J., & Bowers, C. (1997). Loss of aircrew situation awareness: A cross-validation. In *Proceedings of the Human Factors and Ergonomics Society 41st Annual Meeting* (Vol. 2, pp. 1379). Santa Monica, CA: Human Factors and Ergonomics Society.

Jones, D. G. (1997). Reducing situation awareness errors in air traffic control. In *Proceedings of the Human Factors and Ergonomics Society 41st Annual Meeting* (Vol. 1, pp. 230–233). Santa Monica, CA: Human Factors & Ergonomics Society.

Jones, D. G., & Endsley, M. R. (1996). Sources of situation awareness errors in aviation. *Aviation, Space and Environmental Medicine, 67*(6), 507–512.

Kaempf, G. L., Klein, G. A., Thordsen, M. L., & Wolf, S. (1996). Decision making in complex naval command and control environments. *Human Factors, 38*(2), 220–231.

Kaempf, G. L., Wolf, S., & Miller, T. E. (1993). Decision making in the AEGIS combat information center. In *Proceedings of the Human Factors and Ergonomics Society 37th Annual Meeting* (Vol. 2, pp. 1107–1111). Santa Monica, CA: Human Factors & Ergonomics Society.

Klein, G. (1995, November). *Studying situation awareness in the context of decision-making incidents*. Paper presented at the Experimental Analysis and Measurement of Situation Awareness, Daytona Beach, FL.

Klein, G. A. (1989). Recognition-primed decisions. In W. B. Rouse (Ed.), *Advances in man-machine systems research* (Vol. 5, pp. 47–92). Greenwich, CT: JAI.

Klein, G. A. (1993). Sources of error in naturalistic decision making tasks. In *Proceedings of the Human Factors and Ergonomics Society 37th Annual Meeting* (Vol. 1, pp. 368–371). Santa Monica, CA: Human Factors & Ergonomics Society.

Klein, G. A., Calderwood, R., & Clinton-Cirocco, A. (1986). Rapid decision making on the fire ground. In *Proceedings of the Human Factors Society 30th Annual Meeting* (Vol. 1, pp. 576–580). Santa Monica, CA: Human Factors Society.

Lave, J. (1988). *Cognition in practice*. Cambridge, England: Cambridge University Press.

Logan, G. D. (1988). Automaticity, resources, and memory: Theoretical controversies and practical implications. *Human Factors, 30*(5), 583–598.

Marshak, W. P., Kuperman, G., Ramsey, E. G., & Wilson, D. (1987). Situational awareness in map displays. In *Proceedings of the Human Factors Society 31st Annual Meeting* (Vol. 1, pp. 533–535). Santa Monica, CA: Human Factors Society.

Mogford, R. H. (1997). Mental models and situation awareness in air traffic control. *International Journal of Aviation Psychology, 7*(4), 331–342.

Mosier, K. L., & Chidester, T. R. (1991). Situation assessment and situation awareness in a team setting. In Y. Queinnec & F. Daniellou (Eds.), *Designing for everyone* (pp. 798–800). London: Taylor & Francis.

Neisser, U. (1976). *Cognition and reality: Principles and implications of cognitive psychology*. San Francisco: W. H. Freeman.

Nisbett, R. E., & Wilson, T. D. (1977). Telling more than we can know: Verbal reports on mental processes. *Psychological Review, 84*(3), 231–259.

Press, M. (1986). *Situation awareness: Let's get serious about the clue-bird*. Unpublished manuscript.

Sarter, N. B., & Woods, D. D. (1991). Situation awareness: A critical but ill-defined phenomenon. *The International Journal of Aviation Psychology, 1*(1), 45–57.

Sarter, N. B., & Woods, D. D. (1995). How in the world did I ever get into that mode: Mode error and awareness in supervisory control. *Human Factors, 37*(1), 5–19.

Secrist, G. E., & Hartman, B. O. (1993). Situational awareness: The trainability of near-threshold information acquisition dimension. *Aviation, Space and Environmental Medicine, 64*, 885–892.

Smith, K., & Hancock, P. A. (1994). Situation awareness is adaptive, externally-directed consciousness. In R. D. Gilson, D. J. Garland, & J. M. Koonce (Eds.), *Situational awareness in complex systems* (pp. 59–68). Daytona Beach, FL: Embry–Riddle Aeronautical University Press.

Smith, K., & Hancock, P. A. (1995). Situation awareness is adaptive, externally directed consciousness. *Human Factors, 37*(1), 137–148.

Spick, M. (1988). *The ace factor: Air combat and the role of situational awareness.* Annapolis, MD: Naval Institute Press.

Taylor, R. M. (1990). Situational awareness rating technique (SART): The development of a tool for aircrew systems design. In *Situational Awareness in Aerospace Operations* (AGARD–CP–478; pp. 3/1–3/17). Neuilly Sur Seine, France: NATO–AGARD.

Taylor, R. M., Endsley, M. R., & Henderson, S. (1996). Situational awareness workshop report. In B. J. Hayward & A. R. Lowe (Eds.), *Applied aviation psychology: Achievement, change and challenge* (pp. 447–454). Aldershot, England: Ashgate Publishing Ltd.

Taylor, R. M., & Selcon, S. J. (1994). Situation in mind: Theory, application and measurement of situational awareness. In R. D. Gilson, D. J. Garland, & J. M. Koonce (Eds.), *Situational awareness in complex settings* (pp. 69–78). Daytona Beach, FL: Embry–Riddle Aeronautical University Press.

Tenney, Y. T., Adams, M. J., Pew, R. W., Huggins, A. W. F., & Rogers, W. H. (1992). *A principled approach to the measurement of situation awareness in commercial aviation* (NASA Contractor Report 4451). Langely, VA: NASA Langely Research Center.

Wickens, C. D. (1992). *Engineering psychology and human performance* (2nd ed.). New York: HarperCollins.

Wilson, J. R., & Rutherford, A. (1989). Mental models: Theory and application in human factors. *Human Factors, 31*(6), 617–634.

# Chapter 2

# The State of Situation Awareness Measurement: Heading Toward the Next Century

Richard W. Pew
*BBN Technologies*

The term, *situation awareness* (SA), has emerged as a psychological concept similar to such terms as intelligence, vigilance, attention, fatigue, stress, compatibility, or workload. Each began as a word with a multidimensional, but imprecise general meaning in the English language. Each has taken on an importance because it captures a characteristic of human performance that is not directly observable, but that psychologists and human factors specialists have been asked to assess or purposefully manipulate. The key step in the translation of a vague term into a quantitative concept is operational definition and measurement. However, many of these concepts have remained controversial because of the great difficulty in achieving operational definitions that are neither vacuous nor circular. SA is no exception. Defining SA in a way that is susceptible to measurement, different from generalized performance capacity, and usefully distinguishable from other concepts like perception, workload, or attention has proved daunting. Nevertheless, communities of analysts involved with human performance, including military planners and training specialists, military and commercial aviation experts, accident investigators, and power plant engineers, find the concept useful because for them it expresses an important, but separable element of successful performance. The responsibility falls on human factors specialists, represented by the authors of this book, to define the term in ways that lead to useful, valid, and reliable measurement at the same time that it meets the needs of this analyst population. This chapter is my contribution toward this goal.

I begin with a reductionist definition of SA and a discussion of definitional issues. Most of these issues have been raised previously in Pew (1994), Adams, Tenney, and Pew (1995), and Deutsch, Pew, Rogers, and Tenney (1994).

## FORMAL DEFINITION OF SITUATION AWARENESS

Most definitions of SA focus on the awareness part and neglect the definition of a situation. It must be recognized that SA is context dependent. What awareness is appropriate will depend on the specific circumstances that define the situation. On the one hand, it would be useful if situations were stable enough that particular SA requirements remained constant, at least long enough so that it would be possible to assess whether they have been satisfied. On the other hand, it must be recognized that contexts change more or less continuously. It must be expected that SA requirements will shift with changing contexts. I define a "situation" as a set of environmental conditions and system states with which the participant is interacting that can be characterized uniquely by a set of information, knowledge, and response options.

For this definition to be meaningful, I must insist on only a discrete and denumerable set of situations. Awareness requirements must be broken discretely and each associated with a set of system states. More than one requirement may be active at a given time, but each must apply to a situation that meets this definition. This ability to partition situations implies that, although the environment is more or less continuously changing with time, only some of the changes are large or severe enough to create a changed situation from the perspective of the crew member. It must be possible to identify the boundaries at which we wish to say that a situation has changed. Examples of such changes that are severe enough to redefine one or more SA requirements might be: a forest fire that has run out of its firebreak, a ship that enters the range of oncoming traffic, a train that encounters a conflicting train on the same track, a power plant that transitions from start-up to full power, or an aircraft autopilot disengagement, either expectedly or unexpectedly.

The more standard part of the definition requires that a "situation" have associated with it the information and knowledge that I call awareness. I refer to both information and knowledge because it is not enough simply to have current information. Every situation has associated with it specific knowledge required to accomplish a particular task. For example, SA requires immediate access to the procedures required to accomplish a task as well as the information required. Skilled performers will carry the procedures in long-term memory and bring them into working memory when they are specifically needed.

- Current state of the system (including all the relevant variables).
- Predicted state in the "near" future.
- Information and knowledge required in support of the crew's current activities.
- Activity Phase.
- Prioritized list of current goal(s).
  - Currently active goal, subgoal, task.
  - Time.
- Information and knowledge needed to support anticipated "near" future contexts.

FIG. 2.1.　Elements of awareness, given the situation.

Figure 2.1 shows the elements of awareness that need to be included. Although these elements are consistent with Endsley (1988), they generalize her proposal to include awareness of current goals and tasks as well as maintaining a perspective on the temporal status of goal accomplishment. Keeping track of the time course of goal-task completion is important because it plays a critical role in planning and task management. It is typically implicit in SA, but I wish to make it explicit. To achieve SA, the crew member may draw on various information resources, including: sensory information from the environment, visual and auditory displays, decision aids and support systems, extra- and intracrew communication, and crew member background knowledge and experience.

It is also important to note that, although much of the SA literature focuses on spatial awareness, there are many other aspects of systems and their operations that require awareness. In addition to spatial awareness, other elements of awareness that must be considered include: mission-goal, system, resource, and crew.

Spatial awareness is self-explanatory. Mission-goal awareness refers to the need to keep current with respect to the phase of the mission and the currently active goals that are to be satisfied. System awareness is especially important in complex, highly automated systems. The work of Sarter and Woods (1994) identifies, for example, the critical difficulties associated with understanding and tracking the mode of flight management computers. This is a system awareness issue. Resource awareness is needed to keep track of the state of currently available resources, including both physical and human resources. Each crew member must know the current activities of other crew members so that their availability for critical tasks is known. This is different from the crew awareness element that refers to the need for the team of crew members to share their information and interpretation of current system events. They need to be operating in a common framework of information at all times.

To have defined situations and the components of awareness is still not enough because the requirements for SA presented by a given situation must also be defined; otherwise, there exists no standard against which to judge how successful a crew member has been in achieving SA. Our thinking about this has led to the consideration of an *ideal* awareness and the *achievable ideal*, in addition to the actual SA achieved.

The ideal is that SA defined by experts evaluating the requirements at leisure or even after the fact. It includes both the information and knowledge requirements. The achievable ideal is that subset actually available for the crew member to acquire. When defining the obtainable ideal the availability of well-designed information resources is assumed. Also taken into account is the fact that what is practically available is constrained by expected human cognitive abilities. It does not seem "fair" to evaluate a crew member's actual SA achievement with respect to a goal that is not practically achievable. Both the ideal and the achievable ideal are, in principle, assessed independently of crew member performance. However, the *actual* SA must be inferred from measurement or observation. Figure 2.2 illustrates how both achievable and actual SA are subsets of ideal SA.

It is a significant challenge to define the achievable ideal. It involves considerations of information display design as well as that information humans may be expected to absorb and retain. For example, although the ideal for a commercial pilot might be to keep track of the call signs of several aircraft in the immediate vicinity, we know that working memory is limited in this respect. We should not expect achievable SA to overload memory. As a practical matter, expecting to define the achievable ideal entirely independently of actual performance is probably unrealistic. The definition may be seen as an iterative process that will be revised as a result of actual measurement. After seeing what users actually accomplish, we then go back and reflect on whether actual performance was constrained by fundamental cognitive limitations, whether it reflected individual differences in performance, or whether further training should be expected to improve performance.

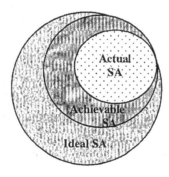

FIG. 2.2. Venn diagram showing the interrelationships among ideal, achievable, and actual SA.

It is the difference between ideal SA and achievable ideal that creates a space for evaluating design alternatives contributing to improved SA. The ideal SA reflects the analyst's front-end analysis to identify information and knowledge requirements; requirements that are formulated with SA in mind. The achievable ideal reflects the success that a designer has had in translating and implementing those requirements to produce useful displays to ensure the success of the tasks to be performed.

The difference between achievable ideal and actual SA creates the space for evaluating individual differences in the ability to achieve SA and for developing training opportunities. If the achievable ideal is not being actualized, the analyst should ask if actual SA could be improved by better training. Of course, this analysis may also conclude that the achievable ideal did not adequately consider human perceptual limitations and may result in a redesign of the information presentation for the system.

With those fundamentals elaborated, I present my view of just how SA plays out in a real-time environment. At any given time the crew members of a system have many tasks on their queue in various stages of completion. Some tasks are waiting to be carried out and some have been attended to and the system is now playing them out, but the tasks must be monitored to assure that they play out as intended. Other tasks are merely anticipated in the near future. This queue has, at best, an explicit, but very dynamic priority structure (at worst that structure is only implicit). The urgency associated with individual tasks changes with the passage of time or with the acquisition of new information. Environmental events intervene, sometimes predictably, but often unpredictably. In either case these events can change the priority structure—some tasks increase in priority while others become moot. The pace with which these changes take place depends on the system time constants. Crew members communicate in order to share awareness, to communicate changing priorities, or to request action from one another.

It is in this dynamic environment that SA plays so important a role and is so hard to assess. Information and status updates are required to support all the tasks currently on the queue as well as those anticipated to be added soon. Some information is itself inherently dynamic and, depending on the bandwidth of variations, is more or less predictable as a function of time since the last update. With constantly changing priorities, information needs are also constantly changing. It is not enough to keep up with the pace of information being delivered. It must be interpreted and related to other information and to the task requirements.

These issues are elaborated in Adams, Tenney, and Pew (1995). I mention them here because understanding measurement alternatives requires understanding the character of the dynamic informational context in which the crew members operate. Describing this generalized scenario helps to articulate the difficulty of defining the end of one "situation" and the beginning of another. It helps to explain the context dependence, scope, and transience of SA requirements.

## A TAXONOMY OF MEASUREMENT METHODS

The subject of measurement encompasses more than just the selection of performance measures. It includes development of the measurement context, including the systems and scenarios of use. In Endsley (1995a), a taxonomy is presented and measurement methods are reviewed. She found weaknesses in every measurement context except job samples of operationally realistic scenarios. Because her work at the time was based largely on SA in air-to-air combat this was not inappropriate; however, I take a more eclectic view. I believe there are many relevant measurement contexts, but each is appropriate to particular purposes. SA measurement methods may be broken down as follows: direct system performance measures, direct experimental measures, verbal protocols, and subjective measures.

Endsley (1995a) argued that only scenarios with full face validity were appropriate for the measurement of SA. I believe that any simulation experiment involves compromises with the realism of the real world. The issue is only one of degree—the face validity required of the scenario depends on the purpose of the assessment, as is illustrated in the discussion that follows. Each category in my taxonomy is discussed in turn.

### Direct System Performance Measures

I agree with Endsley that there are only limited occasions where direct system performance measures per se, are appropriate for assessing SA. Those occur in cases where there is general agreement among an audience of peers that the performance in question is driven solely or largely by SA.

A more common approach is to invest significant planning resources in the careful design of the scenario to specifically create opportunities for using nonobtrusive performance measures to assess specific SA issues. Sometimes it involves introducing subtle or counterintuitive indicators. For example, Busquets, Parrish, Williams, and Nold (1994) studied the usefulness of specific navigation displays during aircraft approach and landing for promoting SA. On a dual runway airport they deliberately introduced a second aircraft apparently intending to land on the second runway. However, just as the pilot-under-study was about to reach final approach, the second aircraft deviated and appeared to be landing on the runway to which the pilot-under-study was committed. They measured the time it took the pilot-under-study to take action to avoid the second aircraft. This, by design, was very clearly a measure of the pilot's SA regarding activity in the airspace around him.

A second method of scenario manipulation is to introduce disruptions intended to disorient the crew or operator and from which they must recover. The measure of SA is the recovery time or success of recovery under the assumption that the amount of time required to recover is proportional to how well the crew could use the system displays for recovery from the disorientation. Busquets et al. (1994) used this method in their experiment by blanking the cockpit displays and, during the period that the displays were blanked, introduced a systematic offset in the aircraft position. The crew had to use the displays to return to their position on the flight path. Both of these methods produced significant differences in performance as a function of the display conditions under study.

A third method, in the spirit of suggestions by Sarter and Woods (1991), is to introduce, without the knowledge of the subjects, anomalous data or instrument readings; that is, readings that create a pattern that could not have been produced by any realistic condition. The typical measure of performance is the time required to detect the anomaly. Hahn and Hansman (1992) used this technique to compare the utility of graphical, aural, and textual communication links. They introduced faulty air traffic control directives and detected the pilots' recognition of them as faulty. For faulty weather and routing clearances, the graphical link provided the most faulty clearances detected immediately upon examination, as contrasted with detection during later study, taking action with respect to them, or never detecting them.

As Endsley (1995a) pointed out, the detection time will depend not only on the time to detect the anomaly, but also on the locus and urgency of the crew members' attentional focus at the time the anomaly was introduced. As a result the variance in detection times may be larger than using some other methods. Nevertheless, this method should be relevant and useful if detection of the anomaly requires an understanding that is central to successful task performance and the goal of the displays that support SA is to facilitate that understanding.

**Direct Experimental Techniques**

The most common measurement method is to use direct experimental techniques such as queries or probes and measures of information seeking. Probes can be introduced during ongoing task performance if the pace of the task is slow or there are many periods of relative inactivity. However, it is more common to suspend the task—to freeze the simulation—and to ask one or more questions about the state of the task or the environment before resuming the action. This method has been formalized by Endsley (1988) as the Situation Awareness Global Assessment Technique (SAGAT) and has been applied in many situations. It and its variants are now widely used.

I like SAGAT. This technique makes use of systematic, preplanned procedures. When preparing to use this technique, the analyst is forced to think through in detail exactly what aspects of SA are going to be assessed. It uses computer administration of the set of SA questions during the simulation freeze to assure control of the question administration. Endsley emphasized three levels of SA: perception of information, integration of data with goals, and projection of the near future. This technique requires that the questions be specific and that the questions asked at a particular sampling point are drawn at random. This makes it very different from the direct system performance measures that attempt to assess some specific aspect of SA through performance assessment based on some predesigned scenario feature. By using SAGAT, one can obtain interesting and rich data about the aggregate levels of specific classes of SA, but cannot answer specific questions about SA at a particular, perhaps critical, point in a scenario. It is useful for assessing aggregate individual differences in SA in the context of a particular scenario and therefore may be diagnostic of selection or training needs or opportunities. It is possible to compare displays designed to support SA, but only at an aggregate level because the randomness of the questions does not allow evaluation of the acquisition of a specific SA feature that might require using a particular display. For example, if it were critical at a particular point in a scenario for a pilot to know aircraft altitude, one could not assure that a question about altitude would be asked at that point. Because the probes are randomized over a given scenario, and often data can come from multiple sources, it is difficult to assign a failure of SA to a specific display or system designed to support it since, without further corroborating information, it is not known whether any particular display was responsible for the failure.

Two controversial issues are often introduced when the probe technique is considered: whether the use of the freeze technique disrupts ongoing task performance thereby placing the subject in an unrealistic setting and producing an unrealistic assessment of SA, and whether the expectation that probes will be presented changes the subjects' behavior. Once the first probe is presented, do they anticipate that additional probes will be presented and prepare for them? If so, then the experimenter obtains a realistic assessment of the level of SA only on the first probe. These issues were addressed in Endsley (1994, 1995a). The performance on 25% of the trials where the subjects received no probes was compared to the performance on the remaining 75% where varying numbers of probes were presented. She found no significant difference in performance among the groups. However, the design of the experiment was such that the same subjects participated in all conditions. They were not told before each trial whether probes should be expected or not. Although this may address the first issue concerning the impact of the freeze technique, it does not address the second issue concerning anticipating additional probes because the subjects were primed to expect probes whether they occurred or not.

Subsequently, Endsley (1995b) reported on an additional experiment. The presence or absence of probes was manipulated between subjects and still no significant differences were found. This is a much sounder design for examining this question; however, both groups were presented with training on how to respond to probes and how to behave when the simulation was frozen. Does the probe response training create such an expectation on the part of the subjects that they do not behave differently whether or not probes are presented?

An alternative design that addresses the issue of surprise was used by Wickens and Prevett (1995). In an experiment involving the design of navigation displays, they interrupted the simulation to ask a series of questions similar to Endsley's procedure. They found that the first probe had a significantly longer reaction time, but that there was no significant difference between the first and subsequent probes in the accuracy of response.

In my opinion, these issues are deserving of somewhat less attention than they have been given. Because most measurement is relative, main effects that appear can still be useful diagnostic indicators although moderated by the experimental techniques. A different level of performance on the first and later trials may occur, but both trials will still accurately reflect the relative differences among treatment conditions. The issue that should be addressed arises when the interruption or surprise creates a statistical interaction between two or more of the other variables of interest. For example, in an experiment by Olson and Sivak (1986), the relation between driver age and reaction time was impacted by introducing differing expectancies in a driving task.[1]

In addition to queries and probes, assessments of information seeking are included, in the category of direct measures. Although some investigators define a separate category of physiological measures, I consider the two primary information seeking measures, namely eye movements and eyeblink response, to be direct experimental measures. The medium by which they are recorded is not really relevant to their application to SA measurement. Eye movements have frequently been used to assess information seeking. It cannot be argued conclusively that simply because an observer's eye is directed toward a specific object or expression, that it is seen. However, the correlation between looking and seeing is likely to be

---

[1]I am indebted to Neil Charness for supplying this example.

quite high. Vidulich, Stratton, Crabtree, and Wilson (1994), in a comprehensive experiment designed to evaluate alternative measures of SA, used eyeblink as an element of a larger set of measures that they studied and found that display condition had a significant effect on eyeblink duration.

Another method for appraising information seeking is to force the user to ask explicitly for the information that is needed. This technique, originally developed by Payne, Bettman, and Johnson (1993), provided an experimental methodology for studying decision making. It was then possible to track the frequency and order in which decision alternatives and their attributes were examined. The subjects were required to select each item they wished to consider with mouse clicks. This approach can provide very useful information for understanding situation assessment (the process of achieving SA).

It has recently been applied to interpreting the way in which aircrew use an approach plate, a map or chart that provides the crew with comprehensive data about the approach and landing at a particular airport. Ricks, Jonsson, and Barry (1996) prepared a computerized version of the approach plate. The labels for each piece of information were available in the location of the actual information, but the crew member had to "button" the location with a mouse in order to actually see it for a brief period of time. Repeated viewing required repeated activation. Detailed data were obtained about what information was sought and the timing and sequence in which it was requested.

### Verbal Protocols

A verbal protocol is information recorded from the observer or crew during or immediately after an exercise or during a video replay of that exercise. The subject may be asked to "think aloud" or to explain the information relied on. This is a technique that is most useful early in the development of an evaluation. It may help to solidify SA concepts that need to be measured more systematically. It is disruptive when the observer is asked to report in real time during the scenario execution, but the information gained may be worth the disruption.

### Subjective Measures

It is so difficult to obtain quantitative objective measures of SA that many investigators instead rely on subjective measures such as self-assessments, expert judgments, peer ratings, and supervisor or instructor ratings.

The most thoughtful and systematic development of a self-administered test of SA to date is reported by Taylor (1990), and is referred to as the Situation Awareness Rating Technique (SART). SART consists of three subscales that are combined in an equation to produce an overall estimate of the subject's SA. The technique is discussed further in Taylor and Selcon (1994). Taylor presents an equation representing the algebraic combination of a set of subscales that serves as an integrated measure of SA. The measure includes, as one of the subscales, the demand on attentional resources. Including attention demand in the integrated equation defining SA confounds SA with workload.

In my opinion, workload and SA should be viewed as different, but of course, they may be correlated in practice. The relation between SA and workload is somewhat analogous to the relation between speed and accuracy of performance. Although one often hears of the speed–accuracy trade-off, a trade-off is only one of the ways that speed and accuracy can be related. First, when speed and accuracy are measured across a set of individuals, we expect a high positive correlation between the two measures. Those subjects who tend to be fast, also tend to be accurate. Similarly, thinking of it as an individual difference variable, subjects with good intrinsic ability to maintain SA would be expected to achieve it with less workload. Second, if we train an individual to improve performance, we expect speed and accuracy to improve more or less together. There is a high positive correlation between speed and accuracy within an individual across a period of practice. The analogous statement with respect to SA and workload is that training can be expected to increase SA for an equivalent or lower level of workload. Finally, analogous to the speed–accuracy trade-off, if we challenge an individual to improve SA at a particular point in practice, we would expect that the workload associated with achieving that improvement to be increased. To include aspects of attentional demand in the formula for the assessment of SA makes it difficult for these relationships to emerge.

Vidulich et al. (1990) evaluated SA in the context of a military air-to-ground attack target detection task and had three display conditions. The condition predicted to provide the most SA used an integrated navigation and targeting display and made the targets a distinctive color as might be provided by an automatic target recognizer. The condition predicted to be next best, from the point of view of SA, used the same display, but made the targets a camouflaged color that was more difficult to detect. The third condition, predicted to be worst, presented separate displays for navigation and targeting, and, as a result, required that information be integrated cognitively in order to assess the relative positions of the various targets. This experiment further illustrates the difficulties of confounding workload and SA. Arguably the first two conditions differ in the amount of workload required for detection. The second and third conditions differ in the cognitive effort associated with assimilating and integrating the information from one or two displays. These manipulations do not seem to me to be varying the same dimensions of SA; the first simply requires more workload to achieve the detection on which SA is based, whereas the second genuinely manipulates the difficulty of achieving an integrated SA picture.

By definition self-ratings can only reflect self-awareness. The operators do not necessarily have a perspective on what they *should be* aware of. The "bad news" is that such measures usually reflect little more than a confidence rating or preference on the part of the operator or crew member. The "good news" is that sometimes that is exactly what is of interest to the investigator.

The best example of supervisor–peer rating is reported in the U.S. Air Force SAINT study (McMillan, 1994). The investigators were interested in assessing SA as an individual difference variable in the combat skill of F-15 fighter aircraft pilots. They developed a series of SA rating scales that were actually inclusive enough that they should be called combat skill ratings rather that strictly SA ratings. They obtained supervisor (squadron commanders) and peer ratings for 200 line combat pilots. Interestingly, the peer ratings (that is pilots' ratings of each other) correlated with their supervisors ratings 0.80.

In a separate study, a representative subset of 40 of these pilots flew a demanding combat simulation and two instructor-pilots, who observed these simulation exercises, used the same scales to produce a set of independent subjective SA ratings. The instructors did not know about the peer or supervisor ratings. When the investigators then correlated the combined supervisor–peer ratings with the instructor ratings, they found that the supervisor–peer ratings correlated 0.56 with the instructors ratings. Said another way, the peer ratings accounted for 31% of the variance in the instructor's ratings. While one might hope for higher correlations, this is quite substantial for these types of rating data.

A second, very different example was provided by a NASA Langley Research Center experiment. The paradigm is described in Schutte and Trujillo (1996). A commercial aircraft scenario, in which several equipment failures occurred, was flown in a generic glass cockpit simulator. The aircrew's task was to detect and manage the failures. Periodically (approximately every 5 minutes), the pilots were asked to report the perceived state of their awareness with respect to three kinds of awareness: spatial (e.g., altitude, aircraft position), mission (e.g., flight planning, fuel, weather), and systems (e.g., operations, status). If they were busy at the time the response was requested, they were allowed to delay reporting until they could comfortably do so.

Although the sample size was small, a significant correlation was observed between reported average level of SA and the time to take actions critical to success of the flight once the failures had occurred. For example, the time at which pilots decided to divert to an alternative airfield was significantly correlated with the value of all three awareness measures obtained just before they made the decision (–.74 = spatial, –.51 = mission, –.50 = systems). The correlations were negative because a lower value of SA led to a longer delay in reporting. However, the time at which pilots declared that an engine was out (the right diagnosis) was only significantly correlated with the average systems awareness of –.80. The difference in SA pattern for these two action items suggests that the measure was responsive to real effects because one would expect a diversion to depend on all three kinds of awareness, whereas diagnosing an engine out is only a system awareness issue.

## CONCLUSION

As illustrated here, there are many SA measurement techniques that are now beginning to be investigated in practical situations to evaluate their relevance and usefulness. Researchers are beginning to understand what measures are good for what purposes. The think-aloud, and other seriously disruptive techniques, are probably best suited to preliminary research to establish measures and understand SA requirements. The scenario manipulation techniques are best for evaluating individuals' ability to meet scenario-specific SA requirements. Techniques such as SEGAT, are best for examining SA across experimental treatment conditions. Finally, the subjective approaches will always be needed for situations where objective quantitative data is either impractical, too costly, or too time-consuming to collect.

To search for the universal measure is to search for the Holy Grail. Rather, an investigator must make a judicious choice of the measurement context together with the appropriate choice of measures. The choices should be made with full understanding of the classes of situations to be measured and some thought about what will index transitions from one situation to another. The investigator must also figure out ahead of time, perhaps in pilot studies with experts, just what the SA requirements of the situation(s) are so that they can assess achieved SA in relation to those requirements.

## ACKNOWLEDGMENTS

I wish to thank my colleagues, Yvette J. Tenney, Marilyn Jager Adams, Stephen Deutsch, and William H. Rogers who were my collaborators on some of the ideas reported in this chapter. Parts of this work were supported under Contract NAS1-18788 with the NASA Langley Research Center, Dr. Raymond Comstock, Technical Monitor.

## REFERENCES

Adams, M. J., Tenney, Y. J., & Pew, R. W. (1995). Situation awareness and the cognitive management of complex systems. *Human Factors, 37*(1), 85–104.

Busquets, A. M., Parrish, R. V., Williams, S. P., & Nold, D. E. (1994). Comparison of pilots' acceptance and spatial awareness when using EFIS vs pictorial display formats for complex, curved landing approaches. In R. D. Gilson, D. J. Garland, & J. M. Koonce (Eds.), *Situational Awareness in complex systems* (pp. 139–170). Daytona Beach, FL: Embry-Riddle Aeronautical University Press.

Deutsch, S. E., Pew, R. W., Rogers, W. H., & Tenney, Y. J. (1994). *Toward a Methodology for Defining Situation Awareness Requirements—A Progress Report.* National Aeronautics and Space Administration, BBN Report No. 7983, Cambridge, MA: BBN Corp.

Endsley, M. R. (1988). Design and evaluation for situation awareness enhancement. In *Proceedings of the Human Factors Society 32nd Annual Meeting* (Vol. 1, pp. 97–101). Santa Monica, CA: Human Factors Society.

Endsley, M. R. (1994). Situation awareness in dynamic human decision making: Measurement. In R. D. Gilson, D. J. Garland, & J. M. Koonce (Eds.), *Situational awareness in complex systems* (pp. 79–98). Daytona Beach, FL: Embry-Riddle Aeronautical University Press.

Endsley, M. R. (1995a). Measurement of situation awareness in dynamic systems. *Human Factors, 37*(1), 65–84.

Endsley, M. R. (1995b). Direct Measurement of situation awareness in simulations of dynamic systems: Validity and use of SAGAT. In D. J. Garland & M. R. Endsley (Eds.), *Proceedings of the international conference on experimental analysis and measurement of situation awareness* (pp. 107–113). Daytona Beach, FL: Embry-Riddle Aeronautical University Press.

Hahn, E. C., & Hansman, R. J. (1992). *Experimental studies of the effect of automation on pilot situational awareness in the datalink ATC environment* (SAE Technical Paper Series 922022). Warrendale, PA: Society of Automotive Engineers.

McMillan, G. R. (1994). Report of the Armstrong Laboratory Situation Awareness Integration (SAINT) team. In M. Vidulich (Ed.), *Situation Awareness: Papers and annotated bibliography* (pp. 37–47). Crew Systems Directorate, Human Engineering Division, Wright-Patterson AFB, OH.

National Aeronautics and Space Administration. (1994). *Toward a methodology for defining situation awareness requirements—A progress report* (BBN Report No. 7983). Cambridge, MA: S. E. Deutsch, R. W. Pew, W. H. Rogers, & Y. J. Tenney.

Olson, P. L., & Sivak, M. (1986). Perception-response time to unexpected roadway hazards. *Human Factors, 28*(1), 91–96.

Payne, J. W., Bettman, J. R., & Johnson, E. J. (1993). *The Adaptive decision maker.* New York: Cambridge University Press.

Pew, R. W. (1994). An introduction to the concept of situation awareness. In R. D. Gilson, D. J. Garland, & J. M. Koonce (Eds.), *Situational Awareness in complex systems* (pp. 17–26). Daytona Beach, FL: Embry-Riddle Aeronautical University Press.

Ricks, W. R. T., Jonsson, J. E., & Barry, J. S. (1996). *Managing approach plate information study (MAPLIST): An information requirements analysis of approach chart use* (NASA Technical Paper 3561), Hampton, VA: NASA Langley Research Center.

Sarter, N. B., & Woods, D. D. (1991). Situation awareness: A critical but ill-defined phenomenon. *The International Journal of Aviation Psychology, 1,* 45–57.

Sarter, N. B., & Woods, D. D. (1994). Pilot interaction with cockpit automation II: An experimental study of pilots' model and awareness of the flight management system. *International Journal of Aviation Psychology, 4*(1), 1–28.

Schutte, P. C., & Trujillo, A. C. (1996). Flight crew task management in non-normal situations. In *Proceedings of the Human Factors and Ergonomics Society 40th Annual Meeting* (Vol. 1, pp. 244–248). Santa Monica, CA: Human Factors Society.

Taylor, R. M. (1990). Situation awareness rating technique (SART): The development of a tool for aircrew systems design. In *Situational Awareness in Aerospace Operations (AGARD-CP-478)*. Neuilly Sur Seine, France: NATO–AGARD.

Taylor, R. M., & Selcon, S. J. (1994). Situation in mind: Theory, application and measurement of situational awareness. In R. D. Gilson, D. J. Garland, & J. M. Koonce (Eds.), *Situational awareness in complex settings* (pp. 69–78). Daytona Beach, FL: Embry-Riddle Aeronautical University Press.

Vidulich, M. A., Stratton, M., Crabtree, M., & Wilson, G. (1994). Performance-based and physiological measures of situational awareness. *Aviation, Space, and Environmental Medicine, 65*, (5 Suppl.), A7–A12.

Wickens, C. D., & Prevett, T. T. (1995). Exploring the dimensions of egocentricity in aircraft navigation displays. *Journal of Experimental Psychology: Applied, 1*(2), 110–135.

# Part II

# MEASUREMENT APPROACHES

# Chapter 3

# Analysis of Situation Awareness from Critical Incident Reports

Gary Klein
*Klein Associates Inc.*

## WHAT IS SITUATION AWARENESS?

I define the phenomenon of *situation awareness* (SA) as the perception of reactions to a set of changing events. This definition emphasizes the affordances in the situation and views a person's understanding in terms of what can be done (even if it is only to gather more data) instead of merely the recalled stimuli. A person's understanding of a context rather than just the ability to accurately recall disconnected data elements is emphasized. The link between SA and task performance, instead of viewing the observer as passively monitoring conditions without a need to respond, is underscored by my definition. The complexity of handling multiple, interacting tasks rather than a single, isolated one is also emphasized.

The ability to remember discrete data elements may be important for establishing SA because inattention to the data results in a weak basis for making interpretations. There are also times when a person is engaged in passive observations as well as occasions when only a single task is performed (e.g., monitoring a track on a radar scope to see if it shows any signs of hostile intent). Nevertheless, the study of SA may be facilitated by establishing an image of the prototypical observer in the most realistic posture instead of simplifying the phenomenon. Researchers are interested in studying SA because of what can be learned about difficult conditions—ones where the observer is faced with overlapping tasks and changing events, and is desperately trying to make sense of the situation.

There are four reasons why the phenomenon of SA is important. First, SA appears to be linked to performance. For example, a fighter pilot performing a complex mission must track many different factors simultaneously, such as weather, his location, the location of friendly resources (e.g., AWACS, tankers, jammers, other aircraft in the formation), and the location and type of enemy resources. Clearly, the more of these a pilot can track, the more adaptive the responses can be. A second related reason is that limitations in SA, perhaps due to working memory capacity or inadequate attentional control, may result in errors. The third reason is that SA may be related to expertise. Classic studies such as Larkin, McDermott, and Simon (1980) and Chi, Feltovich, and Glaser (1981) showed that more experienced physics researchers classified physics problems differently than novices. Finally, the fourth reason (from the perspective of Naturalistic Decision Making) is that SA is the basis for decision making in most cases.

The most widely known account of SA is the one provided by Endsley (1995). She distinguished the three levels of SA as the detection of the environment's relevant elements (Level 1 SA), the comprehension of the elements' meaning (Level 2 SA), and the projection of the elements' status into the future (Level 3 SA).

Level 1 SA is the most straightforward and is concerned with what elements are present, where they are located, and how fast they are moving. This level is the platform for the other two. By itself, it appears to be of minor interest. A novice individual who had perfect recall of the elements, but lacked an understanding of the patterns and implications, is not considered to have good SA. Most of the research on SA has centered on this first level primarily because it is the easiest to investigate and measure. However, without recourse to the other two levels, Level 1 SA is difficult to bound. Potentially, an infinite number of objects in an environment may be identified and characterized. The inclusion of all objects may not be necessary when measuring accuracy of SA, but without recourse to Level 2 SA it is difficult to determine which objects are relevant.

There is also the issue of how to represent context here. Context is not simply the inclusion of more elements; it is the framework for understanding the elements, and that only comes into play in Level 2 SA. That is why Level 2 SA is so critical: to allow us to emphasize context, not to limit it. Measures of Level 1 SA may be misleading if they suggest that SA is only the sum of the elements that are correctly recalled. For a pilot accurate knowledge of altitude is one of the aspects of SA. If a pilot's estimate is wrong then we can claim that the SA is weak. However, a commercial pilot flying at 32,000 feet can be off by 100 feet without any dire consequences, whereas an error of 100 feet is critical for a fighter pilot flying nap-of-the-earth or a pilot preparing to land on an aircraft carrier. What counts as an error is a function of the task being performed.

We see the same process at work when we look at weather forecasters (Pliske, Klinger, Hutton, Crandall, Knight, & Klein, 1997). There are no standard features to sample in order to understand the weather. Rather, the skilled forecasters are actively searching for ways to follow the trajectories of different fronts as they interact with terrain features. The area to be searched more intently, the grain of analysis (e.g., 500 millibars, 200 millibars) is a function of what the forecasters are seeking. Sometimes, they need the finer resolution offered by computer systems that tends to smooth the areas of instability that must be spotlighted. Their SA determines how they will search for the Level 1 SA elements, not vice versa.

The second level is where meaning enters. This is where diagnoses are made (e.g., Endsley & Robertson, 1996) and patterns detected. To achieve Level 2 SA, a person must synthesize a diverse mixture of events. Level 2 SA may be considered as a state or a process. The advantage of considering Level 2 SA as a state is that it leads to attempts to measure the SA for adequacy. The disadvantage is that such measures may not be terribly meaningful. In this chapter I consider Level 2 SA a process of sense-making (e.g., Weick, 1995). Furthermore, by considering Level 2 SA as a process, it can also be linked to attention management as O'Hare (1997) suggested.

If Level 2 SA is concerned with meaning, it may be helpful to consider the suggestion of Olson (1970), who used an information theoretic perspective to define meaning with regard to alternatives. A general's advice to troops to "win the battle" has little or no meaning because there are no relevant alternatives. In contrast, the general's advice to "split the enemy's forces" does have meaning because relevant alternatives do exist. Within this framework, it is valuable to study Level 2 SA as the categorization of a situation from a finite set of potential alternatives. Often, these alternatives are prototypes formed through experience that enable a decision maker to recognize the pattern and judge it as a familiar or typical class of events. In this process, recognition of the prototype carries with it implications for the judgments and decisions that must be made. Similarly, functional, as opposed to structural, categories should be considered. A structural category is synthesized from the elements of the situation. However, if one of the primary purposes of SA is to guide decision making, then relevant categories of SA are based on alternative modes of responding. For example, in research with fireground commanders we found that they were not using categories of structures (e.g., single family residence, apartment building, factory). Instead, the categories they offered were functional: a search and rescue fire, an aggressive interior attack fire, or a fire to be prevented from spreading to other structures. The categories they used were based on the situation and their reactions to it, not on the physical features of the buildings. This suggests that while studying SA researchers should not be limited to considering only the environmental characteristics; the perspective and purposes of the observer must also be examined. That is the reason for defining SA as our perception of the responses that the situation affords us.

Another aspect of Level 2 SA is the detection of leverage points in a situation. In their examination of problem solving, Klein and Wolf (1998) defined leverage points as the opportunities for making dramatic changes in a situation. For a battle commander, the leverage points are the opportunities to seize control of the battlefield. For a weather forecaster seeking to make an accurate forecast, the leverage points are the unsettled areas of the atmosphere where major shifts are likely to occur. For a chess master, the leverage points are the ways of bringing pressure on the adversary to turn the game around. These leverage points are a feature of the environment as well as a function of the expertise of the problem solver. A novice bridge player who doesn't understand how to finesse most likely will not see opportunities to make finesses—the detection of leverage points depends on functional categories for lines of responding. A potential Level 2 SA measurement method is to use the leverage points as a means of capturing the way an observer is understanding the dynamics of the situation. If an observer is communicating his or her SA to another, researchers hypothesize that critical aspects of the communication are about the leverage points.

Although the third level of Endsley's SA definition is concerned with projecting forward into the future, there exist situations in which the projection is backward into the past. For example, a medical diagnostician or a troubleshooter attempts to explain a sequence of events in order to understand the present more fully. Klein and Crandall (1995) described the process of mental simulation whereby stories are constructed to generate predictions of the future and explanations of the past. Klein and Crandall hypothesized that some degree of expertise is needed to construct mental simulations. Klein and Peio (1989) found that skilled chess players were significantly more accurate in predicting move sequences in chess games than were less-skilled players. This ability should be particularly amenable to measurement.

## SITUATION AWARENESS AND THE
## RECOGNITION-PRIMED DECISION MODEL

It may be fruitful to study SA in the context of decision-making incidents (both actual and simulated). Otherwise, everything that a person is or could be aware of may be included in SA. There is no convenient rule to provide guidance as to when the analysis should be terminated. Likewise, there are no basic "elements" that permit us to objectively define the contents of SA. This is especially true of research within a rich context; the context affects the way the aspects of SA are defined. These complications may be avoided by studying SA in an environment with a restricted context, but this will reduce the ability to generalize the findings. For these reasons, it may be useful to study decisions and judgments, and then to examine SA as it contributes to these judgments and decisions. Instead of studying the question of **what**—what is the content of a person's SA, we can study the question of **how**—how the SA affects action. In doing so, we can identify some of the important aspects of SA—those that impact judgments and decisions.

Klein, Calderwood, and Clinton-Cirocco (1986) formulated a Recognition-Primed Decision (RPD) model of how people are able to make decisions in naturalistic settings without having to compare options. The key is that people use expertise to evaluate situations and recognize typical courses of action as the first ones to be considered. Expertise centers around SA. The most recent version of the RPD model (Klein, 1998) is presented in Fig. 3.1 and shows the three variations of the RPD model.

The first variation of the RPD model is the most simplistic: Decision makers use their experience to recognize a situation as familiar or typical and know what the typical reaction is. This is a decision point because other decision makers, with different experiential backgrounds, may carry out different actions. These alternative courses of action are reasonable and all may constitute "good decision making." Even though the decision maker may not actually consider any alternatives, experience is used to select one option instead of another.

The second variation of the RPD model covers cases where there exist alternative interpretations of the situation's nature. Here, the decision maker has difficulty in recognizing a situation as typical. It may be necessary to analyze the evidence to see if one hypothesis is supported more than others. It may be necessary to construct a story to see if each of the alternative explanations or diagnoses is consistent with the evidence.

The third variation of the RPD model describes how decision makers can evaluate a course of action without comparing it to others. The strategy here is to mentally simulate the course of action to see if it will work or whether it needs to be modified. Sometimes, there is no obvious way to save a flawed course of action and it is rejected; the decision maker then considers another option from the response repertoire for that situation.

Figure 3.1 shows that SA is central to all three versions of the RPD model. For the first variation, the recognition of the situation is sufficient to evoke a course of action. The second variation requires effort to determine how to interpret the situation in order to know how to proceed. In the third variation, the SA generates a course of action that is evaluated; sometimes the evaluation will identify aspects of the situation that result in a better understanding of the dynamics.

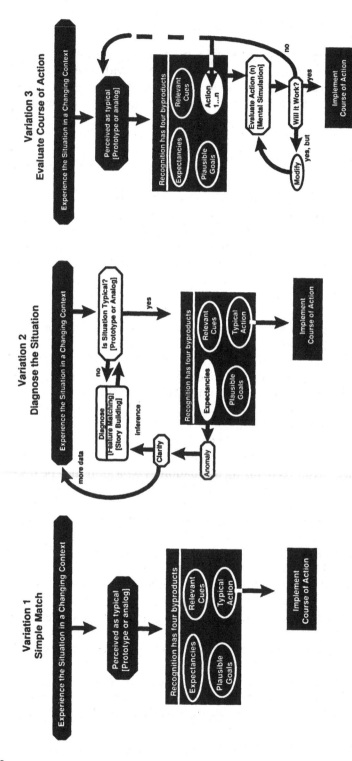

FIG. 3.1. Recognition-primed decision model.

56

The RPD model suggests four key aspects of SA. When a situation is recognized as familiar or typical, this recognition carries with it the following knowledge: the most critical cues, the relevant goals, the events to expect next, and the actions commonly taken. I do not claim that a decision maker determines each of these four components and then infers the nature of the situation; rather, the contrary is true. The recognition of the situation means that the decision maker understands these four aspects as part of that recognition. By knowing the critical cues, the experienced decision maker can focus attention where it will be most efficient and is not as overloaded with information as a less experienced person. Knowing which goals are reasonable allows the decision maker to engage in more efficient problem solving rather than struggling with unattainable or irrelevant goals. Having reasonably accurate expectations facilitates preparation for future events; additionally, the decision maker has a means of assessing the accuracy of the interpretation of situational dynamics. Consistently violated expectancies signal that perhaps the dynamics are different from those that were assumed. Knowing the typical actions frees the decision maker from the necessity of generating large numbers of options that must be compared and analyzed.[1]

The RPD model is consistent with Endsley's SA model. Although her Level 1 SA is not a part of the RPD model, it can be interpreted in Fig. 3.1 as the totality of the events in the environment that could be experienced. Level 2 SA would map on more directly as the recognition of the situation itself, particularly as it leads to a determination of the most important cues, the relevant goals, and the reasonable actions. The concept of leverage points enters here as part of the decision maker's understanding of the dynamics of the situation. Endsley's Level 3 SA, the projection forward into the future, is represented within the RPD model as the expectancies generated once a situation is recognized as typical. Therefore, the processes described by Endsley appear to be relevant for describing some aspects of decision making.

**Critical Decision Method**

This background on the connection between SA and decision making suggests that one way to study and measure SA is within the context of inci-

---

[1]All of these aspects reflect top–down and bottom–up processes. The situation, as it develops over time, generates expectancies but these are also conditioned by the interpretation of the decision maker. The critical cues are a joint product of changes in the environment and the attentional and perceptual skills of the decision maker. The goals are a function of opportunities in the situation and the motivations of the decision maker. The typical reactions are affordances in the situation that depend on the response repertoire of the decision maker.

dents that involve decision making. I have concerns about measuring SA outside the context of task performance. Because SA is so central to naturalistic decision making, it seems reasonable to examine it as it forms and shifts during a decision-making episode.

The classical way to study episodes was described as the **critical incident** method (Flanagan, 1954; see also Shattuck, 1995). Flanagan intended to describe jobs by representing typical events and describing abstract task listings. Klein, Calderwood, and MacGregor (1989) and Hoffman, Crandall and Shadbolt (1998) described a method of examining the critical decisions performed in the context of a task. The Critical Decision Method (CDM) aims to capture the nature of the decision maker's understanding during the course of an incident. This method emphasizes nonroutine incidents and their nonroutine aspects because they usually carry the most information about how expertise is used.

The consistencies between Endsley's work and the RPD model do not suggest that one model can be subsumed under the other, but rather that Endsley's framework can be extended to cover decision making. Her ability to drive dependent variables for assessing SA could have useful parallels in the assessment of decision-making skills.

We can distinguish two ways of studying incidents: using **retrospective memory** for actual incidents and using **process-tracing** methods for simulated or ongoing incidents. Process-tracing methods include think-aloud protocols and other ways of gaining insight into the thought processes linked to task performance.

*Retrospective Accounts.*   The most typical use of the CDM is to study previous incidents. The CDM interview identifies interesting incidents and then probes the decision-making activities during these incidents.

Klein, Calderwood, and Clinton-Cirocco (1986) developed the CDM to understand how fireground commanders made difficult decisions. Because it was impractical to observe fireground commanders and gather verbal protocols during actual incidents, they decided to use retrospective accounts of challenging incidents. This study found that the fireground commanders rarely compared options as to which course of action to pursue. Instead, the fireground commanders were able to use their experience to recognize the typical course of action in a situation. Therefore, most of the information gathered during the study focused on how the fireground commanders understood the situation and how their SA shifted as the conditions changed. At times, the shift was an elaboration of SA as they were able to reduce uncertainty about the details. On occasion the SA shifted radically as the fireground commanders rejected one interpretation in favor of another and at other times the situation changed in accord with their actions.

Table 3.1 shows a record of SA shifts during an actual incident in which a truck carrying petroleum overturned on a highway and caught fire. The markers showing how the incident evolved are represented as the successive interpretations—the shifts in SA. Initially, the fireground commander held one interpretation. Because the situation was dynamic, each SA shift either deepened the existing understanding (referred to in Table 3.1 as a SA elaboration) or it forced a revision of that understanding (referred to in Table 3.1 as a SA shift).

Representing the decision making described by the fireground commanders in this study was difficult until the interview results were analyzed in terms of the shifts in SA. Once the findings were framed using successive SA shifts, it became evident that the adoption of a course of action was usually a straightforward reaction to the interpretation of the situation. The fireground commanders did not struggle with generating and comparing alternative actions. The central role of SA in skilled decision making has been replicated many times in our studies. It is now understood that recognition of the situation is to know "what to do."

TABLE 3.1
SA Record From Tanker Incident

| | |
|---|---|
| **SA-1** | |
| Cues/knowledge | overturned truck on highway; ruptured fuel tank; engulfed in flames; intense heat (highway signs melted); another truck 50 feet away; citizen rescuing driver |
| Expectations | potential explosion; life hazard |
| Goals | complete the rescue; extinguish fire; block traffic |
| Decision point-1 | aid in driver rescue |
| Decision point-2 | call for additional units, rescue unit, police, foam |
| **SA-2 (elaboration)** | |
| Cues/knowledge | additional equipment arrives; fire more involved |
| Expectations | chance of explosion is less |
| Goals | protect firefighters; gain needed resources (water) |
| Decision point-3 | set up covering streams |
| Decision point-4 | hook up pumper hoses |
| **SA-3 (elaboration)** | |
| Cues/knowledge | protective streams functioning; foam trucks arrive |
| Expectations | fire banking down |
| Goals | optimal truck placement; set up foam operations |
| Decision point-5 | directed truck and foam placement (using angles, impact, and wind direction cues) |
| **SA-4 (shift)** | |
| Cues/knowledge | storm drain behind operations blows up; determines jet fuel has leaked into storm sewers |
| Expectations | fire will span out of control |
| Goals | check fire and prevent spread |
| Decision point-6 | call second alarm |

The CDM has been further developed (e.g., Klein, Calderwood, & MacGregor, 1989; Klein, 1993; Hoffman, Crandall, & Shadbolt, 1998). Currently the method centers around a difficult judgment or decision that requires expertise. The method's objective is to uncover the expertise, particularly the cues and patterns, used to identify the dynamics of the situation.

The process requires the interviewer to conduct several "sweeps" through the incident once the subject-matter expert (SME) recalls a demanding event. The first sweep provides the SME's initial account of the incident. The second sweep is a more deliberate recounting of the incident that includes putting it on a time scale and identifying decision points for subsequent probing. In the third sweep, the interviewer uses a variety of probes to examine the judgments and decisions that appear to depend most strongly on expertise. These probes include asking about the following: salient cues, relevant knowledge, goals being pursued, overall understanding of the situation, potential courses of action, and hypothetical issues about the effect on the decision if key features of the situation were different. A fourth sweep is sometimes used to probe how a person with much less experience would have sized up the situation at key points and to identify those cues or patterns an inexperienced decision maker might miss or misinterpret.

The CDM is useful for learning about SA and decision making in context. Researchers can discover important considerations and factors not usually covered in manuals or in global accounts of the task. In a study of anti-air warfare decision making by U.S. Navy officers (Kaempf, Wolf, Thordsen, & Klein, 1992), virtually every one of the 14 incidents studied included a key variable that was context-specific and is typically **not** included in simulations. These variables ranged from judgments of the other officers' competence to the difficulties of waking commanding officers from sleep.

Kaempf et al. (1992) prepared detailed diagrams of the successive stages of SA for the naval officers and the factors that contributed to each SA change. Table 3.2 presents a portion of the analysis of the SA evolution during one of the incidents involving an AEGIS cruiser. Although there are additional categories along the top, the basic concept is the same as in Table 3.1—to document the development of SA as an incident unfolded.

The incident involved Iranian F-4 aircraft that were harassing an AEGIS cruiser. Initially, the aircraft circled closer and closer to the cruiser and then they turned on search and fire control radars. Each new step changed the interpretation of what the F-4s intended to do and, therefore, changed the functional understanding of the situation as the commanding officer shifted his response posture. In the beginning, the commanding officer was intent on defending an attack, but was also careful to avoid any

TABLE 3.2
Successive Stages of SA

| Time | Cues | Factors | SA |
|------|------|---------|-----|
| T | New tracks on the display | Recent increase in Iranian F-4 flights | Two aircraft have launched from nearby Iranian airport |
| | Origination in Iran | Recent hostilities with Iran | They may be a threat |
| T + 1 | Circular course of aircraft | Military aircraft typically do *not* circle the airfield | These aircraft are not flying the usual route |
| | Proximity to U.S. ships | Cruiser is escorting a flagship | They may have designs on the flagship |
| T + 2 | Radar emissions from lead aircraft | Weapons range of Iranian F-4's | Aircraft may be preparing an attack |
| | Circular orbit is expanding | Rules of engagement | Trajectory is bringing aircraft closer to the U.S. ships |
| | | | Situation is getting more threatening |

provocation. Once the F-4s activated their fire control radar systems, the commanding officer was permitted by the rules of engagement to launch missiles against the aircraft. The shift in understanding was due to the increased threat from the F-4s as well as the change in defensive posture.

Figure 3.2 shows another way to represent the successive stages of SA over the entire F-4 incident.

In studying the decision making of AEGIS commanders, Kaempf et al. (1992) confirmed earlier findings that less effort was given to selecting courses of action than to understanding the nature of the situation. They also discovered that the mental simulation process used to evaluate courses of action was also used to explain the situation. The decision makers were often trying to build a story that encompassed the events and were using the plausibility of the story to gauge the adequacy of their explanation.

For example, the incident shown in Table 3.2 and Fig. 3.2 involved story construction. Because the commander was unable to construct a plausible story to explain how the Iranian F-4 pilots might be planning to attack his ship in broad daylight and under obvious scrutiny, he found it easier to interpret the F-4s' actions as harassment rather than an attack. As a result, he decided not to engage the F-4s with his missiles despite their provocation. He maintained preparation for self-defense, but did not escalate the incident. His judgment was vindicated—the F-4s eventually withdrew. Information gathered from incidents such as this indicated the

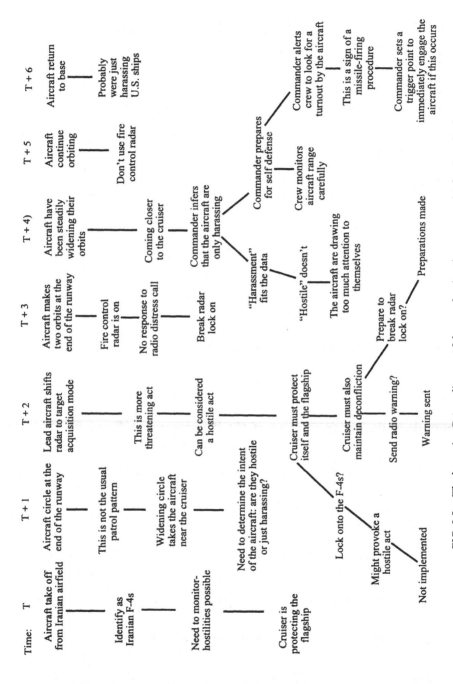

FIG. 3.2.   The harassing F-4s: a diagram of the way the situation was perceived as it evolved.

62

importance of mental simulation for generating explanations and essentially for creating SA. These findings resulted in the addition of Variation 2 to the RPD model, the diagnosis of the situation (Fig. 3.1).

Each of the incidents studied by Kaempf et al. (1992) included contextual factors that are rarely covered in laboratory simulations. Examples are: variations in the credibility of individuals performing different functions that affected how assignments were made; the reputations of various ships that affected the credibility of their reports; and the opportunities to ignore unpopular orders because communications difficulties allowed the crew members to claim they had not received the orders.

Another example of the use of CDM comes from the work of Crandall and Getchell-Reiter (1993). They described how the method was used to uncover the cues and patterns used by nurses in a Neonatal Intensive Care Unit (NICU) to diagnose sepsis. In this study, NICU nurses were interviewed about difficult and nonroutine incidents and nine incidents involving sepsis were collected. (Sepsis is systemic infection and is extremely dangerous for newborns.)

The NICU nurses appeared to be able to detect which newborns were just starting to develop sepsis (often in advance of laboratory tests); however, the nurses were unable to articulate the basis for their judgments. The CDM interviews identified a large range of critical cues and patterns that the nurses were using to spot sepsis in its very early stages. At the time of the study, many of the cues revealed by the CDM had not been published in the medical or nursing literature. Table 3.3 presents the critical cue inventory that was developed from the CDM interviews.

After reviewing this inventory, the NICU nurses confirmed the accuracy of the entries. It was suggested that training would be more successful if the cues were embedded in the context of the stories themselves. Figure 3.3 provides an example of an annotated interview. This short selection illustrates the story that the nurse was telling when describing the incident and the types of expertise filled in through the CDM probes.

*Process Tracing.*   Research has shown that there are some situations and tasks for which the CDM is not well suited. These boundary conditions include those tasks that are highly repetitive, lack clear feedback, or do not result in dramatic or memorable incidents. When researchers encounter such tasks, they rely on alternative methods such as process-tracing.

Process-tracing methods usually are employed with simulated incidents, although they can also be used with actual performance. For example, Kaempf, Klinger, and Wolf (1994) studied baggage screeners by observing them at work in airports and questioning them about the judgments they were making in selecting items to scrutinize more closely.

TABLE 3.3
CDM Critical Cue Inventory

| | Indicator Found In | |
| | --- | --- |
| Indicator | Nursing/Medical Literature | Nurses' Accounts of Septic Infants |
| Unstable temperature | Yes | Yes |
| Hypothermia | Yes | Yes |
| Elevated temperature | Yes | No |
| Feeding abnormality | Yes | Yes |
| Abdominal distension | Yes | Yes |
| Increased residuals | Yes | Yes |
| Vomiting | Yes | No |
| Lethargy | Yes | Yes |
| Irritability | Yes | No |
| Color changes | Yes | Yes |
| Respiratory distress | Yes | Yes |
| Apnea | Yes | Yes |
| Cyanosis | Yes | No |
| Tachypnea | Yes | No |
| Seizures | Yes | No |
| Purpura | Yes | No |
| Jaundice | Yes | No |
| Bradycardia | No | Yes |
| Unresponsive | No | Yes |
| Poor muscle tone | No | Yes |
| Perfusion/mottling | No | Yes |
| Edema | No | Yes |
| "Sick" eyes | No | Yes |
| Clotting problems | No | Yes |

Because process tracing is often used with simulated incidents it offers the potential for greater control of the research. The incident can be developed to tap into the variables of greatest interest. Both experts and novices can be studied in the same tasks to compare their SA. Unobtrusive measures and interventions can be incorporated (e.g., requiring the participants to take actions in order to obtain certain types of information and testing workload by presenting secondary tasks). The simulation can be halted to allow more detailed questioning. Pairing simulations and process-tracing strategies offers a wide variety of data-gathering opportunities.

Process tracing has been found to be highly informative about SA as it relates to judgment and decision making in simulated tasks. As with the CDM, the same types of probes are used (e.g., asking about the cues that were most salient for sizing up the situation and the types of misinterpretations that novices might make). By asking participants what alternative actions were possible, the features of the situation that precluded these ac-

## CASE ACCOUNT    ELICITED KNOWLEDGE

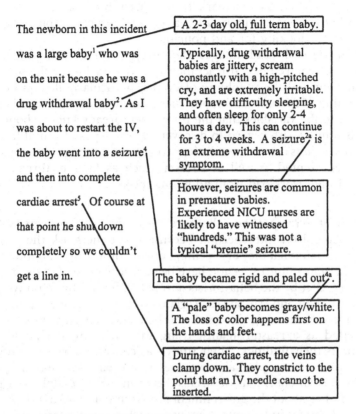

The newborn in this incident

was a large baby[1] who was

on the unit because he was a

drug withdrawal baby[2]. As I

was about to restart the IV,

the baby went into a seizure[4]

and then into complete

cardiac arrest[5]. Of course at

that point he shut down

completely so we couldn't

get a line in.

A 2-3 day old, full term baby.

Typically, drug withdrawal babies are jittery, scream constantly with a high-pitched cry, and are extremely irritable. They have difficulty sleeping, and often sleep for only 2-4 hours a day. This can continue for 3 to 4 weeks. A seizure[2a] is an extreme withdrawal symptom.

However, seizures are common in premature babies. Experienced NICU nurses are likely to have witnessed "hundreds." This was not a typical "premie" seizure.

The baby became rigid and paled out[4a].

A "pale" baby becomes gray/white. The loss of color happens first on the hands and feet.

During cardiac arrest, the veins clamp down. They constrict to the point that an IV needle cannot be inserted.

FIG. 3.3.    NICU nurse interview.

tions may be elicited. Asking what would have to happen to make the alternatives more likely to be adopted reveals more about the interplay between different factors. If a person is asked whether a certain course of action makes sense and responds with "It depends," that is a natural opening to explore the features of the task and situation upon which it depends. Wolf, Hutton, Miller, and Klein (1995) used this type of process tracing to examine the SA of Patriot missile battery officers.

One very powerful pairing of simulations and process tracing is to contrast the SA of experts to novices when faced with the same circumstances. DeGroot (1946/1965) made these contrasts using chess positions and uncovered a number of interesting differences between strong and weak players in their assessments of the positions' dynamics. Charness (1989) did the same using bridge hands. It is worth noting that these simulations involved relatively straightforward stimuli—pictures of chess boards and of bridge hands. Wolf et al. (1995) used simulations built from paper maps and transparent overlays. In a market research study of consumer decision making, photographs of supermarket aisles were used to probe what the consumers were looking for and noticing.

There are several disadvantages when using process-tracing methods. Effort may be required to construct the simulation including the scenario of the incident and the unfolding of events, particularly if high-fidelity representation of dynamic and subtle cues is needed. Because simulations are limited to the variables that the researcher already knows about, the method is limited with regard to discovering new types of causal factors. Finally, simulations can be unrealistic, particularly with regard to the ways cues are represented as well as stressors such as fatigue, threat, high stakes, and high workload from additional tasks.

*The Use of Introspection.* Both retrospective and process-tracing methods depend on introspection because researchers ask the participants to articulate what they are seeing, noticing, inferring, and interpreting. The dangers of introspection have been clearly presented during the past century, first by the behaviorists and most recently by cognitive psychologists. Nisbett and Wilson (1977) claimed that people inaccurately report their own reasoning processes. Nisbett and Wilson's study is frequently cited as a reason against using introspection at all and this misinterprets their findings. Their study was concerned with identifying the inaccuracies in peoples' ability to describe their own thinking processes. There is an enormous difference between asking people to report what they notice about the environmental surround and asking them why they formed certain judgments. Morever, much of the research reviewed and critiqued by Nisbett and Wilson concerned social judgments (e.g., the thought processes involved in evaluating the trustworthiness of a peer vs. a stranger). This research focus is quite different from people's reports of what they notice and how they act to accomplish tasks in a work setting.

Clearly, researchers must be aware of the problems with uncritical acceptance of self-reports, particularly reports about the reasons why people performed certain actions or made certain judgments. Research done using CDM avoids questions about WHY judgments and decisions were made. Instead, the interview probes emphasize the cues and patterns that were noticed and do not require introspection about the inferential processes themselves.

Despite these cautions, researchers should not feel that they have to abandon self-reports. Howard (1994) argued that self-reports are important and useful forms of information. He noted that even behavioral measures of performance are susceptible to biases and distortion. The selection of stimulus materials and dependent variables can limit the generalization of findings. Some studies (e.g., Cole, Howard, & Maxwell, 1981; Cole, Lazarick, & Howard, 1987; and Howard, Conway, & Maxwell, 1985) found that self-report data have **higher** validity than do behavioral measures. Researchers must be careful to understand the limitations of any methods used and should document methods to allow the replication of their findings by others. Nevertheless, self-reports are an important and useful form of data that can have higher validity and utility for SA researchers than behavioral measures.

I emphasize this point for two reasons. First, it is important to defend the use of self-report methods as a way of learning about SA. The other concern is that researchers are too quick to adopt empirical paradigms that emphasize quantitative data for the study of SA and as a result learn little about it. Some researchers may be uncomfortable or unfamiliar with analyzing self-report data; they then adopt paradigms to study phenomena such as SA by defining categories and counting frequencies. The strength of such an approach is to provide rigor. The weakness is to spend a great deal of energy for relatively little return.

The most rigorous way to study SA is to confine the investigation to Endsley's Level 1 SA—the features of the environment that a person can recall; however, Endsley's Levels 2 and 3 SA may be more informative. These levels deal with the decision maker's interpretation of the situation, not the quantity of cues recalled correctly. Furthermore, as researchers better understand Levels 2 and 3 SA, they may be able to conduct better studies of Level 1 SA. For example, if a person asserts that an aircraft is at an altitude of 9,500 feet when it is really at 8,500 feet, does this count as a correct response or an error? Although criteria may be defined, perhaps in terms of standard deviations, what really matters is whether the discrepancy leads to functionally different reactions. That takes us to Level 2 SA where the decision maker interprets the meaning of the situation as a whole and interprets the meaning of the features. The functional categories can help establish parameters for Level 1 SA. An exhaustive analysis of Level 1 SA data will not enable us to understand the inferences at Level 2 SA. Rather, an understanding of Levels 2 and 3 SA facilitates comprehension of the Level 1 SA data.

## SUMMARY AND CONCLUSIONS

One way to study SA is through critical incidents. Critical incidents can be presented in the form of realistic simulations, observed in a workplace setting, or elicited and probed through interviews that generate retrospective accounts. One advantage of studying retrospective accounts of actual incidents is that they typically include details and patterns of observation that reflect various aspects of context rarely incorporated into laboratory simulations.

Researchers can use Cognitive Task Analysis (CTA) methods to examine SA in critical incidents. CTA methods have been developed to elicit information about the way a person perceives a situation, comprehends the situation's components, and makes decisions and plans. CTA includes elicitation and analysis of the data and the representation of the findings. Many variations of CTA rely on interview techniques. For a study of retrospective accounts of interviews, the interview might cover the successive stages of SA as the incident unfolded, the interpretation of each stage, and alternative interpretations that were considered but rejected. Crandall and Getchell-Reiter (1993) found that skilled NICU nurses were able to diagnose cases of sepsis using a wide variety of cues. Many of these cues had not previously been listed in medical or nursing literature. If the study had taken the literature as a point of departure, these new cues and patterns would not have been discovered. Only by looking at critical incidents was it possible to detect their use.

The use of critical incidents allows researchers to perform qualitative as well as quantitative studies. The danger in overemphasizing quantitative studies is that these studies assume the meaning of a situation can be decomposed into basic elements and that these elements can be adequately defined and measured. However, in context-rich settings there may not be any basic elements. The nature of the elements depends on the interpretation given to the situation. Therefore, investigations must begin with the overall interpretations made by the participants, rather than trying to induce the meaning from the basic elements. CTA of actual incidents can be very helpful in understanding how decision makers comprehend situations. One hypotheses is that decision makers do not categorize situations by structural qualities—the features of the environment; instead, they rely on functional categories that are the types of reactions afforded by the situation. If researchers place more emphasis on qualitative aspects of SA, less time will be spent asking **what** are the contents of SA. The focus will be on **how** the situation's interpretation affects judgments and decisions.

Klein, Calderwood, and Clinton-Cirocco (1986) found that the SA of fireground commanders changed as the situational dynamics changed. The investigation of how SA is shifted, and how this affects decision making, depends on qualitative research rather than on a specification of features accurately recalled. Kaempf et al. (1992) used a qualitative study of critical incidents to identify a story-building strategy used by AEGIS commanders to diagnose complex situations. Merely measuring the cues recalled, or the confidence placed in the SA, would not have identified the strategies used to formulate an interpretation of what was happening as a situation unfolded.

SA is not composed of elements. SA defines the elements. It is possible to interpret the three stages of Endsley's model as pinning down the data, deriving inferences, and then making predictions. Although Endsley does not state this prescription, others might interpret her work in this way. Such an interpretation can be misleading. I believe there are no basic, context-free elements of a situation. The elements that are important at a given point in time depend on the task context and the interpretation the person already brings. Studies of skilled decision makers in a number of different domains indicate that the way they interpret situations is based on their ability to spot violations of expectancies, inconsistencies, leverage points, sudden changes, and areas of instability. These are not context-free data elements. Considerable expertise is needed to detect these cues. A military commander can show insight (e.g., the coup d'oeil) when peering at a map and quickly noticing the key terrain to be attacked. A skilled weather forecaster shows the same phenomenon when looking at a weather map and spotting the area of instability. These interpretations guide the types of data that will need to be monitored.

The development of SA is an active process of guided information seeking, rather than a passive receipt and storage of details. We may assume that experts work with more data than novices—the information use hypothesis—but Shanteau (1995) obtained data that reject this hypothesis. Experts don't have more data. They have the ability to identify which cues to use, whereas novices tend to place too much emphasis on cues that are less diagnostic.

Skilled decision makers appear to use the same line of questions that Wiener (1989) described for pilots wrestling with clumsy automation: What is it doing; why is it doing that; what is it going to do next; and what can I do about it.

When decision makers have a sense of the ways in which they can react, they have achieved SA. Their SA is their perception of the reactions they can make to a set of changing events.

Researchers may feel impelled to conduct controlled empirical studies of phenomena and to test hypotheses. However, there may be value in first trying to learn more about the nature of a complex phenomenon such as SA. Then it becomes easier to comprehend the process of understanding situations, shifting that understanding to reflect new events and disconfirmations, and seeking new information and managing attention. Qualitative research using critical incidents can help us gain a richer perspective on SA and thereby to improve our research questions.

## ACKNOWLEDGMENT

I would like to thank Beth Crandall for her helpful review and recommendations.

## REFERENCES

Charness, N. (1989). Expertise in chess and bridge. In D. Klahr & K. Kotovsky (Eds.), *Complex information processing: The impact of Herbert A. Simon* (pp. 183–208). Hillsdale, NJ: Lawrence Erlbaum Associates.

Chi, M. T. H., Feltovich, P. J., & Glaser, R. (1981). Categorization and representation of physics problems by experts and novices. *Cognitive Science, 5*, 121–152.

Cole, D. A., Howard, G. S., & Maxwell, S. E. (1981). The effects of mono versus multiple operationalization in construct validation efforts. *Journal of Consulting and Clinical Psychology, 49*, 395–405.

Cole, D. A., Lazarick, D. M., & Howard, G. S. (1987). Construct validity and the relation between depression and social skill. *Journal of Counseling Psychology, 34*, 315–321.

Crandall, B., & Getchell-Reiter, K. (1993). Critical decision method: A technique for eliciting concrete assessment indicators from the "intuition" of NICU nurses. *Advances in Nursing Sciences, 16*(1), 42–51.

DeGroot, A. D. (1946/1965). *Thought and choice in chess.* New York: Mouton.

Endsley, M. R. (1995). Towards a theory of situation awareness. *Human Factors, 37*(1), 32–64.

Endsley, M. R., & Robertson, M. M. (1996). Team situation awareness in aviation maintenance. In *Proceedings of the Human Factors and Ergonomics Society 40th Annual Meeting* (Vol. 2, pp. 1077–1081). Santa Monica, CA: HFES.

Flanagan, J. C. (1954). The critical incident technique. *Psychological Bulletin, 51*, 327–358.

Hoffman, R. R., Crandall, B. E., & Shadbolt, N. R. (1998). A case study in cognitive task analysis methodology: The critical decision method for the elicitation of expert knowledge. *Human Factors & Ergonomics Society, 40*(2), 254–276.

Howard, G. S. (1994). Why do people say nasty things about self-reports? *Journal of Organizational Behavior, 15*, 399–404.

Howard, G. S., Conway, C. G., & Maxwell, S. F. (1985). Construct validity of measures of college teaching effectiveness. *Journal of Educational Psychology, 77*, 187–196.

Kaempf, G. L., Klinger, D., & Wolf, S. (1994). *Development of decision-centered interventions for airport security checkpoints* (Contract No. DTRS–57–93–C–00129). Fairborn, OH: Klein Associates Inc.

Kaempf, G. L., Wolf, S., Thordsen, M. L., & Klein, G. (1992). *Decision making in the AEGIS combat information center* (Contract No. N66001–90–C–6023). Fairborn, OH: Klein Associates Inc.

Klein, G. (1998). *Sources of power: How people make decisions.* Cambridge, MA: MIT Press.

Klein, G. (1993). *Naturalistic decision making—Implications for design.* Dayton, OH: CSERIAC.

Klein, G. A., Calderwood, R., & Clinton-Cirocco, A. (1986). Rapid decision making on the fire ground. In *Proceedings of the Human Factors Society 30th Annual Meeting* (Vol. 1, pp. 576–580). Santa Monica, CA: Human Factors Society.

Klein, G. A., Calderwood, R., & MacGregor, D. (1989). Critical decision method for eliciting knowledge. *IEEE Transactions on Systems, Man, and Cybernetics, 19*(3), 462–472.

Klein, G. A., & Crandall, B. W. (1995). The role of mental simulation in naturalistic decision making. In P. Hancock, J. Flach, J. Caird, & K. Vicente (Eds.), *Local applications of the ecological approach to human-machine systems, Volume 2* (pp. 324–358). Hillsdale, NJ: Lawrence Erlbaum Associates.

Klein, G. A., & Peio, K. J. (1989). The use of a prediction paradigm to evaluate proficient decision making. *American Journal of Psychology, 102*(3), 321–331.

Klein, G. A., & Wolf, S. (1998). The role of leverage points in option generation. *IEEE Transactions on Systems, Man, and Cybernetics, 28*(1), 157–160.

Larkin, J., McDermott, D. P., & Simon, H. A. (1980, June). Expert and novice performance in solving physics problems. *Science, 208*, 1335–1342.

Nisbett, R. E., & Wilson, T. D. (1977). Telling more than we can know: Verbal reports on mental processes. *Psychological Review, 84*, 231–159.

O'Hare D. (1997). Cognitive ability determinants of elite pilot performance. *Human Factors, 39*(4), 540–552.

Olson, D. (1970). Language and thought: Aspects of a cognitive theory of semantics. *Psychological Review, 77*(4), 257–273.

Pliske, R., Klinger, D., Hutton, R., Crandall, B., Knight, B., & Klein, G. (1997). *Understanding skilled weather forecasting: Implications for training and the design of forecasting tools* (Technical Report No. AL/HR–CR–1997–0003). Fairborn, OH: Klein Associates Inc.

Shanteau, J. (1995). Expert judgment and financial decision making. In B. Green (Ed.), Risk behavior and risk management. *Proceedings of the First International Stockholm Seminar on Risk Behavior and Risk Management* Stockholm, Sweden.

Shattuck, L. (1995). *Communication of intent in distributed supervisory control systems.* Doctoral dissertation in preparation. Columbus: The Ohio State University.

Weick, K. (1995). *Sensemaking in organizations.* Thousand Oaks, CA: Sage.

Wiener, E. L. (1989). *Human factors of advanced technology ("glass cockpit") transport aircraft* (NASA Report 177528). Moffett Field, CA: Ames Research Center.

Wolf, S., Hutton, R., Miller, T., & Klein, G. (1995). *Identification of the decision requirements for air defense planning* (PO #TC392277). Fairborn, OH: Klein Associates Inc.

# Chapter 4

# Post Hoc Assessment of Situation Awareness in Air Traffic Control Incidents and Major Aircraft Accidents

Mark D. Rodgers
*Federal Aviation Administration*

Richard H. Mogford
*National Aeronautics & Space Administration*

Barry Strauch
*National Transportation Safety Board*

Post hoc assessment of the relevance of *situation awareness* (SA) to incident and accident causation has become an important area of study for human factors engineers, system engineers, and equipment designers. It is usually impractical or unsafe to measure operator SA in real settings, so researchers must rely on post hoc data. Researchers can draw conclusions regarding the SA of controllers or pilots by using data that originate from a database, videotape, analysis of accident data, or other sources. However, determining the relevance of SA to the incident or accident situation may be difficult.

Occupations requiring interaction with complex, dynamic systems make the post hoc assessment of SA problematic due to the lack of available information that directly relates to the cognitive processes of relevant personnel. As a result, it may be difficult to utilize the results to prevent future incidents and accidents. Nonetheless, valuable information regarding SA can be obtained through investigating various data sources and may provide insight into potential life saving modifications of procedures, training methodologies, and system designs.

Differences in information requirements associated with various occupations make the assessment of SA domain-specific. For example, the examination of SA in air traffic control (ATC) incidents and major aircraft accidents has been initiated only recently. With the increasing focus on human factors during the post hoc review of these events, the importance of operator SA has been recognized. These investigations and studies offer insight into the procedures and methods available to researchers studying SA. Therefore, this chapter concentrates on the post hoc analysis of SA as it relates to relevant personnel in the occurrence of ATC incidents and major aircraft accidents.

ATC incident and major aircraft accident investigations have two primary objectives: to determine the cause of the incident or accident and to make recommendations to prevent a recurrence. To meet these objectives, investigators often attempt to determine the SA of the air traffic controller(s) involved in the incident, crewmembers piloting the accident aircraft, or other persons potentially relevant to the cause of the event. The processes involved in determining the cause of an incident or accident are similar. However, incident investigations lack the detail and amount of information typically collected during a major aircraft accident investigation.

The Federal Aviation Administration (FAA) documents ATC incidents that involve the loss of separation between two or more aircraft in Operational Error (OE) final reports. The National Transportation Safety Board (NTSB) investigates major aircraft accidents through the compilation and analysis of the information in the reports of the various accident investigation subgroups into a single final report, that is then made available to the public. Incident and accident investigation data from these two sources are assimilated into independent databases for subsequent analysis. This chapter describes the procedures associated with the collection of incident and accident data, issues and concerns associated with using databases for assessing SA, and post hoc analysis techniques suitable for assessing SA in ATC incident and major aircraft accident occurrence.

The chapter starts with a brief discussion of post hoc assessment and SA. Discussions of the methods used to collect information during incident and accident investigations are provided to illuminate the limitations of incident and accident data. Two exploratory studies of air traffic controller awareness of OE occurrence are presented as examples of the type of information that can be gleaned from the analysis of databases associated with ATC incidents, and to illustrate how that information was used to augment the current OE report form. The first study addresses OE severity and SA, the second study addresses OEs, Sector Characteristics, and SA. This is followed by a discussion of major aircraft accident investigation findings related to SA. The chapter closes with a review of the difficulties associated with applying post hoc findings related to SA to the prevention of incidents and accidents.

## POST HOC ASSESSMENT

Post hoc, or after the fact, techniques range from a statistical analysis of experimental data and various databases to the identification, retrieval, and analysis of selected information from an aircraft accident. Unfortunately, not all phenomena of potential value to the researcher are easily observable, measurable, or accessible through the application of post hoc methods. Much depends on the nature of the data, the design of the record keeping or data capture systems, and the procedures established to consolidate relevant data. A prior analysis of information required for effective post hoc studies may result in sufficient data being collected. However, important items may be overlooked and practical or economic considerations may constrain the resources available. As a result, the information that is accessible to the researcher after the incident or accident may not be comprehensive and may limit the possible analyses. In general, data that address cognitive task components, such as SA, are difficult to retrieve and have not been included in the procedures established for retrospective data analysis. Post hoc reviews must make the best possible use of the available data and provide feedback on missing information so that recording systems can be improved to better meet research needs.

## SA

Endsley (1995) defined SA as "the perception of the elements in the environment within a volume of time and space, the comprehension of their meaning, and the projection of their status in the near future" (p. 36). As applied to piloting an airplane, for example, SA refers to flight crew recognition and comprehension of the present and near term state of the aircraft systems and components and the present and near term airplane attitude and flight path. Because the loss of pilot SA of any of these elements can be potentially significant for the safety of flight, it can be assumed that the maintenance of SA is critical to operator performance in complex systems. A crucial factor in understanding SA in any environment rests on the elucidation of the elements in Endsley's definition. Each setting (i.e., flight deck, tower cab, terminal radar, en route radar, and maintenance operations) requires the identification of the underlying components of SA. For example, Endsley and Rodgers (1994) analyzed the SA information requirements for en route ATC under the current National Airspace System (NAS) concept of operations. This was conducted as a goal-directed task analysis in which the major goals, subgoals, decisions, and associated SA information requirements were delineated based on interviews with eight experienced air traffic controllers. A determination of the major SA information requirements for en route ATC was developed. This provided a basis for the creation of SA measures for ATC and a foundation for future system development.

The importance of SA to aviation safety derives largely from its influence on decision making (Klein, 1993; Orasanu, 1993). SA forms the critical inputs to, but is separate from, decision making that is the basis for all operator actions. Proper implementation of rules and procedures depends on the quality of the operator's SA. Even the best trained, most experienced controllers or pilots, using effective decision making techniques, can make poor decisions if they have incomplete or inaccurate SA. Conversely, an inexperienced operator may accurately understand what is occurring in the environment, yet not know the correct action to take. For this reason, it is important that SA be considered separately from decision making and performance.

Effective decision making is founded upon good situation assessment and the resulting SA. Even with the numerous comprehensive standard operating procedures, specific phraseology, and other rules and guidance, air transport pilots, air traffic controllers, and other participants in the aviation system routinely must make decisions that have profound effects on the safety of flight. Their ability to maintain accurate SA, therefore, is critical to the quality of the decisions that result and to the safety of the aviation system.

Repeatedly, accidents have resulted from flaws in the flight crew's decision making that was based on inadequate SA. For example, in 1994 a warning light illuminated in the cockpit of a regional airliner as the airplane was on final approach, at night and in marginal visibility conditions (NTSB, 1995b). Although this was a relatively benign occurrence, the captain misdiagnosed the warning light by interpreting it as an engine failure. Consequently, he failed to assign appropriate responsibility to the first officer, incorrectly decided to break off the approach and attempt a go-around, and improperly applied engine power. The aircraft crashed about five miles short of the runway.

Although this accident illustrates a rather extreme example of the potential consequences of incorrect SA, the fact remains that aviation safety relies on accurate SA on the part of all personnel critical to the operation of the airspace system including pilots, air traffic controllers, dispatchers, maintenance personnel, and flight attendants. Later in this chapter we will describe the methods used by researchers and aircraft accident investigators to assess the SA of relevant personnel in circumstances where SA deficiencies are believed to have played a part in the incident or accident.

## AIR TRAFFIC CONTROL INCIDENT
## INVESTIGATION

In the history of the FAA, no aircraft have collided while under positive control in en route airspace. However, such collisions have occurred in terminal and other radar equipped air traffic environments, and aircraft have violated prescribed separation minima and approached in close proximity. Violations of separation minima can occur as a result of a pilot deviation or an OE. An OE takes place when an air traffic controller allows less than applicable minimum separation criteria between aircraft (or an aircraft and an obstruction). Standards for separation minima are described in the ATC Handbook (7110.65F, and supplemental instructions). At flight levels between 29,000 ft and 45,000 ft, controllers at en route facilities are required to maintain either a 2,000 ft vertical separation or a 5 mile horizontal separation between aircraft. At flight levels below 29,000 ft, with aircraft being operated under instrument flight rules (IFR), en route controllers are required to maintain either a 1,000 ft vertical separation or a 5 mile horizontal separation.

Immediately after the detection of an OE, a detailed investigation is conducted in an attempt to fully describe the events associated with the error's occurrence. This includes removing the controller from the operating position, obtaining a statement from each of the involved specialists, gathering the relevant data (voice and computer tapes), and reviewing in detail the events associated with the error's occurrence. In each of the NAS Air Route Traffic Control Centers (ARTCCs), the Systematic Air Traffic Operations Research Initiative (SATORI) system is used in the investigation process to re-create the error situation in a format much like the one originally displayed to the controller (Rodgers & Duke, 1993). SATORI allows for a more accurate determination of the factors involved in the incident. Once the OE has been thoroughly investigated, an OE final report is filed. This report, FAA 7210-3, contains detailed information obtained during the investigation process and includes an item on the controller's awareness of the developing error.

Specifically, FAA 7210-3 requires a yes or no answer to the question: Was the employee aware that an OE was developing? Previous research has suggested that controller awareness of error development is related to the severity of OEs (Rodgers & Nye, 1993; Durso, Truitt, Hackworth, Ohrt, Hamic, Crutchfield, & Manning, 1998). Rodgers and Nye (1993) found that controller awareness of error development was associated with less severe errors. Durso et al. (1998), in an analysis of data for a 1 year period not covered in the earlier Rodgers and Nye study, confirmed that controller awareness of error development resulted in significantly less severe errors. Given the relationship of controller awareness to OE severity, further study of the awareness variable was warranted.

This section reviews two studies conducted by the FAA on OEs that occurred in ARTCCs. In each study, available data were reviewed for several purposes, including the assessment of controller SA at the time of the incident. The data illustrate the kind of information that can be gained from post hoc analysis of FAA databases and by using systems such as SATORI that recreate the radar and radio communication information available during the error occurrence.

## OE Severity and SA

The first study, conducted by Rodgers and Nye (1993), had several purposes. The first was to examine the relationship between the level of error severity (degree of loss of vertical and horizontal separation between aircraft) and air traffic controller workload as measured by the number of aircraft being worked and the traffic complexity level reported at the time of the occurrence. Earlier studies (Kershner, 1968; Kinney, Spahn, & Amato, 1977; Schroeder, 1982; Schroeder & Nye, 1993; Stager & Hameluck, 1990) found a general trend for an increased percentage of errors under light and moderate as compared to heavy workloads. Schroeder and Nye (1993) also found two measures traditionally used to assess controller workload (number of aircraft being worked and situation complexity) to be highly correlated. Although OEs tend to occur more often under light and moderate workloads, it was hypothesized that the more severe OEs happen under heavy workload conditions.

The second purpose of the Rodgers and Nye (1993) study was to determine which of the causal factors attributed to the air traffic controller in the OE final report tended to be associated with OE severity. The general categories of causal factors include: data posting, radar display, coordination, communication, and position relief briefing (see Appendix). Each of these factors is subdivided into 5–15 more specific categories. It was hypothesized that the OEs involving the causal factor categories more closely related to the controller's SA were more likely to result in greater severity (i.e., OEs involving less separation). Another variable (not defined as a causal factor) reported on the OE form addresses whether the controller was aware the error was developing. This factor was included in the analyses because it relates directly to the topic of study.

The Rodgers and Nye (1993) study's third purpose was to determine the relation between the severity of an OE and the flight profile and altitude of the aircraft involved in the errors. It was hypothesized that aircraft in a climbing or descending profile are more likely to be involved in the more severe OEs due to the rapidly changing dynamics of the control situation. It was also hypothesized that aircraft at lower altitudes, where less separation is required, are more likely to be involved in the more severe OEs.

The fourth purpose of this study was to determine how these variables are distributed among individual ARTCCs. It was hypothesized that clusters of facilities could be determined in regard to certain factors related to OE severity. The study's fifth purpose was to determine what factors operated during a major OE. Because there were only 15 major OEs in the database during the time frame of this study, only frequencies of the factor categories occurring during a major error were reviewed.

## Method

### OE Data Base

Since 1985, the FAA has operated an OE database to track the OEs reported within the nation's ATC system. It uses data collected by quality assurance (QA) personnel at each ARTCC who are responsible for completing a specific report, FAA 7210-3 (U.S. Department of Transportation, 1991a). Therefore, for the purposes of the analysis, the sample included error reports for the years 1988, 1989, 1990 plus approximately two thirds of the OEs from 1991 and yielded a total of 1,053 errors. The distribution of errors across each of the years was relatively constant.

### Error Severity

The severity of an OE is categorized and reported by the FAA according to the closest proximity of the involved aircraft in terms of both horizontal and vertical distances. Table 4.1 shows the separation parameters and the corresponding point assignments; the total point number determines whether the severity was major, moderate, or minor. This calculation is made by a subroutine of the OE database. Of the 1,053 errors in our sample, only 15 were coded as major errors. Some of the key points regarding the separation standards can be summarized as follows:

1. A major error is defined as less than 0.5 mi horizontal separation and less than 500 ft vertical separation between aircraft at the time of occurrence.

2. The designation of a moderate or minor error is based on the determination of the altitude of occurrence:

TABLE 4.1
Definition of Severity Categories

| Vertical Separation | Points | Horizontal Separation | Points |
|---|---|---|---|
| If occurrence was at or below | | Less than ½ mi | 10 |
| altitude of 29,000 ft (FL 290) | | ½ mi to 1 mi | 8 |
| Less than 500 ft | 10 | 1 mi to 1½ mi | 7 |
| 500 ft to 600 ft | 9 | 1½ mi to 2 mi | 6 |
| 600 ft to 700 ft | 8 | 2 mi to 2½ mi | 5 |
| 700 ft to 800 ft | 6 | 2½ mi to 3 mi | 4 |
| 800 ft to 900 ft | 4 | 3 mi to 3½ mi | 3 |
| 900 ft to 1,000 ft | 2 | 3½ mi to 4 mi | 2 |
| | | 4 mi to 4½ mi | 1 |
| If occurrence was above | | | |
| altitude of 29,000 ft (FL 290) | | | |
| Less than 500 ft | 10 | | |
| 500 ft to 700 ft | 9 | | |
| 700 ft to 1,000 ft | 8 | | |
| 1,000 ft to 1,500 ft | 3 | | |

*Note.* Severity
20 points = Major
14–19 points = Moderate
13 or less points = Minor

    a. at 29,000 ft (Flight Level [FL] 290) or below, a moderate error can involve separation parameters ranging from less than 0.5 mi horizontal combined with up to 900 ft vertical to almost 3 mi horizontal combined with less than 500 ft vertical. A minor error can involve less than 0.5 mi horizontal separation only if the vertical separation is 900 ft or greater. Also, an error can be classified as minor with less than 500 ft vertical separation only if the aircraft were 3 mi or farther apart.

    b. above FL 290, a moderate error can involve less than 0.5 mi horizontal separation combined with up to 1,000 ft vertical separation. Although severity points are added for vertical separation less than 2,000 ft, at this flight level, all errors occurring with vertical separation of 1,000 ft or greater are classified as minor.

Of the 15 major errors, 7 resulted in the filing of a near midair collision report by at least one of the involved pilots. Because the FAA has categorized these errors as qualitatively different from other errors, and because there were so few of them, they were analyzed separately with only frequency data presented in this chapter.

## Air Traffic Controller Workload and Causal Factors

The two measures of workload reported in the OE database are the number of aircraft being worked by the air traffic controller at the time of the error and the air traffic complexity (as estimated by the QA specialist and based on factors such as weather, airspace restrictions, and variety of duties). The rating is made on a 5-point scale with anchors of 1 = *easy* and 5 = *complex*. The causal factors attributed to the controller comprise a hierarchy of specific elements within more general categories as shown in Table 4.2. For example, if a computer entry had been incorrectly updated (causal factor 1a.1), then that error also involved the less specific category of data posting. An OE can involve multiple causal factors. For this analysis, an OE was coded as 1 if a factor was recorded and 0 if it was not. Table 4.2 also includes the percentage of the 1,053 errors that involved each factor. Sometimes the more global categories were not recorded, whereas at other times only the global categories were recorded.

## Aircraft Profile Characteristics

One of six possible combinations of aircraft flight characteristics was identified for each error in the sample. The profiles were as follows: (a) all aircraft were climbing, (b) all aircraft were descending, (c) all aircraft were at level flight, (d) one (or more) aircraft was descending and one (or more) aircraft climbing, (e) one (or more) aircraft was at level flight and one (or more) aircraft climbing, or (f) one (or more) aircraft was at level flight and one (or more) aircraft descending.

## Results[1]

## Factor Relationships to OE Severity

Since OE severity is defined differently for aircraft above or below FL 290, the following results are presented in terms of that distinction. As shown in Table 4.3, the average number of aircraft being worked at the time of the OEs ranged from 8.1 for minor errors at FL 290 or below to 10.6 for moderate errors above FL 290. The number of aircraft being worked was not significantly different between severity levels, $F(1,1048) = 2.2$, NS, whereas the number of aircraft being worked was significantly

---

[1]This section is intended to illustrate the application of post hoc data analysis techniques. Therefore, not all of the data in the following tables are discussed in the text. Please refer to the original report for more information.

TABLE 4.2
Percentages of "Minor" or "Moderate" Errors at En Route
Facilities that Involved Each Causal Factor

| | | | | |
|---|---|---:|---|---:|
| 1. | Data posting | 13 | b. Transposition | 5 |
| | a. Computer entry | 6 | c. Misunderstanding | ·4 |
| | (1) Incorrect input | 2 | d. Readback | 20 |
| | (2) Incorrect update | 1 | (1) Altitude | 14 |
| | (3) Premature termination of data | 1 | (2) Clearance | 3 |
| | (4) Other | 2 | (3) Identification | 4 |
| | b. Flight progress strip | 9 | e. Acknowledgment | 5 |
| | (1) Not prepared | 0 | f. Other | 8 |
| | (2) Not updated | 3 | | |
| | (3) Posted incorrectly | 0 | 4. Coordination | 15 |
| | (4) Reposted incorrectly | 0 | a. Area of occurrence | |
| | (5) Updated incorrectly | 2 | (1) Interposition | 2 |
| | (6) Sequenced incorrectly | 0 | (2) Intraposition | 3 |
| | (7) Resequenced incorrectly | 0 | (3) Intersector | 4 |
| | (8) Interpreted incorrectly | 2 | (4) Interfacility | 2 |
| | (9) Premature removal | 0 | b. An aircraft penetrated designated air- | |
| | (10) Other | 3 | space of another position of operation | |
| | | | or facility without prior approval | 4 |
| 2. | Radar display | 59 | c. Coordination was effected and | |
| | a. Misidentification | 14 | controller(s) did not utilize in- | |
| | (1) Overlapping data blocks | 3 | formation exchanged | 6 |
| | (2) Acceptance of incomplete or | | (1) Aircraft identification | 0 |
| | difficult to correlate position info | 1 | (2) Altitudes/flight level | 2 |
| | (3) Improper use of identifying turn | 0 | (3) Route of flight | 1 |
| | (4) Failure to reidentify aircraft when | | (4) Clearance limit | 0 |
| | accepted target identity becomes | | (5) Speeds | 0 |
| | questionable | 1 | (6) APREQS | 0 |
| | (5) Failure to confirm aircraft | | (7) Special instructions | 0 |
| | identity after accepting a radar | | (8) Other | 2 |
| | hand-off | 0 | | |
| | (6) Other | 11 | 5. Position relief briefing deficiencies | |
| | b. Inappropriate use of displayed data | 47 | noted | 3 |
| | (1) Conflict alert | 2 | a. Employee did not use position | |
| | (2) Quick look | 0 | relief checklist | 1 |
| | (3) Mode C | 12 | b. Employee being relieved gave | |
| | (4) MSAW/EMSAW | 0 | incomplete briefing | 1 |
| | (5) Other | 37 | c. Relieving employee did not make | |
| | | | use of pertinent data exchanged | |
| 3. | Communications error | 36 | at briefing | 1 |
| | a. Phraseology | 3 | d. Other | 1 |

*Note.* 0 indicates <.5%

TABLE 4.3
Air Traffic Controller Workload and Severity of OEs

| | Severity | | | |
| --- | --- | --- | --- | --- |
| | Minor | n | Moderate | n |
| # of Aircraft Worked | 8.35 | 737 | 8.50 | 315 | |
| FL ≤ 290 | 8.10 | 487 | 8.25 | 281 | FL $F_{1,1048}$ = 13.9; $p$ < .001 |
| FL > 290 | 8.83 | 250 | 10.59 | 34 | Severity $F_{1,1048}$ = 2.2; NS |
| Traffic Complexity Rating | 3.02 | 737 | 3.09 | 315 | |
| FL ≤ 290 | 3.03 | 487 | 3.01 | 281 | FL $F_{1,1048}$ = 0.06; NS |
| FL > 290 | 3.06 | 250 | 3.29 | 34 | Severity $F_{1,1048}$ = 0.93; NS |

*Note.* FL = flight level, 290 = 29,000 ft

TABLE 4.4
Severity Classification Within Each Complexity Rating

| Traffic Complexity | Moderate Severity % | Minor Severity % | n |
| --- | --- | --- | --- |
| Easy | 27.2 | 72.8 | 103 |
| Below average | 29.9 | 70.1 | 197 |
| Average | 28.5 | 71.5 | 376 |
| Above average | 32.7 | 67.3 | 306 |
| Complex | 31.0 | 69.0 | 71 |
| Overall | 30.0 | 70.0 | 1,053 |

*Note.* $\chi^2_{(4)}$ = 1.89; NS. The *n* represents the number of OEs that occurred in each complexity category.

greater, $F(1,1048)$ = 13.9, $p$ < .001, at FL 290 or higher. The average air traffic complexity rating was not significantly different between the severity categories, $F(1,1048)$ = .93, NS, or flight levels, $F(1,1048)$ = .06, NS.

As indicated in Table 4.4, the percentages of errors that were moderate (30%[2] overall), when compared to minor errors, were relatively consistent regardless of the traffic complexity. For example, 27% of the OEs that occurred under the easy complexity rating were classified as moderate in severity whereas 31% of the OEs that were evaluated as occurring under complex conditions resulted in a moderately severe OE.

The error causal factors that were differentially related to OE severity are shown in Table 4.5. The number of cases in each category in Table 4.5 equals the total number of OEs (both minor and moderate in severity) that involved either a given causal factor or controller awareness that the error was developing. The percentages represent the proportion of the OEs that were moderately severe for each causal factor. The factors that were

---

[2]For this discussion all percentages are rounded to the nearest whole number.

TABLE 4.5
Percentages of Errors That Were Moderately
Severe by Causal Factor and ATCS Awareness

| | Overall | | FL ≤ 290 | | FL > 290 | |
|---|---|---|---|---|---|---|
| | % Moderate | N | % Moderate | N | % Moderate | N |
| Factors involved with greater severity | | | | | | |
| Misuse—conflict alert | 52.0 | 25 | 60.0 | 20 | 20.0 | 5 |
| Communications | 36.9 | 379 | 45.0 | 300 | 6.3 | 79 |
| Readback | 40.8 | 206 | 47.9 | 169 | 8.1 | 37 |
| Readback—altitude | 41.1 | 151 | 46.5 | 127 | 12.5 | 24 |
| Coordination | 38.5 | 161 | 42.7 | 110 | 28.0 | 51 |
| Interfacility coordination | 53.8 | 26 | 60.0 | 20 | 33.3 | 6 |
| Factors involved with lesser severity | | | | | | |
| Misuse of displayed data (excluding conflict alert) | 22.0 | 473 | 27.9 | 323 | 9.3 | 150 |
| ATCS aware error was developing | 21.0 | 267 | 26.0 | 177 | 11.1 | 90 |

Note.  All of the causal factors were significant at $p < .01$ on chi-square tests.
$N$ = the number of OEs that were related to each factor.
% = the percentage of each $N$ that was moderate (compared to minor) in severity.
$FL$ = flight level, 290 = 29,000 ft.

associated with a lower percentage of moderate errors included the misuse of displayed data (excluding use of conflict alert) and awareness by the controller that the error was developing. Twenty-two percent of the 473 OEs that involved misuse of displayed data and 21% of the 267 OEs involving controller awareness of the developing situation resulted in errors of moderate severity. By contrast, other causal factors were more likely to result in moderate severity. In particular, over 50% of errors that involved misuse of conflict alert or interfacility coordination were classified as moderate. Another factor related to a greater loss of separation was "readback" (a communication during which the controller fails to detect a pilot's incorrect response to a clearance provided by the controller) and, more specifically, readback involving altitude information. When readback was a causal factor over 40% of those errors were classified as moderately severe.

Table 4.6 lists the differences in average horizontal separation (overall, below FL 290, and above FL 290) between those errors that involved each of the causal factors in Table 4.5 versus those that did not. For example, for the errors in which misuse of displayed data (excluding use of conflict

TABLE 4.6
Differences in Horizontal Separation Between Aircraft When
OE Involved Certain Causal Factors and ATCS Awareness

| | Horizontal Separation | | | | | |
| --- | --- | --- | --- | --- | --- | --- |
| | Overall | | FL ≤ 290 | | FL > 290 | |
| | Miles | n | Miles | n | Miles | n |
| Factors involved with greater severity | | | | | | |
| Misuse—conflict alert | −.31 | 25 | −.34 | 20 | −.15 | 5 |
| Communications | −.25 | 379 | −.28 | 300 | −.16 | 79 |
| Readback | −.30 | 206 | −.27 | 169 | −.39 | 37 |
| Readback—altitude | −.29 | 151 | −.25 | 127 | −.43 | 24 |
| Coordination | −.25 | 161 | −.20 | 110 | −.32 | 51 |
| Interfacility coordination | −.42 | 26 | −.43 | 20 | −.41 | 6 |
| Factors involved with lesser severity | | | | | | |
| Misuse of displayed data (excluding conflict alert) | 0.34 | 473 | 0.33 | 323 | 0.37 | 150 |
| ATCS aware error was developing | 0.28 | 267 | 0.33 | 177 | 0.13 | 90 |

Note. The values presented are mean differences in horizontal separation in miles for errors that involved a given causal factor versus all other errors. Negative values represent less separation at occurrence while positive values indicate greater separation.
$n$ = the number of OEs that were related to each factor.

alert) was involved, the average horizontal separation between aircraft was 0.34 mi greater than the OEs that did not involve this causal factor. By contrast, in the OEs in which readback was a factor, the resultant horizontal separation was 0.30 mi less than the OEs that did not involve readback. We found that vertical separation was not significantly related to any of the factors. The impact of these factors in terms of OE severity category, as illustrated in Table 4.5, was found to be related primarily to the horizontal separation parameter.

Table 4.7 examines the relation between aircraft profile in conjunction with flight levels and OE severity. Tests of significance were not conducted due to low expected values in some cells; however, the results are presented for descriptive purposes. Most OEs in the overall sample occurred when one or more aircraft was level and either others were climbing ($N = 323$, 31%) or others were descending ($N = 469$, 45%). Surprisingly, the greatest likelihood for moderate severity occurred when all aircraft were in level flight (43% overall, 53% at FL 290 or less, and 29% at FL 290 or more). Also, flight level was related to severity for each aircraft profile (i.e., a greater percentage of OEs were moderately severe at FL 290 or less than those at FL 290 or more).

TABLE 4.7
Percentages of Errors That Were Moderately
Severe by Profile of Aircraft

| | Overall | | FL ≤ 290 | | FL > 290 | |
|---|---|---|---|---|---|---|
| Profile of Aircraft | % Moderate | n | % Moderate | n | % Moderate | n |
| All Climbing | 21.9 | 32 | 23.3 | 30 | 0.0 | 2 |
| Level & Climbing | 26.3 | 323 | 34.4 | 218 | 9.5 | 105 |
| Descending & Climbing | 28.0 | 75 | 31.1 | 61 | 7.7 | 13 |
| Level & Descending | 30.3 | 469 | 37.6 | 346 | 9.8 | 123 |
| All Descending | 34.5 | 58 | 36.5 | 52 | 16.7 | 6 |
| All Level | 43.0 | 86 | 52.9 | 51 | 28.6 | 35 |

*Note.* $n$ = the number of OEs that occurred in each aircraft profile.
% = the percentage of each $n$ that was classified as moderately severe.
$FL$ = flight level, 290 = 29,000 ft.

## Facility-Level Characteristics of Moderately Severe OEs

Tables 4.8 through 4.10 illustrate the results of analyses by facility for the moderately severe errors at ARTCC facilities located in the continental U.S. (excluding ZSU, ZHN, and ZAN). Only moderately severe errors are reported here given that so few errors fall in the severe category. Table 4.8 lists the average number of aircraft being worked and average air traffic complexity rating. The results of a series of cluster analyses produced two interpretable groups of facilities based on reported controller workload. The relatively low workload facilities (ZAU, ZBW, ZJX, ZLA, ZNY, ZOA, and ZSE) were characterized by average air traffic complexity ratings of less than 3 combined with averages of 8.2 or fewer aircraft being worked. It should be noted that, given potential differences in reporting standards and practices, some degree of variability could be expected between facilities as well as within a facility. Table 4.8 also illustrates facility differences in the percentages of moderately severe errors involving controller awareness that the error was developing. Specifically, ZAU and ZMP reported that no controller was aware of the developing error in any of their OEs. By contrast, several facilities (ZDV, ZKC, ZLC, and ZOA) reported that the controller was cognizant of the situation prior to loss of separation in one-third or more of their moderately severe OEs.

To examine controller workload together with an awareness that the error was developing, multidimensional scaling (MDS) was applied with the results shown in Fig. 4.1. First, the two workload measures and controller awareness data were aggregated at the facility level. Then the $z$ score values for the number of aircraft being worked and traffic complexity were summed to compute a composite measure representing the average con-

TABLE 4.8
Average ATCS Workload During Moderately
Severe OEs at En Route Facilities

| Facility | Number of OEs | Average Number of Aircraft Worked | Average Traffic Complexity | Facility Workload Category | Percent ATCS Awareness |
|---|---|---|---|---|---|
| ZAB | 9 | 10.00 | 3.33 | (2) | 22 |
| ZAU | 31 | 7.68 | 2.77 | (1) | 0 |
| ZBW | 19 | 5.79 | 2.74 | (1) | 11 |
| ZDC | 35 | 9.09 | 3.71 | (2) | 23 |
| ZDV | 11 | 9.27 | 3.73 | (2) | 36 |
| ZFW | 16 | 9.31 | 3.31 | (2) | 6 |
| ZHU | 13 | 8.92 | 3.31 | (2) | 15 |
| ZID | 23 | 9.30 | 3.13 | (2) | 4 |
| ZJX | 20 | 8.20 | 2.50 | (1) | 20 |
| ZKC | 9 | 10.00 | 3.78 | (2) | 33 |
| ZLA | 15 | 7.00 | 2.60 | (1) | 20 |
| ZLC | 4 | 11.25 | 3.25 | (2) | 75 |
| ZMA | 11 | 8.64 | 3.09 | (2) | 18 |
| ZME | 10 | 10.60 | 3.30 | (2) | 10 |
| ZMP | 5 | 8.60 | 3.20 | (2) | 0 |
| ZNY | 26 | 6.88 | 2.77 | (1) | 31 |
| ZOA | 11 | 7.45 | 2.82 | (1) | 36 |
| ZOB | 19 | 9.26 | 3.00 | (2) | 21 |
| ZSE | 6 | 7.33 | 2.67 | (1) | 17 |
| ZTL | 9 | 8.33 | 3.78 | (2) | 11 |
| Overall | | | | | |
| Mean | | 8.65 | 3.14 | | 21 |
| SD | | 1.35 | .40 | | |

*Note.* The number in ( ) represents a classification of each facility as either relatively high ATCS workload (2) or low workload (1) reported at the time of the moderately severe OEs. Awareness = the percentage of each facility's moderately severe OEs in which the ATCS was aware that the error was developing.

troller workload at the time of OE occurrence for each facility. A dissimilarity matrix among facilities was created based on Euclidean distances using the composite workload and error awareness variables. The matrix was then analyzed using the classical MDS approach that plotted the cases (en route facilities) in two dimensions. Thus, MDS provided a geometric representation of both the similarity and dissimilarity among facilities in terms of workload and controller awareness. Dimension 1 geometrically represents controller workload, while dimension 2 is controller awareness that the error was developing.

The intersection of the dimensions produced four quadrants into which facilities were grouped based on their relative similarity and dissimilarity. Quadrant 1 was defined by relatively high controller workload combined with greater than average error awareness. This high workload, greater

TABLE 4.9
Causal Factors With Moderately Severe
OEs at En Route Facilities (% of Errors)

| Facility | Displayed Data | Conflict Alert | Readback | Readback Altitude | Interfacility Coordination |
|---|---|---|---|---|---|
| ZAB | 22 | 11 | 33 | 22 | 00 |
| ZAU | 32 | 03 | 32 | 26 | 00 |
| ZBW | 32 | 16 | 21 | 11 | 05 |
| ZDC | 43 | 00 | 20 | 14 | 00 |
| ZDV | 27 | 00 | 27 | 27 | 00 |
| ZFW | 38 | 06 | 25 | 25 | 13 |
| ZHU | 23 | 08 | 46 | 31 | 00 |
| ZID | 35 | 00 | 30 | 30 | 09 |
| ZJX | 25 | 15 | 20 | 20 | 05 |
| ZKC | 11 | 00 | 44 | 22 | 00 |
| ZLA | 33 | 00 | 27 | 13 | 07 |
| ZLC | 100 | 00 | 00 | 00 | 00 |
| ZMA | 27 | 09 | 09 | 09 | 00 |
| ZME | 20 | 00 | 30 | 20 | 20 |
| ZMP | 40 | 20 | 00 | 00 | 00 |
| ZNY | 23 | 04 | 19 | 15 | 12 |
| ZOA | 27 | 00 | 45 | 09 | 18 |
| ZOB | 26 | 00 | 32 | 26 | 00 |
| ZSE | 33 | 00 | 33 | 33 | 00 |
| ZTL | 44 | 00 | 33 | 11 | 00 |
| Overall | 33 | 05 | 26 | 18 | 04 |

*Note.* The data are the percentages of a facility's moderately severe OEs that involved these causal factors.

awareness condition was illustrated best by ZDV, ZLC, and ZKC. The high workload, less awareness facilities (Quadrant 2) included ZFW, ZTL, and ZME. ZAU and ZBW were characterized by relatively low controller workload combined with less awareness (Quadrant 3), whereas ZNY and ZOA reported relatively low controller workload combined with greater awareness (Quadrant 4).

There was also considerable variability in the frequency of the causal factors previously found to be related to severity. For example, the percentages of OEs that involved misuse of displayed data ranged from 11% to 44% across facilities (see Table 4.9). Similarly, although readback was involved in 26% of all moderately severe OEs, between 0% and 46% of the OEs involved readback as a factor at the facility level.

Table 4.10 illustrates the percentages of moderately severe OEs that occurred under the various aircraft profile categories described previously. The most frequent profile, aircraft level and aircraft descending, characterized over 40% of the OEs with a range of 14% to 78% across facilities.

TABLE 4.10
Profile of Aircraft in Moderately Severe OEs at En Route Facilities

| Facility | All Climb | Descend Climb | All Descend | Level Climb | Level Descend | All Level |
|---|---|---|---|---|---|---|
| ZAB | .0 | .0 | 22.2 | .0 | 77.8 | .0 |
| ZAU | .0 | 6.5 | .0 | 22.6 | 61.3 | 9.7 |
| ZBW | .0 | 15.8 | 5.3 | 21.1 | 52.6 | 5.3 |
| ZDC | 2.9 | 2.9 | 5.9 | 20.6 | 58.8 | 8.8 |
| ZDV | .0 | .0 | 9.1 | 36.4 | 45.5 | 9.1 |
| ZFW | .0 | .0 | .0 | 37.5 | 43.8 | 18.8 |
| ZHU | 7.7 | .0 | 23.1 | 30.8 | 30.8 | 7.7 |
| ZID | 4.5 | 13.6 | 4.5 | 31.8 | 13.6 | 31.8 |
| ZJX | .0 | 5.0 | 5.0 | 20.0 | 60.0 | 10.0 |
| ZKC | .0 | .0 | .0 | 55.6 | 33.3 | 11.1 |
| ZLA | .0 | .0 | .0 | 33.3 | 46.7 | 20.0 |
| ZLC | .0 | 25.0 | 25.0 | .0 | 25.0 | 25.0 |
| ZMA | 9.1 | 18.2 | .0 | 27.3 | 36.4 | 9.1 |
| ZME | .0 | .0 | .0 | 50.0 | 50.0 | .0 |
| ZMP | .0 | .0 | 20.0 | 40.0 | 20.0 | 20.0 |
| ZNY | 3.8 | 7.7 | 7.7 | 30.8 | 46.2 | 3.8 |
| ZOA | .0 | 20.0 | .0 | 20.0 | 40.0 | 20.0 |
| ZOB | 5.3 | 5.3 | 5.3 | 31.6 | 42.1 | 10.5 |
| ZSE | .0 | .0 | 16.7 | 33.3 | 50.0 | .0 |
| ZTL | .0 | 11.1 | 11.1 | 22.2 | 44.4 | 11.1 |
| Overall | 1.7 | 6.6 | 8.0 | 28.2 | 43.9 | 11.6 |

Note. The data are percentages of a facility's moderately severe OEs that occurred under the various aircraft profiles.

The all level aircraft profile was present between 0% and 32% depending on facility.

These results illustrate why future research regarding OEs should measure facility-level differences to estimate the extent to which findings are generalizable among ARTCCs. It would be interesting to investigate the relations that may exist between these differences and facility variations in traffic density, local procedures, and airspace configuration.

## Characteristics of OEs of Major Severity

Six of the 15 OEs categorized as major had a complexity rating of above average, four were classified as average, two as below average, and three had an easy rating. This produced a mean complexity rating of 2.9. The average number of aircraft being worked was 8.5 with the range between 2 and 15. These findings illustrate the sizable variability in controller workload conditions found in the major OEs.

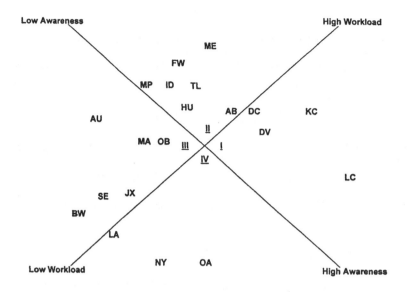

Note: Quadrant 1   =  high workload - high awareness
Quadrant II   =  high workload - low awareness
Quadrant III  =  low workload - low awareness
Quadrant IV  =  low workload - high awareness
Workload     =  standardized levels for number of aircraft being worked plus air traffic
                          complexity at the time of the OEs
Awareness   =  ATCS awareness that the OE was developing

FIG. 4.1. Multidimensional scaling of en route facilities based on workload and SA characteristics at the time of OEs.

As to flight characteristics of the aircraft in the major OEs, three of the
OEs occurred at FL 290 or above. Other characteristics found included:
five cases with all aircraft in a level profile, seven with aircraft level and descending, one with aircraft level and climbing, and two with aircraft descending and climbing.

Table 4.11 shows the frequencies of the causal factors that resulted in
OEs that were rated as major in severity. The data posting factor was involved in five OEs, radar display in nine, communications in six, coordination in five, and relief briefing in one OE. (As noted previously, an OE can
involve multiple causal factors.)

## Discussion

### General Findings

A counterintuitive finding of this study was that neither a quantitative measure of controller workload nor a rating of air traffic complexity was found to

TABLE 4.11
Frequency of Causal Factor Categories that
Resulted in "Major" Errors at En Route Facilities

| | | | | |
|---|---|---|---|---|
| 1. Data posting | 5 | b. Transposition | | 1 |
| a. Computer entry | 2 | c. Misunderstanding | | 1 |
| (1) Incorrect input | 2 | d. Readback | | 2 |
| (2) Incorrect update | 1 | (1) Altitude | | 1 |
| (3) Premature termination of | | (2) Clearance | | 3 |
| data | 1 | (3) Identification | | 1 |
| (4) Other | 2 | e. Acknowledgment | | 2 |
| b. Flight progress strip | 4 | f. Other | | 1 |
| (1) Not prepared | 0 | | | |
| (2) Not updated | 3 | 4. Coordination | | 6 |
| (3) Posted incorrectly | 0 | a. Area of occurrence | | |
| (4) Reposted incorrectly | 0 | (1) Interposition | | 1 |
| (5) Updated incorrectly | 2 | (2) Intraposition | | 3 |
| (6) Sequenced incorrectly | 0 | (3) Intersector | | 2 |
| (7) Resequenced incorrectly | 0 | (4) Interfacility | | 1 |
| (8) Interpreted incorrectly | 1 | b. An aircraft penetrated designated | | |
| (9) Premature removal | 0 | airspace of another position of | | |
| (10) Other | 1 | operation or facility without prior | | |
| | | approval | | 1 |
| 2. Radar display | 9 | c. Coordination was effected and | | |
| a. Misidentification | 2 | controller(s) did not utilize in- | | |
| (1) Overlapping data blocks | 3 | formation exchanged | | 5 |
| (2) Acceptance of incomplete or | | (1) Aircraft identification | | 1 |
| difficult to correlate position | | (2) Altitudes/flight level | | 1 |
| info | 1 | (3) Route of flight | | 1 |
| (3) Improper use of identifying | | (4) Clearance limit | | 0 |
| turn | 0 | (5) Speeds | | 0 |
| (4) Failure to reidentify aircraft | | (6) APREQS | | 0 |
| when accepted target identity | | (7) Special instructions | | 1 |
| becomes questionable | 1 | (8) Other | | 2 |
| (5) Failure to confirm aircraft | | | | |
| identity after accepting a radar | | 5. Position relief briefing deficiencies | | |
| hand-off | 0 | noted | | 1 |
| (6) Other | 2 | a. Employee did not use position | | |
| b. Inappropriate use of displayed | | relief checklist | | 1 |
| data | 7 | b. Employee being relieved gave | | |
| (1) Conflict alert | 2 | incomplete briefing | | 1 |
| (2) Quick look | 0 | c. Relieving employee did not make | | |
| (3) Mode C | 1 | use of pertinent data exchanged | | |
| (4) MSAW/EMSAW | 0 | at briefing | | 1 |
| (5) Other | 7 | d. Other | | 1 |
| | | | | |
| 3. Communications error | 6 | | | |
| a. Phraseology | 3 | | | |

be related to OE severity. Future research could examine the correspondence between these workload measures and workload assessments made with instruments such as the National Aeronautics and Space Administration Task Load Index (Hart & Staveland, 1988) in order to expand the understanding of the workload conditions of air traffic controllers.

Causal factors that relate to controller awareness appear likely to include those that are not easily corrected if not caught immediately, thereby allowing the error situation to build until it is too late to avoid. In general, the causal factors that resulted in a greater loss of separation were those likely to reduce awareness of the situation by the air traffic controller (i.e., interfacility coordination, incorrect readback of altitude by pilot). This assertion is supported by the finding that if the controller was aware the error was developing, the errors tended to be less severe. Additionally, that the misuse of displayed data (excluding conflict alert) was associated with less severity suggests that the factors associated with more severe errors have an insidious nature not characteristic of errors involving the misuse of displayed data.

It is interesting to note that only the horizontal separation parameter (not the vertical) was significantly affected as a function of some of the causal factors and severity categories. This may be because altitude, used to provide vertical separation, is reported numerically in the data block. Thus, altitude information is likely to be much more salient than that information used to judge the extent of horizontal separation between aircraft. Horizontal separation is presented graphically (or pictorially) and hence requires interpretation by the controller to recognize relative position. Determination of horizontal separation is based on the visually-judged distance between two targets on the plan view display and evaluations of their relative speeds.

Although most OEs occur when at least one aircraft is in level flight and at least one aircraft is either descending or ascending, the all level aircraft profile was found to be associated with a greater percentage of moderately severe OEs. Perhaps less attention is directed to aircraft in the all level profile thus resulting in a greater likelihood of severe errors. Additionally, moderately severe errors were more likely to occur at or below FL 290.

These data are national averages and the values may be different for certain facilities. Differences in traffic density, local procedures, and airspace configuration may limit the ability to generalize these results across different facilities. This is evidenced by the results of the facility-level analyses presented earlier. Furthermore, without additional information concerning the percentage of time controllers spend controlling traffic under the various complexity or workload conditions, it is difficult to determine the primary factors associated with these outcomes.

## Facility-Level Findings

Although neither the number of aircraft being worked nor a rating of air traffic complexity was found to be related to OE severity, a series of cluster analyses produced two interpretable groups of facilities based on reported controller workload. This measure, that for low controller workload was characterized by an average air traffic complexity rating of less than 3 combined with 8.2 or fewer aircraft being worked, was used to examine controller workload in conjunction with SA.

Although there are few definite patterns in the data, several hypotheses of the effect of workload on SA are offered. The first suggests that awareness decreases as workload increases. This might be due to difficulty in maintaining an accurate mental picture of the air traffic situation as its complexity or information load increases. This phenomena is more likely to operate under very high workload conditions. Another possibility is that as workload shifts from a high to low level awareness decreases. This situation is likely to exist after a busy traffic push. Additionally, awareness may decrease if low workload conditions are sustained for an extended period of time. Furthermore, a fatigue effect could be operating under sustained periods of high workload and result in a decrease in awareness during later periods of a high workload watch.

It appears that reduced SA might occur as an effect of different workload-related causes. In this research, several clusters of facilities were distinguished based on differences along the dimensions of controller workload and SA. Facilities were classified as follows: high workload/high awareness, high workload/low awareness, low workload/high awareness, or low workload/low awareness. The data were not available to determine which of the four classifications might have been operating during each of the errors analyzed in this study.

Two facilities best exemplified the relatively low workload/low awareness combination (ZAU, ZBW). They accounted for 17% of all the moderately severe errors. However, it was not possible to determine the extent to which these errors were a result of traffic density, airspace configuration, local operating procedures, workstation design, poor technique, or other factors. In fact, since normative information is lacking on variables such as traffic density, only a description of the factors associated with operational irregularities is available.

## Major OE Severity Findings

Over the 3½ years of data analyzed and presented in this chapter, only 15 of the 1,053 OEs had a major severity rating. Causal factor frequencies for the major errors proportionately matched the frequencies of the minor and moderate causal factor categories. Additionally, the average number of aircraft being worked did not differ between severity categories. The profile in which the majority of minor and moderate errors occurred (level and descending) was also the profile of the majority of the major errors. Although 40% of the major errors were rated above average in complexity, the mean complexity for the major errors (2.9) was less than that for the minor and moderate errors (3.1). However, given the small sample of major errors, this finding was not statistically significant.

## Conclusion

The types of analyses presented allow for the identification of factors more likely to precipitate air traffic control OEs. However, due to the lack of normative data, one must settle for a description of the factors associated with these operational irregularities. Additionally, the reporting process, including the reporting reliability of the investigators, may affect the extent to which these relationships can be determined. For the investigative process to result in a more definitive determination of the factors involved in operational irregularities, investigators must be able to review the dynamics associated with the air traffic situation.

Previously, QA specialists reviewed printouts of data that detailed aircraft locations, controller and host computer system interactions, audio recordings of the pilot and controller, and interphone communications to determine events surrounding the occurrence of an operational irregularity. The task of piecing together the "big picture" from multiple sources of data is a difficult and time-consuming part of the QA specialist's job. The reliability of the QA investigation findings affects the quality of the report filed for each irregularity. To allow the QA team to spend more time identifying the sources of problems associated with an incident and less time piecing together the dynamics of the ATC situation, a method was needed that permitted the re-creation of incidents and integrated the various data sources into a coherent picture. SATORI was developed for the purpose of re-creating ATC incidents and collecting data to assist in determining more definitively the location and likelihood of breakdowns in the human-machine system (Rodgers & Duke, 1993).

It may be possible to quantify characteristics of the ATC situation such as workload, complexity, and severity as a by-product of incident re-creation using SATORI data. This would reduce the variability that appears to exist between investigators and facilities. Standardization of the investigation process in this manner may allow for an improved determination of the factors involved in operational irregularities. Additionally, SATORI may provide the capability to collect normative information and potentially provide more than just a descriptive view of these occurrences. The second study reviewed in this chapter used data from several sources, including SATORI, to investigate the relationship between ATC complexity factors (including SA) and OEs.

### OEs, Sector Characteristics, and SA

This study had two goals. The first was to examine the characteristics associated with those OEs involving controller awareness of the developing error as well as those OEs involving controller unawareness. It was hypothesized that awareness of error development differed as a function of sector type, number of aircraft, number of people on position, amount of aircraft separation, time of day, and sector complexity. The second goal was to identify those sectors in the ARTCC that demonstrated a relation to the awareness variable. It was hypothesized that certain sectors could be identified or characterized as low awareness sectors. Further, it was hypothesized that sectors identified as low awareness sectors would differ with regard to various ATC complexity measures.

### Method

### OE Data

QA personnel at each facility are responsible for gathering data and completing a specific report, FAA 7210-3 (U.S. Department of Transportation, 1991a). For this study, a number of fields from 103 OE reports from the Atlanta ARTCC were coded and entered into a data file. Those OEs where more than one sector was involved (13), no final report was available (4), or the error was attributed to an equipment failure (1) were not included in this analysis. This left a sample of 85 OEs covering a 3 year period from June 1992 to June 1995 (Rodgers, Mogford, & Mogford, 1998).

### Controller Awareness

As mentioned earlier, one of the items contained in the OE final report requires an assessment of the involved employee's awareness of the developing error. This item has been on the OE final report form for the past 14 years. After interviewing the involved controller and reviewing the error with SATORI, QA specialists make a determination as to the controller's awareness. Although SATORI assists in the formulation of this judgment, most QA specialists find the answer relatively easy to ascertain. Typically, if the control action to provide separation was not issued in a timely manner or no control action was initiated, the controller was judged to be unaware of the developing error. However, if the controller actively attempted to provide separation to the involved aircraft, although the control action was either inappropriate or inadequate, the controller was judged to be aware of the developing error.

## Sector Complexity Measures

Two measures of sector complexity were used in this study. The first was a facility average complexity estimate calculated by ARTCC personnel during the sector validation conducted each year. This assessment involves estimating the sector complexity workload using a formula that weights various ATC functions (U.S. Department of Transportation, 1991b). Variables include: number of departures, number of arrivals, number of radar vectored arrivals, number of en route aircraft requiring control actions, number of en route aircraft not requiring control actions, number of emergencies, number of special flights, and number of required coordinations. These 7 factors are evaluated, weighted, and totaled to derive the sector complexity workload value. The sector complexity workload data from 1995 were used in this study.

The second measure of sector complexity was obtained using a 16 factor survey developed by Mogford, Murphy, and Guttman (1994). The complexity factors (CF) include:

1. Amount of climbing or descending traffic.
2. Degree of aircraft mix (VFR, IFR, Props, Turboprops, Jets).
3. Number of intersecting flight paths.
4. Number of multiple functions the controllers must perform (terminal feed, in-trail spacing).
5. Number of required procedures that must be performed.
6. Number of military flights.
7. Amount of coordination required.
8. Extent to which the controller is affected by airline hubbing.
9. Extent to which weather related factors affect ATC operations.
10. Number of complex aircraft routings.
11. Extent to which the controller's work is affected by restricted areas, warning areas, military operating areas, and their associated activities.
12. The size of the sector airspace.
13. The requirement for longitudinal sequencing and spacing.
14. Adequacy and reliability of radio and radar coverage.

15. Amount of radio frequency congestion.
16. Average traffic volume.

There are seven areas of specialization at Atlanta Center with approximately seven sectors per area. Each of the above 16 CFs was evaluated for each sector in each area by the airspace and procedures specialist assigned to that area of specialization using a 7 point scale.

## Results

Of the 85 OEs analyzed, the controller charged with the error was judged not aware in 72% (62) of the cases and aware in 28% (23) of the cases. Chi-square tests of awareness (not aware vs. aware) by sector type (ultra high, high, and low), number of aircraft being controlled, number of persons on position at the sector, sector complexity (facility estimation and 16 CF survey total), and time of day did not yield significant relations. Greater separation existed for both the vertical and horizontal dimensions for controller aware errors; however, only the horizontal parameter achieved statistical significance: $t(54.69) = -3.06, p < .003$.

In an effort to describe the association of the awareness variable to specific sectors, two groups of sectors with 1 or more errors were created. Those sectors with 1 to 3 errors were categorized as low error sectors, and those with 4 or more errors were categorized as high error sectors. (Because the mean of OEs for all sectors in which they occurred was 1.9 with a *SD* of 2.24, the cut between low and high error sectors was set between 3 and 4.) Since there was no variance associated with the awareness variable for the single error sectors (all errors were no awareness), they were excluded from this analysis.

Sectors were ordered by number of OEs that occurred in each sector (low to high) and plotted as a function of the number of non-aware vs. aware errors (see Fig. 4.2). There is an apparent trend in the graph toward less awareness of OEs as total errors in a sector increases. This was demonstrated in a significant overall correlation between total OEs in a sector and the number of OEs without awareness. This result was true when including all of the sectors with 2 or more errors, $r = 0.84, p < .001$, and also when focusing on the high error sectors alone, $r = 0.86, p < .01$. Correlations of total sector errors with errors when there was awareness were not statistically significant in either group.

Further analyses were conducted in an effort to describe sector complexity differences as a function of sector error categorization. Sectors were categorized as either non-error sectors (0 errors), low error sectors (1 to 3), or high error sectors (4 or more errors). Analyses of variance for sector complexity were significant for both the facility average complexity es-

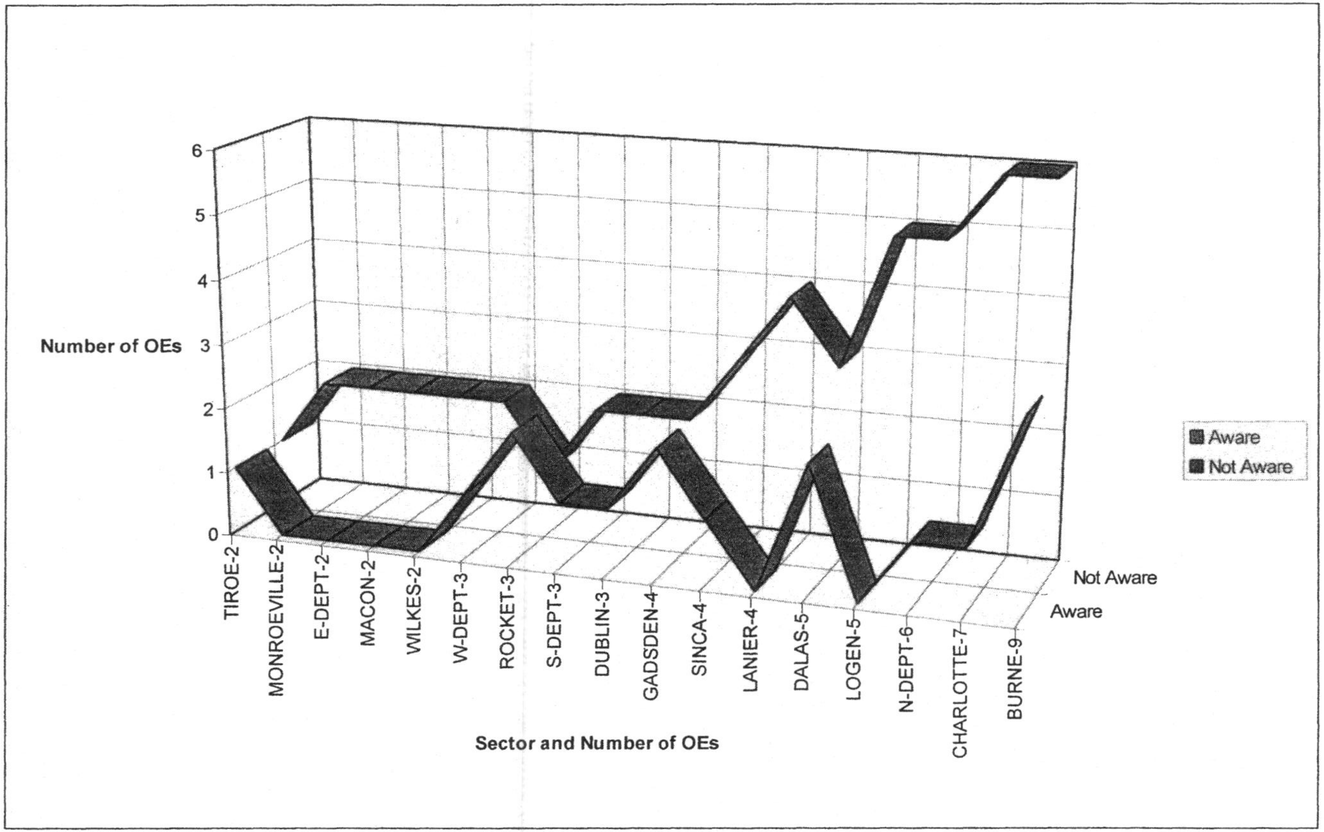

FIG. 4.2  SA for sectors with two or more OEs.

timates, $F(2, 45) = 5.45, p < .007$, and the 16 CF total, $F(2,45) = 3.12, p < .05$. Analysis of the individual 16 CFs yielded significant one-way ANOVAs for 2 of the factors. Frequency congestion, $F(2, 45) = 6.19, p < .004$, and weather, $F(2,45) = 3.75, p < .03$, significantly differed as a function of sector error categorization. Tukey's Honestly Significant Difference post hoc test was used to determine which groups contributed to the significant differences. For both of the global complexity measures and the two survey factors, the difference between the zero OE group and the four or more OE group was significant at $p < .05$.

## Discussion

This study found that error development awareness was significantly related to sector error rates. In general, controllers involved in errors at sectors in the Atlanta ARTCC that have greater error frequency tend to demonstrate less awareness of the developing errors. With further study, it may be possible to reliably define sector characteristics that distinguish high error-low awareness sectors. Low awareness, presumably generated by sector complexity, may lead to higher OE incidence. In support of this hypothesis is the finding that both of the global measures of sector complexity were significantly related to high error rate sectors.

It is interesting to note for the awareness variable that only the horizontal separation parameter, not vertical separation, was significantly affected. This finding was first suggested by Siddiqee (1974), discussed earlier in the review of Rodgers and Nye (1993), and confirmed by Durso et al. (1995).

Much work is required to explain how the loss of SA relates to OE occurrence. In an attempt to provide more detailed information about the nature of controller awareness of error development, the OE final report was modified to allow for a more detailed description of the loss of awareness. Three causal factor categories that directly assess the nature of the loss of awareness were added to the final report: failure to detect displayed information, failure to comprehend displayed information, and failure to project the future status of displayed data. These three factors directly relate to the Endsley (1995) model of SA. Initial analysis indicates frequent use of these causal factor categories (see Table 4.12). Additionally, since their inclusion, there has been a corresponding decrease in the use of the causal factor category "inappropriate use of displayed data—other." It is hoped that through the use of SATORI and revised OE causal factor categories, it will be possible to more accurately characterize the nature of SA losses during the occurrence of OEs.

TABLE 4.12

Top 10 Causal Factors: 3rd Quarter Fiscal Year 1998
NASDAC Data Period: January 1, 1997 through December 31, 1997

| Terminal facilities | |
|---|---|
| *Causal factor* | *N* |
| 1. Aircraft observation; actual observation of aircraft | 57 |
| 2. Radar display; inappropriate use of displayed data; failure to project future status of displayed data | 53 |
| 3. Aircraft observation; improper use of visual data; taking off | 38 |
| 4. Coordination; area of incident; inter-facility | 38 |
| 5. Radar display; inappropriate use of displayed data; failure to detect displayed data | 31 |
| 6. Coordination; failure to utilize/comply with precoordination information | 30 |
| 7. Aircraft observation; improper use of visual data; landing | 29 |
| 8. Communications error; misunder-standing | 29 |
| 9. Aircraft observation; improper use of visual data; ground operation; taxiing across runway | 25 |
| 10. Radar display; inappropriate use of displayed data; failure to comprehend displayed data | 24 |

| TRACON facilities | |
|---|---|
| *Causal factor* | *N* |
| 1. Radar display; inappropriate use of displayed data; failure to project future status of displayed data | 89 |
| 2. Radar display; inappropriate use of displayed data; failure to detect displayed data | 54 |
| 3. Coordination; area of incident; inter-facility | 44 |
| 4. Radar display; inappropriate use of displayed data; failure to compre-hend displayed data | 36 |
| 5. Coordination; area of incident; inter-sector/position | 16 |

| | |
|---|---|
| 6. Coordination; area of incident; intra-sector/position | 16 |
| 7. Data posting; flight progress strip; not updated | 13 |
| 8. Communications error; readback; altitude | 12 |
| 9. Coordination; failure to utilize/comply with precoordination information | 12 |
| 10. Communications error; other | 11 |

| Enroute facilities | |
|---|---|
| *Causal factor* | *N* |
| 1. Radar display; inappropriate use of displayed data; failure to project future status of displayed data | 204 |
| 2. Radar display; inappropriate use of displayed data; failure to detect displayed data | 122 |
| 3. Radar display; inappropriate use of displayed data; failure to comprehend displayed data | 104 |
| 4. Coordination; area of incident; inter-facility | 60 |
| 5. Data posting; flight progress strip; interpreted incorrectly | 34 |
| 6. Communication error; readback; altitude | 29 |
| 7. Coordination; area of incident; intersector/position | 29 |
| 8. Radar display; inappropriate use of displayed data; other | 25 |
| 9. Data posting; computer entry; input/update not made | 23 |
| 10. Communications error; transposition | 21 |

| AFSS/FSS facilities | |
|---|---|
| *Causal factor* | *N* |
| 1. Communications error; other | 1 |
| 2. Data posting; computer entry; incorrect input | 1 |
| 3. Data posting; flight progress strip; other | 1 |

**Conclusion**

The preceding two studies and the information presented regarding the modification of the OE causal factor categories are examples of how FAA researchers make use of databases to perform post hoc analyses of ATC incidents and improve the quality of the information available in these databases. Many kinds of analyses can be performed to investigate the conditions leading to OEs. With SA, however, much depends upon the post-error QA investigation and a judgment as to whether the controller involved was aware of the developing error. With the addition of more detailed queries to the OE final report, it will be important to focus on the interview process immediately following an OE occurrence. QA personnel should be trained not only to carefully question the involved controller about SA issues, but to observe the SATORI re-creation of the incident to seek independent confirmation of the SA situation. Although such observations are usually not fully conclusive, it is possible to note that the failure of a controller to take any observable action indicates a lack of SA regarding a developing error. Alternately, if SA was present, it is possible to notice events during a SATORI re-creation (such as communications to aircraft) that suggest awareness. In any case, it is evident that the combination of a comprehensive reporting procedure regarding ATC OEs combined with the capability for recreating the synchronized communication and radar display data associated with the incident provide many avenues for investigating controller SA.

## SA AND MAJOR AIRCRAFT ACCIDENT INVESTIGATION

The next section of this chapter considers the process associated with determining the events leading up to major aircraft accidents. Given the differences between the investigation processes associated with ATC incidents and major aircraft accidents, this section is organized quite differently. This section initially considers the data, people, equipment, and environmental issues associated with accident causation and then develops a broad discussion of major aircraft accident post hoc investigation processes and SA.

It is widely accepted that good flight crew SA is critical for safe aircraft operation. As with ATC, there are several data sources provided, but some may be compromised or absent due to the traumatic nature of the event that often involves loss of life. Much of the investigation may involve incident reconstruction from available information. There may be no direct data on flight crew SA and inferences must be made from available ATC data and flight data recorders (including cockpit voice recorders).

*Data.* Aircraft accident investigators typically obtain data pertaining to the persons considered critical to the cause of the accident, the equipment used, and the environment in which the critical personnel and the equipment operated. As in empirical research, investigators sample the universe of data to collect those necessary to meet the objectives of the investigation. The data, in addition to meeting measures of quality, must be internally consistent, logical, and sequentially correct. That is, data obtained from the various sources in an accident investigation must manifest the application of comparable logic and the description of identical events, in the same sequence, leading up to the accident itself.

*The Person.* The majority of aircraft accidents result from an error or series of errors that a person critical to the flight has committed. The data of interest describe the quality of the person's performance of the critical tasks and the state of his or her behavioral and physical health at the time of the accident. The sources of data include medical, personnel, and training records and the characterizations of those who were familiar with the person. When available, reports by flight crew and others involved in the accident are also important.

*The Equipment.* The accident aircraft and other equipment provide considerable data about the events preceding the accident. Items such as control surface positions, instrument readings, non-volatile memory contents, switch and circuit breaker positions, and site damage offer substantial information about the state of the machine before the accident and can be critical to understanding the cause of the accident. For example, preimpact fire damage exhibits different smoke patterns than postimpact fire. Determining when the fire started is critical to learning the accident's cause. Similarly, aircraft wreckage that is concentrated in a small area results from a different flight path than wreckage that is dispersed, thus describing substantially different types of accident sequences and causes.

The most critical equipment-provided information derives from the two recorders required on air transport aircraft that continually sample data on the status and operation of the aircraft. Digital Flight Data Recorders (DFDRs) contain anywhere from 11 to over 100 parameters regarding the airplane's flight surfaces, engines, systems and pilot controls, and flight path from the time of the accident back through the preceding 25 hours. Cockpit Voice Recorders (CVRs) record sounds, conversation, alerts, and warnings within the cockpit, including sounds accompanying changes in aircraft flight status, from the time of the accident back through the preceding 30 minutes.

Additional sources of data are supplied by ATC facilities that record communications between pilots and air traffic controllers and most communications among controllers themselves. Further data are provided by ATC facilities that record radar information revealing the precise location, airspeed, and altitude of all aircraft in the airspace.

*The Environment.* Environmental data include information from radar and other devices that regularly measure and record weather parameters in the relevant airspace. These sources can provide information on the direction and velocity of the winds, visibility, temperature, and precipitation level at different points in time. In addition, dispatch records contain basic information about the flight, including the planned flight path, weight and balance, fuel requirements, and other data relevant to the investigation.

## Major Aircraft Accident Post hoc Analysis

Often, within days of an accident, patterns emerge within the data obtained that suggest significant issues for further exploration. For example, the absence of preexisting hardware failures often leads investigators to examine the actions and decisions of critical personnel involved in the accident flight. In that event, the SA of the individual or individuals may be essential for determining the error or errors that caused or contributed to the accident. As noted, the SA examination follows the collection of data from a variety of sources and the assurance that the data meets the requisite logical and statistical requirements of accident investigation.

For example, on August 16, 1987, Northwest Airlines MD–80 flight 255 crashed shortly after departure from Detroit Metropolitan Airport killing 154 persons (NTSB, 1988). Initial witness reports described flames being emitted from the engines. As a result, much of the early activities focused on collecting data that could corroborate possible power plant anomalies. However, DFDR data subsequently showed that the flaps and slats had been retracted during takeoff. This indicated that without substantial additional airspeed the airfoils were incapable of providing the necessary lift to initiate and sustain the initial climbout. In addition, because the airplane's attitude at rotation was considerably higher than normal (to maintain the desired airspeed with retracted flaps or slats), the airflow into the engines was reduced. The reduced airflow led to compressor stalls within the engines that then produced the flames described by the witnesses.

Physical evidence obtained from the aircraft wreckage corroborated the DFDR data on the flap and slat positions. Other data included the calcu-

lated climbout performance of an MD–80 that matched the actual performance of the accident aircraft as recorded on ATC radar and on the DFDR
with the flaps and slats retracted. Finally, information from the CVR revealed that the pilots failed to check the status of the flaps and slats after
their taxi checklist procedures had been interrupted by an ATC clearance.
This further corroborated the findings that the flaps and slats had not been
extended during the taxi. In this manner, the data from a variety of sources
were consistent in describing the actions of the crew in setting the configuration of the airplane, the airplane's performance in that configuration,
and the effects of that configuration on engine performance.

These efforts established not only the configuration of the airplane at the
time of the accident but, more importantly, provided the data necessary to
determine the sequence of events leading up to the accident. With this
knowledge, the cause of the accident could be determined. By establishing
the sequence of events, investigators could then attempt to reconstruct the
SA of the crew involved. In this accident, the airspeed the crew selected was
appropriate for an airplane with extended flaps and slats. This information, together with the information from the CVR, supported the conclusion that the crew believed, almost up to the point of impact, that the difficulty in the climbout was due to weather factors they had discussed and
prepared for before takeoff, and not to an improper aircraft configuration.

On July 2, 1994, a USAir DC–9 crashed near Charlotte, North Carolina, during a severe thunderstorm (NTSB, 1995a). Investigators determined that the aircraft had traversed an area of intense rain. It then encountered a severe microburst and downburst with a change in wind
direction and velocity from a 35-knot headwind to a 26-knot tailwind with
a vertical velocity of 30 feet per second. The accident occurred during a
go-around after the pilots had attempted to discontinue the approach.
The airplane was destroyed, 37 passengers were killed, and the remaining
20 passengers and crew were injured in the accident. Both pilots, who
were experienced in the DC–9 and had unblemished records with the airline, survived. On-site examination of the wreckage provided no evidence
for a preexisting malfunction of the airplane or its components.

Investigators sought to determine the SA of the two pilots particularly
because of the extreme weather conditions. The assessment of the crew's
SA was critical to understanding their attempt to continue the flight into
the adverse weather because no rational pilot would deliberately endanger
the safety of the flight by attempting to traverse such weather.

The flight sequence began that afternoon in Charlotte, about 1½ hours
before the accident, when the crew flew the accident airplane to Columbia,
South Carolina. This flight segment provided them with a firsthand encounter with the prevailing weather conditions at Charlotte and those that
had been forecast for the time of the accident flight. At that time, visual

conditions dominated the Charlotte area, but late afternoon thunderstorms were predicted.

The flight between Charlotte and Columbia lasted about 30 minutes. On nearing Charlotte, the airborne radar on the accident airplane indicated the presence of storm cells in the area. On the Automated Traffic Information System (ATIS), a continuous tape loop that broadcasts field conditions and other weather information to pilots, controllers reported visual conditions over the area. As late as about 7 minutes before the accident, air traffic controllers told the crew to expect a visual approach to the field, an indication that weather conditions in Charlotte were good.

As the flight neared Charlotte, the weather over the field began to change. The crew observed rain over the airport and used their airborne radar to monitor the presence of a storm cell nearby. Just over a minute after being told to expect a clearance for a visual approach, the approach controller informed the crew that they "may get some rain just south of the field." The crew was then told to expect an instrument landing system (ILS) approach, an indication that conditions had deteriorated. Although the crew acknowledged this communication, there was no indication, either from their conversations on the CVR or with ATC, that they understood its implications. The CVR showed that the pilots were attempting to locate the cell on their airborne radar, but they did not discuss the deterioration of the Charlotte weather. They had initiated their descent and approach check, a time of considerable activity in the cockpit. As a result, although they addressed the possibility of a go-around, the CVR conversation described their attempt to avoid the storm cell identified on the airborne radar rather than a formal review of the missed approach procedures associated with the ILS approach as was required of a crew executing such an approach.

Although Charlotte air traffic controllers acknowledged the deteriorating weather, they did not convey all the critical weather information to the flight crew. For example, tower controllers observed lightning over the airport, a key indicator of thunderstorm activity, and discussed it among themselves, but did not notify the flight crew. As the rain intensified, controllers updated the ATIS, as required after a change in the conditions, to note the rain over the airport. However, the crew missed being alerted to this update because they had changed radio frequencies to communicate with the local controller that would issue them landing clearance. Finally, the crew received incomplete, and thus misleading information regarding the location of windshear that had been detected on the airport surface. Charlotte airport was equipped with a low level windshear alerting system (LLWAS) that detected rapid changes in wind direction and velocity in and around the airport. However, the LLWAS was alerting at all locations around the airport and not in one area exclusively. The crew was aware that they were to traverse the northwest area of the airport and the local controller later told investigators that he was aware of only the northeast boundary alert.

As with all passenger-carrying turbojet aircraft, the accident airplane was equipped with an airborne windshear detection device. This system, like the ground-based LLWAS, detected significant changes in the direction and velocity of the wind the airplane was encountering and warned the crew when it noted a substantial change in either. However, the system had been designed to inhibit a warning if it detected a windshear when the aircraft flaps were in transition, that is, being extended or retracted. During the go-around, the airborne windshear detection system recognized a windshear encounter, but because the flaps were being retracted to an intermediate position at that time, the system did not signal its detection of a windshear.

Although unaware of much of the weather information the controllers had available, the CVR showed that the crew repeatedly discussed the weather conditions during the approach into Charlotte. They observed and commented on the heavy rain over the field. They identified the storm cell on their airborne radar and attempted to locate it. They even discussed the potential presence of windshear and were prepared to execute a go-around if the weather conditions so warranted. Finally, they requested the tower controller to provide them with the airport surface winds and the ride conditions experienced by the pilots of the aircraft just ahead of them. The controller responded by informing the crew of the direction and velocity of the steady and fairly strong crosswind and, after querying the pilots of the USAir jet that was just ahead, conveyed to them the crew's report of a "smooth ride."

In summary, the crew had diligently attempted to obtain information regarding the weather. Yet it was clear by both their statements on the CVR and by their initiation of apparently routine go-around procedures that, until just prior to impact, they were unaware of the magnitude of the microburst they were encountering. The fact that they had attempted to traverse a severe microburst in itself demonstrates that their SA regarding the weather conditions along the final approach path was deficient and this deficiency led to their decision to continue the approach.

Nevertheless, the evidence also indicated that the crew had obtained, but did not perceive or comprehend, considerable information that would have supported an alternative assessment regarding the severity of the weather. They were given the predicted weather of possible thunderstorm activity and visually identified heavy rain over the field. They noted the radar portrayal of a storm cell near the field. Yet, despite obtaining necessary information regarding Charlotte weather information, their SA did not change sufficiently to suggest alternative actions. Specifically, the absence of an alert from the airborne windshear detection system, the incomplete information on the location of the windshear on the airport surface, and the lack of information regarding both the lightning and heavy rain over the airport served to convey to the crew an incorrect assessment of the severity of the weather ahead.

Although the pilots of the accident aircraft had been given incomplete and inaccurate information, it is likely that the ride report from the aircraft crew just ahead was most influential in the crew's SA regarding the weather. As a crew from the same airline as the accident crew and flying turbojet aircraft, the crew of the plane ahead was well versed in the kind of weather that would exceed the capabilities of the accident airplane and was knowledgeable on the company's guidance regarding weather conditions its pilots could not safely traverse. Further, because they were just ahead of the accident crew and closest to the airspace conditions the accident crew was about to enter, their report contained the most timely information on weather. The pilots of both aircraft did not know that the conditions were so dynamic that in the 2 to 3 minutes separating the two flights the airspace conditions had deteriorated to the severe level encountered by the accident flight.

The incorrect SA of the pilots led them, in addition to continuing an approach beyond the point that it should have been discontinued, to fail to anticipate a microburst of the severity that they encountered. In fact, the CVR revealed that just before impact, the captain, who was not flying the airplane, commanded the first officer to lower the nose, an action contrary to the guidance airlines provide their flight crews in a windshear escape maneuver. In response to this call, the first officer did lower the nose and the airplane ceased its climb. The crash occurred within 15 seconds after this action.

### Summary and Conclusions

Assessing SA in aircraft accident investigations requires the collection of evidence from a variety of diverse sources including the aircraft wreckage, the accident site, crew records, ATC radar recordings, the CVR, the DFDR, and various training and personnel records. Analysis of that evidence, as in the Northwest and USAir accidents, illustrates how the SA of a highly qualified and trained crew can be faulty and how that deficient SA could cause the crew to make decisions and take actions that are unsafe.

Pilot SA in a dynamic environment can and should be changeable as more current or more accurate information is obtained. The Detroit accident showed how flight crew SA of aircraft status can be faulty even though the relevant information is available and checklists are provided to ensure attention to critical controls and displays. The Charlotte accident demonstrated how susceptible SA can be to the information from numerous participants in the airspace. For example, the Charlotte air traffic controllers provided information that adversely influenced the SA of the crew. Yet, air traffic controllers are trained to prioritize their tasks so that information considered "informative" or "advisory" is secondary to that considered critical (i.e., the separation of air traffic). Although the air traffic controllers did not violate their own procedures, this accident demonstrated that their actions directly contributed to the accident crew's deficient SA. The NTSB investigation determined that the failure of the Charlotte controllers to provide complete information to the pilots contributed to the accident.

Moreover, in examining the evidence, the USAir pilots of the aircraft just ahead of the accident flight likely inadvertently influenced the deficient SA of the accident flight by providing information on the quality of their flight along the final approach path. Certainly, regardless of the ride report, the accident pilots should have been more sensitive to the possibility of an encounter with a severe microburst. Their lack of awareness likely also led to their failure to anticipate and respond immediately to the microburst encounter.

This accident also illustrates the criticality of SA to flight operations. Although maintaining SA requires obtaining and assimilating considerable information in dynamic environments, safe flight operations also require pilots and others operating in the airspace to anticipate hazards and be prepared for a quick reassessment of the situation when warranted. The failure of the accident crew in Charlotte to do that is perhaps the most significant error.

## CONCLUSION

SA is a cognitive construct that is important when considering the tasks of air traffic controllers and pilots. However, being an internal mental state, it is difficult to measure even during routine task activities. The examination of SA issues is especially relevant when considering the causes of ATC incidents and aircraft accidents. In such situations, an accurate determination of the role of SA in the causation of the problem could lead to improvements in training, procedures, or equipment design that might prevent future errors.

Typical SA assessment techniques in laboratory research settings rely on operator self-reports or experimenter observations (Endsley, 1995). The research and post hoc investigation settings have something in common with regard to gathering data on SA. The initial design of the data collection procedure is critical to the quality and applicability of the SA evaluation that can be completed. By the time a research study is completed or an incident or accident has been investigated, it is too late to modify data collection techniques. A good understanding of the role of operator SA in the system will lead to the design of questionnaires, interview forms, data recording systems, or event re-creation equipment that will permit a useful analysis. If possible, approaches should include both subjective reports of controllers and pilots (when available) as well as objective data that would help to corroborate these statements.

If the data sources supporting a post hoc analysis of an ATC incident or aircraft accident are adequate, it may be possible to determine the role of faulty controller or pilot SA. However, it is often problematic to discover why an operator was not aware of the critical factors signaling the developing problem. The focus of some of the ATC research discussed in this chapter was on defining airspace or traffic flow characteristics that may lead to low SA and errors. Workload and time pressure may also affect awareness level, although clear relations are not always evident. Poor training or equipment design could have resulted in misinformation leading to some of the aircraft accidents mentioned here. In any case, it is not enough just to determine that controller or flight crew SA was lacking. Further investigation of the causes of this problem must be undertaken.

In some cases, it is possible to easily discover why a controller or pilot was not aware of a critical piece of information. Perhaps it was not presented or communicated at the correct time. In other cases, the situation is not so simple. The relevant data may have been available, but was not understood or its future status was not projected accurately and this prevented the operator from properly assessing or recognizing a threatening situation. The reasons for this lack of awareness may not be clear. Other events may have been distracting, or expectations and preconceptions may have been operating to cause the operator to ignore the information or interpret it incorrectly. When post hoc investigations reveal an SA problem of this type, it is important to review these findings so that research can be focused on investigating the conditions under which they occur and recommend interventions. This chapter has reviewed post hoc techniques that have succeeded in identifying possible causes of SA problems. Research to investigate the more subtle issues associated with deficiencies in SA is incomplete and warrants further effort.

**APPENDIX**

Causal Factor Definitions (from FAA Form 7210–3)

1. Data posting—a data posting error is any error of calculation, omission, incomplete data, erroneous entries, handling, or subsequent revisions to this data. This includes errors in posting and recording data. It does not include errors involved in receiving, transmitting, coordinating, or otherwise forwarding this information.
2. Radar display
   a. misidentification—radar misidentification means a failure to properly identify the correct target and includes subsequent errors committed after the original identification was properly accomplished. Indicate the listed item(s) most closely describing the reason for misidentification.
   b. inappropriate use of displayed data—a data or display information error occurs due to a failure to maintain constant surveillance of a flight data display or traffic situation and to properly present or utilize the information presented by the display or situation.
3. Communications—a communications error is a causal factor associated with the exchange of information between two or more people (e.g., pilots and specialists). It refers to the failure of human communication not communications equipment.
   a. phraseology—use of incorrect or improper phraseology.
   b. transposition—errors due to transposition of words, numbers, or symbols by either oral or written means. This involves writing or saying one thing while thinking or hearing something else.
   c. misunderstanding—the failure to communicate clearly and concisely so that no misunderstanding exists for any actions contemplated or agreed upon.
   d. readback—the failure to identify improper or incorrect readback of information.
   e. acknowledgment—the failure to obtain an acknowledgment for the receipt of information.
4. Coordination—any factor associated with a failure to exchange requirement information. This includes coordination between individuals, positions of operation, and facilities for exchange of information such as APREQ's, position reports, and forwarding of flight data.
5. Position relief briefing—relief briefing errors are special errors of both communication and coordination that occur as the result of po-

sition relief. They include such things as failure to give a relief briefing, failure to request a briefing, and an incomplete or erroneous briefing.

# REFERENCES

Durso, F. T., Truitt, T. R., Hackworth, C. A., Crutchfield, J. M., & Manning, C. A. (1998). En Route Operational Errors and Situation Awareness. *The International Journal of Aviation Psychology, 8,* 177–192.

Endsley, M. R. (1995). Toward a theory of situation awareness. *Human Factors, 37*(1), 32–64.

Endsley, M. R., & Rodgers, M. D. (1994). *Situation awareness information requirements for en route Air Traffic Control* (Report No. DOT/FAA/AM–94/27). Washington, DC: Federal Aviation Administration.

Hart, S. G., & Staveland, L. (1988). Development of the National Aeronautics and Space Administration (NASA) Task Load Index (TLX): Results of empirical and theoretical research. In P. A. Hancock & N. Meshkati (Eds.), *Human mental workload* (pp. 139–183). Amsterdam: North Holland.

Kershner, A. M. (1968). *Air traffic control system error data for 1965 and 1966 as related to age, workload, and time-on-shift of involved controller personnel* (Report No. NA-68-32). Atlantic City, NJ: Federal Aviation Administration.

Kinney, G. C., Spahn, J., & Amato, R. A. (1977). *The human element in air traffic control: Observations and analyses of the performance of controllers and supervisors in providing ATC separation services* (Report No. MTR-7655). McLean, VA: METREK Division of the MITRE Corporation.

Klein, G. (1993). *Naturalistic decision making: Implications for design.* Wright-Patterson Air Force Base, OH: Crew System Ergonomics Information Analysis Center.

Mogford, R. H., Murphy, E. D., & Guttman, J. A. (1994). Using knowledge exploration tools to study airspace complexity in air traffic control. *The International Journal of Aviation Psychology, 4,* 29–45.

National Transportation Safety Board. (1988). *Northwest Airlines, Inc., McDonnell Douglas DC-9-82, N312RC, Detroit Metropolitan Wayne County Airport, Romulus, Michigan, August 16, 1987.* Washington, DC: Author.

National Transportation Safety Board. (1995a). *Controlled collision with terrain, Flagship Airlines, Inc., dba American Eagle, Flight 3379, BAE Jetstream 3201, N981AE, Morrisville, North Carolina, December 13, 1994.* Washington, DC: Author.

National Transportation Safety Board. (1995b). *Flight into terrain during missed approach, USAir Flight 1016, DC-9-31, N954VJ, Charlotte/Douglas International Airport, Charlotte, North Carolina, July 2, 1994.* Washington, DC: Author.

Orasanu, J. M. (1993). Decision making in the cockpit. In E. L. Wiener, R. L. Helmreich, and B. G. Kanki (Eds.), *Cockpit resources management* (pp. 137–172). New York: Academic Press.

Rodgers, M. D., & Duke, D. A. (1993). *SATORI: Situation assessment through re-creation of incidents* (Report No. DOT/FAA/AM–93/12). Washington, DC: Federal Aviation Administration.

Rodgers, M. D., Mogford, R. H., & Mogford, L. S. (1998). The relationship of sector characteristics to operational errors. *ATC Quarterly, 5*(4), 241–263.

Rodgers, M. D., & Nye, L. G. (1993). Factors associated with the severity of operational errors at air route traffic control centers. In M. D. Rodgers (Ed.), *An examination of the opera-*

*tional error database for air route traffic control centers* (Report No. DOT/FAA/AM–93/22). Washington, DC: Federal Aviation Administration.

Schroeder, D. J. (1982). The loss of prescribed separation between aircraft: How does it occur? In *Proceedings (P-114), Behavioral Objectives in Aviation Automated Systems Symposium* (pp. 257–269). Washington, DC: Society of Automotive Engineers.

Schroeder, D. J., & Nye, L. G. (1993). An examination of the workload conditions associated with operational errors/deviations at Air Route Traffic Control Centers. In M. D. Rodgers (Ed.), *An examination of the operational error database for air route traffic control centers* (Report No. DOT/FAA/AM–93/22). Washington, DC: Federal Aviation Administration.

Siddiqee, W. (1973). A Mathematical model for predicting the number of potential conflict situations at intersecting air routes. *Transportation Science, 7*(2), 158–167.

Stager, P., & Hameluck, D. (1990). Ergonomics in air traffic control. *Ergonomics, 33*, 493–9.

U.S. Department of Transportation (1991a). *Facility operation and administration* (FAA Order 7210.3). Washington, DC: Federal Aviation Administration.

U.S. Department of Transportation (1991b). *Facility operation and administration* (FAA Order 7210.46). Washington, DC: Federal Aviation Administration.

# Chapter 5

# Subjective Measures of Situation Awareness

Debra G. Jones
*SA Technologies*

Subjectively assessing Situation Awareness (SA) typically involves assigning a numerical value to the quality of SA during a particular period or event (Fracker & Vidulich, 1991). The most common subjective rating techniques utilize a linear scale with verbal descriptors at the endpoints, and many researchers have employed multiple subscales within the rating based on the theory that SA is a multidimensional construct. Subjective assessments of SA are popular because these techniques are fairly inexpensive, easy to administer, and nonintrusive (Endsley, 1996; Fracker & Vidulich, 1991; Taylor & Selcon, 1991). Furthermore, unlike other types of SA metrics, subjective estimations can be collected in controlled real-world settings, as well as during the evaluation of design concepts in simulation studies (Endsley, 1996). Subjective estimations of SA may be made by individual operators regarding their own SA (i.e., self-ratings) or by experienced observers regarding another operator's SA (i.e., observer ratings).

## Self-Ratings

Self-rating techniques involve operators subjectively assessing their own SA and reporting it on a rating scale. One of the main criticisms of self-rating techniques is that operators cannot be aware of their own lack of SA. However, multidimensional rating scales break SA down into its components

**113**

that are, arguably, available for self-rating. This assertion is supported by the validation of techniques across a broad range of settings (S. J. Selcon, personal communication, June 3, 1998). Utilizing self-rating techniques to evaluate SA affords other benefits as well. First, self-rating measures provide a source of insight into the nature of the underlying cognitive process involved in the reported subjective SA experience (Taylor & Selcon, 1991). Second, even though these measures may not provide an actual indication of the operator's true SA, they can provide an indication of the operator's confidence level regarding his or her SA (Endsley, 1994).

Problems exist, however, with the use of self-rating metrics to assess SA and these problems limit the conclusions that can be drawn from such measures. First, if operators are queried about their SA during the session, they report what they perceive. However, the operators' knowledge may not be correct and the reality of the situation may be quite different from what the operators believe (Endsley, 1994, 1995). For example, in a study comparing several cockpit displays designed to facilitate spatial orientation, Fracker and Vidulich (1991) found that the display that produced the best subjective ratings of SA also resulted in the greatest percentage of inverted recoveries; that is, pilots believed they were upright when actually they were inverted. This vivid illustration is a reminder that subjective SA ratings should not be used in isolation but should be interpreted in light of other data such as performance data.

Second, SA may be highly influenced by self-assessments of performance and thus become biased by issues that are beyond the SA construct (Endsley, 1996). Venturino, Hamilton, and Dvorchak (1989) found a high correlation between post-trial subjective measures of SA and performance. That is, operators rated their SA as good if the trial had a positive outcome regardless of whether good SA, luck, or other factors influenced performance. Similarly, the operators reported their SA as poor if the outcome was less than favorable regardless of where the error occurred (e.g., poor SA, insufficient skill or knowledge selection, poor action selection) (Endsley, 1994, 1995). Thus, once a situation has unfolded, a person's memory of what their SA was earlier in the session can be influenced by the outcome, thereby limiting the usefulness of posttrial subjective measures.

Finally, self-assessments of SA assume that individuals can judge and accurately report feelings, thoughts, and facts from experience (Taylor & Selcon, 1991). Errors and systematic bias in human judgment and recall (e.g., anchoring, availability, recency, and saliency), as well as limits on working memory, hinder the levels of accuracy and sensitivity that can be achieved with these measures (Taylor & Selcon, 1991). Furthermore, self-ratings may not provide an accurate quantification of SA since operators may not realize inaccuracies exist in their SA or that information exists of which they are unaware (Endsley, 1996). This issue must be carefully addressed in scale development because finding out what is not known requires knowledge about the questions that need to be asked (Taylor & Selcon, 1991).

### Observer Ratings

Observer ratings involve a trained observer assessing an operator's SA. Observer ratings are appealing since the trained observer typically has more information than the operator about what is really happening in a given simulation, and thus their knowledge of reality may be more complete (Endsley, 1996). However, the observers will have only limited knowledge about what the operator's concept of the situation is and must rely on overt indications to determine the operator's SA. Operator actions and verbalizations may provide useful information regarding explicit SA problems such as misperceptions or lack of knowledge and they may provide an indication that certain information is known (Endsley, 1996). However, actions and verbalizations most likely do not provide a complete representation of an operator's SA; the operator may have a much greater store of information held internally that is not verbalized (Endsley, 1994). Thus, observer ratings provide only a partial indicator of an operator's SA. A variation of this method employs a confederate in the study acting as an associate of the operator (e.g., another crew member or an air traffic controller) and who requests certain information from the operator to encourage more verbalization (Sarter & Woods, 1991). These types of efforts to elicit more information may augment natural verbalizations, but they also may alter the operator's distribution of attention and thus alter SA (Endsley, 1996).

A variety of subjective scales have been proposed and range from simple linear scales (e.g., rating SA on a scale from 1 to 10) to more complex scales that consider the effect of more complex cognitive constructs. For example, in an early attempt to subjectively measure SA, a scale was developed that probed six aspects of SA particularly relevant to tactical air combat (Arbak, Schwartz, & Kuperman, 1987). To complete this scale, subjects rated six statements on a scale from 1 (completely disagree) to 7 (completely agree). The scale reflected consistently higher SA in the manner expected in the study, but the lack of testing of the scale prevents making any conclusion regarding the scale's effectiveness (Arbak, Schwartz, & Kuperman, 1987).

When evaluating the effectiveness of a subjective SA assessment technique, two important issues must be considered: scale validity and scale sensitivity. Validity concerns the extent to which a procedure measures what it purports to measure, whereas sensitivity (a function of scale construction) refers to the capability of a technique to discriminate significant variations in SA (Taylor & Selcon, 1991). An assessment technique that effectively measures SA will possess a high degree of validity and sensitivity. Numerous scales have been developed, with varying degrees of success, to subjectively evaluate SA, including unidimensional scales, Situation Awareness Rating Technique (SART), Situation Awareness-Subjective Workload Dominance (SA-SWORD), and Situational Awareness Rating Scale (SARS).

## UNIDIMENSIONAL SCALES

The use of unidimensional scales to assess SA is attractive because they are easily implemented and interpreted. Unidimensional techniques typically require the operator to rate their SA on a linear scale, either on a line of a certain length or on a numerical scale.

### Validity and Sensitivity

Several studies have investigated the effectiveness of unidimensional scales to assess SA. One study examined the use of an unidimensional scale while investigating the effectiveness of SART on tasks involving skill-based behavior. This study found a lack of diagnostic power in the unidimensional SA scale and supported the assertion that a multidimensional scale is needed to accurately represent the complexity of SA (Selcon & Taylor, 1990).

Another study (Vidulich, Crabtree, & McCoy, 1993) investigated the construct validity and reliability of an unidimensional subjective scale. Construct validity was evaluated by assessing the scale's sensitivity to the experimental manipulations and reliability was examined by calculating the test-retest correlation. On this scale, a single 7-point scale designated the Overall-SA scale, subjects were asked to rate their SA from low to high.

The experimental task examined the effectiveness of the scale in detecting differences in SA in an air-to-ground attack simulation, and it sought to manipulate SA by altering the aircraft display conditions in three ways: optimum, moderate, and degraded display conditions. Subjective SA ratings were collected at designated points in each of two sessions. The theory in this study was that good SA should have produced better performance. Analysis of the performance measures indicated a significant effect of display type and suggested that the intended SA manipulation through differences in display conditions was successful (Vidulich, Crabtree, & McCoy, 1993).

The Overall-SA ratings for each of the experimental conditions from each of the two sessions were averaged for each subject and then analyzed.

No statistically significant effects were found, indicating that the SA ratings were not affected by experimental conditions as were the performance measures. Based on the study's premise, however, improvements in performance should coincide with improvements in SA and these improvements should have been reflected in the subjective assessment of SA. Because the unidimensional rating scale did not reflect any differences in SA, this scale does not appear to provide an useful measure of SA. Finally, the subjects' ratings from the two sessions were correlated and the mean test-retest correlation was calculated and analyzed. Because this correlation was not statistically different from zero, an absence of scale reliability was indicated. The lack of any display effect in the Overall-SA ratings implied that a simple unidimensional SA scale was not an effective SA measure, and the failure of the unidimensional Overall-SA scale should be a warning against relying on such a simplistic approach for evaluating SA (Vidulich, Crabtree, & McCoy, 1993).

**Advantages and Disadvantages**

The primary advantage of an unidimensional scale is the ease of administering the scale and interpreting the data (i.e., the results can be directly read from the scale). Some researchers feel that when the research question is simple, a single scale might be adequate (Fracker & Vidulich, 1991), but the general consensus is that a single scale is not sufficient to capture the richness and complexity of the SA construct (Selcon & Taylor, 1990; Vidulich, Crabtree, & McCoy, 1993). So far, no evidence has been found to support the validity and sensitivity of a unidimensional scale for subjectively assessing SA.

**SART**

SART is another example of a subjective metric of SA and this self-rating scale is one of the best known (Endsley, 1996) and most thoroughly tested subjective technique. SART assumes that operators "use some understanding of situations in making decisions, that this understanding is available to consciousness and that it can readily be made explicit and quantifiable" (Taylor, 1990, chap. 3, p. 2).

To avoid the circularity of measuring only what was previously defined as SA, an a priori definition of SA was not used in scale development. Instead, the definition was developed from information gathered from experienced aircrew (Selcon & Taylor, 1990; Taylor & Selcon, 1991). For scale development, knowledge elicitation techniques were utilized to determine which elements the aircrew considered essential for good SA

| Construct | Definition |
|-----------|------------|
| Instability of situation | Likeliness of situation to change suddenly |
| Variability of situation | Number of variables which require one's attention |
| Complexity of situation | Degree of complication (number of closely connected parts) of situation |
| Arousal | Degree to which one is ready for activity (sensory excitability) |
| Spare Mental Capacity | Amount of mental ability available to apply to new variables |
| Concentration | Degree to which one's thoughts are brought to bear on the situation |
| Division of Attention | Amount of division of attention in the situation |
| Information Quantity | Amount of knowledge received and understood |
| Information Quality | Degree if goodness or value of knowledge communicated |
| Familiarity | Degree of acquaintance with situation experience |

FIG. 5.1  SART construct definitions. *Note.* From "Subjective measurement of situational awareness" by R. M. Taylor and S. J. Selcon. In *Designing for everyone, Proceeds of the 11th Congress of the International Ergonomics Association,* by Queinnec and Daniellou (Eds), (1991). London: Taylor & Francis. Copyright 1991. Crown copyright is reproduced with the permission of the Controller of Her Magesty's Stationery Office.

(Taylor, 1990). From these interviews, 10 generic SA constructs emerged (Selcon & Taylor, 1990; Taylor, 1990; Taylor & Selcon, 1991): instability of situation, variability of situation, complexity of situation, arousal, spare mental capacity, concentration, division of attention, information quantity, information quality, and familiarity (see Fig. 5.1 for definitions). Although the choice of constructs had a degree of arbitrariness, delineation of these constructs was governed by factors such as elicitation frequency, principal coordinate clustering, and intercorrelations, as well as by the goals of simplification, parsimony, and theoretical consistency (Taylor & Selcon, 1991).

The 10 generic constructs were found to cluster into three broad domains (Taylor, 1990). *Attentional demand* encompasses the constructs of instability of situation, variability of situation, and complexity of situation; *attentional supply* includes the constructs of arousal, spare mental capacity, concentration, and division of attention; and *understanding* incorporates information quantity, information quality, and familiarity. These three domains constitute an abbreviated 3-dimensional SART scale that can be used when a shorter scale is advantageous.

Thus, subjective estimates of SA can be collected using either the 10-dimensional or the 3-dimensional SART. The given application and the degree of intrusiveness permitted by the measured task indicates the most appropriate scale to use (Taylor, 1990). When utilizing the 3-D SART, raters typically mark their rating on a continuous 100-millimeter line with the endpoints *low* (0 mm) to *high* (100 mm) (see Fig. 5.2). Current implementations of the 10-dimensional SART employ a 7-point rating scale for each of the 10 generic constructs on which raters quantify their qualitative observations (Taylor & Selcon, 1991) (see Fig. 5.3). The choice of the number of intervals on the scale was drawn from the findings in psychophysics that an untrained subject can absolutely identify between five and nine stimuli, depending on the continuum. Based on these guidelines, the choice of a 7-interval scale was reasonable (Taylor & Selcon, 1991).

The ratings from the SART scale are assumed to have the properties of interval data and therefore to be amenable to factor analysis. Since concerns exist as to whether the ratings should be treated as parametric data with interval rather than ordinal properties, the SART data should be tested for satisfaction of interval data requirements and any necessary statistical transformations applied before interpreting the data (Taylor & Selcon, 1991).

The ratings on the 3-D SART scale can be combined to create a single value for SA by using the algorithm SA(calc) = Understanding – (Demand – Supply). Since this formula was derived from theoretical considerations of how the three domains interact rather than empirical or statistical evaluation, it contains an element of arbitrariness; consequently, undue emphasis should not be placed on the single calculated value (Selcon, Taylor, & Shadrake, 1992). Furthermore, although a single number rating for SA

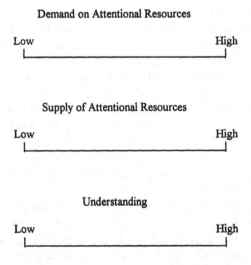

FIG. 5.1   3-D SART scale. *Note.* From "Evaluation of the situational awareness rating technique" by S. J. Selcon and R. M. Taylor. In *Situational Awareness in Aerospace Operations* (p. 2), AGARD-CP-478 (1990), Neuilly Sur Seine, France: NATO–AGARD. Copyright 1990. Crown copyright is reproduced with the permission of the Controller of Her Majesty's Stationery Office.

| | | Low | | | | High | | |
|---|---|---|---|---|---|---|---|---|
| | | 1 | 2 | 3 | 4 | 5 | 6 | 7 |
| DEMAND | Instability of Situation | | | | | | | |
| | Variability of Situation | | | | | | | |
| | Complexity of Situation | | | | | | | |
| SUPPLY | Arousal | | | | | | | |
| | Spare Mental Capacity | | | | | | | |
| | Concentration | | | | | | | |
| | Division of Attention | | | | | | | |
| UNDERS. | Information Quantity | | | | | | | |
| | Information Quality | | | | | | | |
| | Familiarity | | | | | | | |

FIG. 5.3. 10–D SART scale. *Note.* From "Evaluation of the situational awareness rating technique" by S. J. Selcon and R. M. Taylor. In *Situational Awareness in Aerospace Operations* (p. 2), AGARD–CP–478, (1990), Neuilly Sur Seine,
France: NATO–AGARD. Copyright 1990 by Crown copyright is reproduced
with the permission of the Controller of Her Majesty's Stationery Office.

can be useful in comparative system design evaluation, absolute SART
scores should not be specified as system specific design objectives (Taylor
& Selcon, 1991).

## Validity and Sensitivity

Numerous studies have been performed to examine the validity and sensitivity of the SART scale. In one study, RAF pilots rated a videotaped sequence of air combat (as if they had been the pilot in the sequence) at
three levels of difficulty and SART showed sensitivity to task difficulty and
operator experience (Selcon, Taylor, & Koritsas, 1991). In another study,
the structures of both the 3-D and 10-D SART were tested using decision-making tasks by navigators and pilots and found to be reasonably robust (Taylor, 1990; Taylor & Selcon, 1991). Another study examined the
effectiveness of SART in assessing SA in tasks characterized by skill-based
behavior (a multiple-task compatibility study), rule-based behavior (an aircraft attitude recovery study), and knowledge-based behavior (a warnings
comprehension study) (Selcon & Taylor, 1990). In each case, support was
provided for the internal structure of SART and the results indicated a
sensitivity to the design variables (Selcon & Taylor, 1990).

In the study by Vidulich, Crabtree, and McCoy (1993), in which subjective SA metrics were evaluated in an air-to-ground attack simulation, an
Overall-SART rating was examined for validity and sensitivity. The Overall-SART rating was calculated from the 10-D SART scale by utilizing a
variation of the formula SA(calc) = Understanding – (Supply – Demand).
For this Overall-SART score, the mean value of the subdimensions associ-

ated with the three main categories (Demand, Supply, and Understanding) were first calculated, and then the SA(calc) algorithm was applied to these means to obtain the single numerical Overall-SART score (Vidulich, Crabtree & McCoy, 1993).

The Overall-SART ratings for each of the experimental conditions from the two sessions were averaged for each subject and analyzed, and a significant main effect of display was reflected in the Overall-SART scores. This finding suggested that the technique could discriminate the changes in SA that occurred when the different displays were in use. However, analysis of the SART subscale showed that the subjects perceived the display manipulation as a change in demand rather than understanding and that called into question what was being manipulated—SA or workload (Vidulich, Stratton, & Wilson, 1994). Furthermore, the mean test-retest correlation failed to show statistical significance. Thus, the Overall-SART rating produced a measure that appears to be sensitive to changes in SA, but the lack of test-retest reliability raises cause for concern (Vidulich, Crabtree, & McCoy, 1993).

## Advantages and Disadvantages

SART has several advantages in addition to the general advantages of subjective measures. First, SART has high ecological validity since its dimensions were derived directly from operational aircrew (Selcon & Taylor, 1990). Second, the constructs are general in nature and therefore have the potential to be applicable to nonaircrew domains (Taylor & Selcon, 1991). Next, SART provides a certain level of diagnosticity. Diagnosticity refers to a technique's ability to discriminate the causes of differences in the construct it is measuring (in this case SA) and to generate predictions regarding the construct. Although only limited causal and predictive inferences can be drawn (because SA is a state of knowledge rather than a causal process), resolving SA into the individual SART dimensions provides some diagnostic and predictive indicators for delineating the strengths and weakness associated with SA as measured by the scale (Taylor & Selcon, 1991).

Finally, proponents suggest that workload is an integral part of the multidimensional SA construct (Taylor, 1990) and that understanding the relation between SA and workload is essential in order for SA to be an effective design evaluation measure (Selcon, Taylor, & Koritsas, 1991). Thus, since SART takes into account the supply and demand of attentional resources (generally considered workload constructs), it should provide some measure of how changes in workload affect SA (Selcon & Taylor, 1990).

However, like other subjective measures, SART data should always be interpreted in light of performance data since the operators' assessment of SA does not always match the reality of the situation. Additionally, although SART proponents see the inclusion of workload aspects within the measurement scale as an advantage, critics argue that including workload elements within the SA scale confound the measure of SA by confusing it with workload. For example, SART has been shown to be correlated with performance measures, but it is unclear whether this is due to the workload component or the understanding (i.e., SA) component of the scale (Endsley, 1996).

Finally, the necessity and sufficiency of the 10 general constructs within the 3 domains has not been clearly established (Taylor & Selcon, 1991). Developers of the scale acknowledge that considerable scope for scale development remains (e.g., through description improvement, interval justification, and the use of conjoining scaling techniques to condense multidimensional ratings into a single SA score) (Taylor & Selcon, 1991).

## SA-SWORD

Another scale designed to subjectively assess SA was derived from a subjective workload assessment tool called the SWORD technique (Vidulich & Hughes, 1991). This technique was developed by a mathematician from a decision-making procedure called the Analytic Hierarchy Process (AHP). The AHP was found to possess a high level of sensitivity and test-retest reliability, so SWORD was developed to standardize this procedure for use as a workload metric. SWORD uses pair-wise comparisons for data collection and then utilizes a geometric means rating calculation algorithm.

Due to the success of SWORD as a workload metric, this technique was adapted for use as a SA metric (Vidulich & Hughes, 1991). This adaptation required no changes in the data collection process or in the analysis procedures; rather it simply necessitated a change in the instructions to the operators. When applied to the SA domain, this rating scale is called SA-SWORD to distinguish it from the workload measure.

### Validity and Sensitivity

SA-SWORD was tested in a study evaluating two types of displays (Vidulich & Hughes, 1991): the Fire Control Radar (FCR) display that provided relatively raw information from the forward returns of the aircraft's own radar (as is common in the current generation of air-to-air radar displays), and the Horizontal Situation Format (HSF) display that was a map-like display utilizing data-linked information from outside sources (e.g., from an AWACS aircraft) and data provided by the aircraft's own instruments to provide a bird's-eye view of the entire area. Since the HSF display contained vital information that was missing from the FCR display (e.g., information regarding the approach of threats from behind), the study assumed that a potent manipulation of SA was achieved—that is, the HSF display provided more information in an integrated format so it should afford better SA. Thus, according to the study's premise, the sensitivity of a SA metric to the display manipulation would constitute a test of the metric's construct validity (Vidulich & Hughes, 1991).

The study evaluated the two displays during two segments of flight (ingress and engagement) at two levels of enemy threat (low and high). Following the data collection trials, SA-SWORD evaluations were collected. Pilots rated their SA on all possible combinations of displays (HSF vs. FCR), segment (ingress vs. engagement), and threat (low vs. high). Additionally, during a structured interview, subjects were queried regarding the acceptability of the SA-SWORD metric in terms of perceived validity and the ease of use. A three-way analysis of variance found only a main effect of display—SA ratings were higher when the HSF display was in use.

Based on this study's philosophy, the SA-SWORD results supported the assumption that the HSF display improved SA by providing more of the information required to maintain a higher level of SA. However, critics suggest that these differences may have been due to display preferences as opposed to sensitivity to true SA (Endsley, 1995). Furthermore, the lack of sensitivity to differences in flight segments (i.e., ingress and engagement) is disturbing and counterintuitive since SA was expected to be different for the relatively uneventful ingress compared with combat engagement. Further analysis revealed a potential explanation for the unexpected results based on the subjects' definitions of SA. Approximately half of the subjects associated SA with the amount of information they were attempting to track. Thus, this group rated their SA as higher during combat engagement. The other subgroup appeared to associate SA with how much potentially relevant information they thought might be missing and rated SA as lower during the combat engagement phase. The SA-SWORD metric can be strengthened by providing a more detailed definition of SA during the instructions to minimize discrepancies caused by differences in SA definitions. Despite rating differences in flight segments, both groups were uniform in rating the HSF display as superior to the FCR display.

Because each pilot performed the SA-SWORD evaluation only once, a test-retest correlation calculation was not possible. As an alternative, an interrater correlation was calculated. The logic behind this calculation was that if the SA-SWORD ratings are reliable, then multiple raters should experience similar reactions to the experimental conditions. Nine out of 10 pilots positively correlated with the others and the interrater correlation suggested that the SA-SWORD ratings were reliably related to the task conditions.

The patterns of results found by the SA-SWORD ratings were distinctly different than those observed in the workload data that found workload measures were generally very sensitive to segment and generally insensitive to display. According to the researchers, these patterns of results demonstrated that workload and SA metric complement each other as separate parts of a complete evaluation.

### Advantages and Disadvantages

The sensitivity and interrater reliability demonstrated by the SA-SWORD scale in the study by Vidulich and Hughes (1991) suggested that the technique holds promise as an easily implementable subjective measure of SA. Additionally, this metric possesses face validity and appears to have user acceptance. However, because this study was the first attempt to utilize SA-SWORD to assess SA, no final conclusion can be drawn about its effectiveness. Furthermore, proponents acknowledge that SA-SWORD is not a solution to the problem of SA metrics because it is likely that any single metric of SA (particularly subjective metrics) is prone to provide only a part of the picture regarding SA in complex tasks. Nonetheless, proponents feel the current results suggest that SA-SWORD can be a useful adjunct to workload metrics and a useful tool for assessing subjective SA (Vidulich & Hughes, 1991).

### SARS

SARS represents another method for subjectively measuring SA (Bell & Waag, 1995; Waag & Houck, 1994). Three objectives motivated development of this scale (based in the tactical air environment): to define SA, to determine the degree to which pilots can judge other pilots in terms of SA, and to examine the relation between such judgments and performance. In defining SA, this scale emphasizes "perceiving what is important" in the environment and then "using that perception to guide the selection and performance of appropriate behaviors" (Bell & Waag, 1995). The SARS scale was developed through interviews with experienced F-15 pilots, and it consists of a list of 31 behavior elements of SA considered important to mission success. The 31 behavior elements represent 8 categories of mission performance (See Fig. 5.4). When completing the SARS scale, the subjects rate each of the 31 elements on a 6-point scale with the anchors *acceptable* and *outstanding*. (*Acceptable* was chosen as the low end of the scale because all the pilots in the study were mission ready.)

1. GENERAL TRAITS
   Discipline
   Decisiveness
   Tactical knowledge
   Time-sharing ability
   Spatial ability
   Reasoning ability
   Flight management
2. TACTICAL GAME PLAN
   Developing plan
   Executing plan
   Adjusting plan on-the-fly
3. SYSTEM OPERATION
   Radar
   Tactical electronic warfare system
   Overall weapons system proficiency
4. COMMUNICATION
   Quality (brevity, accuracy, timeliness)
   Ability to effectively use information

5. INFORMATION INTERPRETATION
   Interpreting vertical situation display
   Interpreting threat warning system
   Ability to use controller information
   Integrating overall information
   Radar sorting
   Analyzing engagement geometry
   Threat prioritization
6. TACTICAL EMPLOYMENT-BVR
   Targeting decisions
   Fire-point selection
7. TACTICAL EMPLOYMENT-VISUAL
   Maintain track of bogeys/friendlies
   Threat evaluation
   Weapons employment
8. TACTICAL EMPLOYMENT-
   GENERAL
   Assessing offensiveness/defensiveness
   Lookout
   Defensive reaction
   Mutual support

FIG. 5.4.  Items and categories used in SARS. *Note.* From "Tools for assessing situational awareness in an operational fighter environment," by W. L. Waag and M. R. Houck, 1994, *Aviation, Space, and Environmental Medicine, ?,* p. A15. Copyright 1994 by Aerospace Medical Association. Reprinted with permission.

## Validity and Sensitivity

To test SARS, subjects were asked to rate themselves and their peers using the SARS scale. Additionally, subjects were asked to rate the other mission-ready pilots in their squadron based on their general fighter pilot ability as well as their SA ability, and then to rank order them according to their SA ability. Supervisory SARS were also completed by squadron leaders for their subordinates. For the peer and the supervisory SARS, the raters were given the option to omit a pilot if they did not have enough information to accurately rate that particular pilot.

In order to analyze the data, summary scores were created for each of the three types of SARS (Waag & Houck, 1994). For self-report SARS, 9 summary scores were created by averaging the scores on each item within the 8 categories, and an overall score was calculated from the means of all 31 items. For the supervisory SARS the same nine scores were calculated for each rater's assessment of each ratee and then these scores were averaged across all raters to create a final nine summary scores. For the peer SARS, three summary scores were created for each ratee by each rater: ratings of fighter pilot ability, ratings of SA ability, and rank order. The means of these three scores were calculated across all raters and used as the final peer SA scores. Analysis of this data showed that the reliability estimates were quite high in all cases, indicating that the scale consistently measured "whatever" it was measuring (Waag & Houck, 1994). An analysis of the data generated by the peer and supervisory SARS suggested that the pilots could reliably classify their fellow pilots in terms of SA ability. The scale was also found to have high internal consistency and interrater reliability for both peer SARS and supervisory SARS (Bell & Waag, 1995).

In an attempt to determine the relation between SA and mission performance, a composite SA score was calculated from the three peer summary scores and the eight categorical supervisory summary scores. The SA score was calculated using a principal components analysis and then transformed into a distribution with a mean of 100 and a standard deviation of 20 (Waag & Houck, 1994). Based on this composite score, 40 mission-ready flight leads (covering the range of SA scores) were selected to participate in a series of simulated air-to-air combat missions. Imbedded in each of the simulations were events designed to trigger specific goal-directed behaviors necessary for mission accomplishment; SA was inferred from the subjects' responses to these events.

Since measuring air combat skills is difficult, the method of analysis chosen at this point was a behavioral observation by two subject-matter experts (SMEs) who were unaware of the pilots' SA ratings. After each simulator session, the SMEs discussed each engagement and completed a consensus performance rating scale containing 24 behavior indicators based on the SARS.

Analysis of the data indicated a significant relation between squadron ratings of SA and performance in simulated air combat missions. The conclusions from this study were that the results indicated that "SA is a construct that has meaning and can be used by both peers and supervisors to classify mission ready pilots," and that the squadron ratings of SA are correlated with mission success in simulated air combat missions (Bell & Waag, 1995). Thus, this scale appears to measure those factors it was designed to measure.

### Advantages and Disadvantages

The SARS scale appears to assess what it was developed to assess; however, it does not provide a subjective measure of SA in the same sense as the other scales. In the case of SARS, SA is regarded more as an innate ability rather than as a changeable state of knowledge. Additionally, the scale combines assessments on many dimensions besides SA, including decision-making abilities, flight skills, performance, and the subjective impressions of a person's personality traits (Endsley, 1996). Moreover, the scale is closely tied to the particular aircraft type and mission, so the applicability of this measure to other domains is doubtful. Likewise, the usefulness of this scale to evaluate systems instead of individuals is unclear (Endsley, 1996). Finally, because SARS is a fairly new technique that has received only limited testing, more studies are needed to validate the scale before its true utility can be judged.

## CONCLUSION

The ease of implementation, ease of administration, low cost, and non-intrusiveness, as well as the ability to utilize these scales in controlled real-world settings, make subjective SA assessment techniques a popular choice. Although numerous subjective scales have been proposed, few of these scales have been subjected to extensive testing and evaluation for validity and sensitivity. Scales that take into account the multidimensionality and complexity of SA are more likely than single factor scales to provide some insight regarding the SA construct. Nonetheless, the problems inherent in subjective metrics demand that caution be employed when interpreting the data gained from these scales. Furthermore, even under the best circumstances, subjective scales only provide a partial picture of operator SA. For example, the very nature of a "subjective" assessment prevents the scale from being trusted to have complete accuracy—that is, the operator's self-assessment may not reflect the true situation. For this reason, subjective SA data may most effectively be used in conjunction with more objective data such as performance measures.

Proponents of subjective measures suggest that utilizing subjective measures is advantageous because these measures are more closely related to higher order psychological constructs than are other forms of measures (Bell & Waag, 1995). Undoubtedly, utilizing subjective measures provides insight into the operator's subjective experience that cannot be gained from other mediums. Thus, the strength in utilizing subjective assessment metrics is that of a complement rather than a replacement to other forms of SA metrics.

## REFERENCES

Arbak, C. J., Schwartz, N., & Kuperman, G. (1987). Evaluating the panoramic cockpit controls and displays system. In *Proceedings of the Fourth International Symposium on Aviation Psychology*, *1*, 30–36. Columbus: The Ohio State University.
Bell, H. H., & Waag, W. L. (1995). Using observer ratings to assess situational awareness in tactical air environments. In D. J. Garland & M. R. Endsley (Eds.), *Experimental analysis*

*and measurement of situation awareness* (pp. 93–99). Daytona Beach, FL: Embry-Riddle Aeronautical University Press.

Endsley, M. R. (1994). Situation awareness in dynamic human decision making: Measurement. In R. D. Gilson, D. J. Garland, & J. M. Koonce (Eds.), *Situational awareness in complex systems* (pp. 79–97). Daytona Beach, FL: Embry-Riddle Aeronautical University Press.

Endsley, M. R. (1995). Measurement of situation awareness in dynamic systems. *Human Factors, 37*(1), 65–84.

Endsley, M. R. (1996). Situation awareness measurement in test and evaluation. In T. G. O'Brien & S. G. Charlton (Eds.), *Handbook of human factors testing and evaluation* (pp. 159–180). Mahwah, NJ: Lawrence Erlbaum Associates.

Fracker, M. L., & Vidulich, M. A. (1991). Measurement of situation awareness: A brief review. In Y. Queinnec & F. Daniellou (Eds.), *Designing for everyone, Proceedings of the 11th Congress of the International Ergonomics Association* (pp. 795–797). London: Taylor & Francis.

Sarter, N. B., & Woods, D. D. (1991). Situation awareness: A critical but ill-defined phenomenon. *The International Journal of Aviation Psychology, 1*(1), 45–57.

Selcon, S. J., & Taylor, R. M. (1990). Evaluation of the situational awareness rating technique (SART) as a tool for aircrew systems design. In *Situational Awareness in Aerospace Operations* (AGARD–CP–478; pp. 5/1–5/8). Neuilly Sur Seine, France: NATO–AGARD.

Selcon, S. J., Taylor, R. M., & Koritsas, E. (1991). Workload or situational awareness?: TLX vs. SART for aerospace systems design evaluation. In *Proceedings of the Human Factors Society, 35th Annual Meeting* (Vol. 1, pp. 62–66). Santa Monica, CA: Human Factors Society.

Selcon, S. J., Taylor, R. M., & Shadrake, R. A. (1992). Multi-modal cockpit warnings: Pictures, words, or both? In *Proceedings of the Human Factors Society 36th Annual Meeting* (Vol. 1, pp. 57–61). Santa Monica, CA: Human Factors Society.

Taylor, R. M. (1990). Situational awareness rating technique (SART): The development of a tool for aircrew systems design. In *Situational Awareness in Aerospace Operations* (AGARD–CP–478; pp. 3/1–3/17). Neuilly Sur Seine, France: NATO–AGARD.

Taylor, R. M., & Selcon, S. J. (1991). Subjective measurement of situational awareness. In Queinnec & Daniellou (Eds.), *Designing for everyone, Proceedings of the 11th Congress of the International Ergonomics Association* (pp. 789–791). London: Taylor & Francis.

Venturino, M., Hamilton, W. L., & Dvorchak, S. R. (1989). Performance-based measures of merit for tactical situation awareness. In *Situational Awareness in Aerospace Operations* (AGARD–CP–478; pp. 4/1–4/5). Neuilly Sur Seine, France: NATO–AGARD.

Vidulich, M. A., Crabtree, M. S., & McCoy, A. L. (1993). Developing subjective and objective metrics of pilot situation awareness. In R. S. Jensen & D. Neumeister (Eds.), *Proceedings of the 7th International Symposium on Aviation Psychology* (pp. 896–900). Columbus: The Ohio State University.

Vidulich, M. A., & Hughes, E. R. (1991). Testing a subjective metric of situation awareness. In *Proceedings of the Human Factors Society 35th Annual Meeting* (pp. 1307–1311). Santa Monica, CA: Human Factors Society.

Vidulich, M. A., Stratton, M., Crabtree, M., & Wilson, G. (1994). Performance-based and physiological measures of situational awareness. *Aviation, Space, and Environmental Medicine, 65*(5 Suppl.), A7–A12.

Waag, W. L., & Houck, M. R. (1994). Tools for assessing situational awareness in an operational fighter environment. *Aviation, Space, and Environmental Medicine, 65*(5 Suppl.), A13–A19.

# Chapter 6

# Using Observer Ratings to Assess Situation Awareness[1]

Herbert H. Bell
*Air Force Research Laboratory*
*Human Effectiveness Directorate*

Don R. Lyon
*Link Simulation and Training*

This chapter reviews the development and use of observational measures to assess *situation awareness* (SA) among fighter pilots. The chapter begins with a general description of air combat and SA. The next two sections summarize the general approach and the results of this effort. The chapter concludes with a discussion of the advantages and disadvantages of this approach and some general comments on the problems involved in measuring concepts that are as ill-defined as SA.

Uncertainty is characteristic of air combat. This uncertainty places enormous demands on the pilot's cognitive resources (Houck, Whitaker, & Kendall, 1993). The pilot must execute multiple tasks under extreme time pressure. At the same time, the pilot must deal with a variety of data sources, each of which may present only limited information about the current environment. For example, data regarding the location of enemy aircraft may come from on-aircraft systems that are controlled by the pilot. That data may also come from radio calls made by other members of the flight or an air weapons controller. Much of this data often corresponds to what the pilot already knows. Other pieces of data, however, provide new or conflicting information. Consequently, the pilot must filter, analyze, and interpret this data to estimate its timeliness and accuracy. The pilot must synthesize this data, assess the situation, and select a course of action.

---

[1]The views expressed in this chapter do not necessarily reflect those of the United States Air Force or the Department of Defense.

The necessity of making decisions based on the perception and interpretation of incomplete and sometimes conflicting data is one reason that training and experience are so important to the development of SA.

Endsley (1995a) views SA as an hierarchically organized construct with three interrelated levels. At its most basic level, SA involves the perception of current environmental information. At Level 1 SA, a pilot knows factual data such as the aircraft's energy state and the locations of other aircraft. The next higher level of SA reflects the pilot's interpretation of the current environmental data in terms of its immediate significance to the pilot's goals and objectives. An example of Level 2 SA is recognizing relative offensiveness or defensiveness during an air combat engagement. At the third and highest level of SA, the pilot not only recognizes and comprehends the current situation but also uses that information to anticipate future environmental states. At Level 3 SA, the pilot in an air combat engagement knows that both aircraft are currently neutral. In addition, the pilot knows that, based on the current flight paths and energy states, the enemy will soon have an offensive advantage.

In 1991, the Air Force Chief of Staff asked a series of questions about SA. These questions included: what is SA; can we measure SA; can we select individuals for pilot training based on their SA potential; and what impact does training have on SA. In response to these questions, Armstrong Laboratory (now the Air Force Research Laboratory) initiated a SA research program. This chapter summarizes our initial attempts to measure SA in operational fighter squadrons and in multiship air combat simulations.

Our initial efforts focused on three issues: the definition of SA; the degree to which pilots can reliably judge their fellow pilots in terms of SA; and whether there is a relation between such judgments and mission performance.

In response to the question, what is SA, the Air Staff provided a working definition that links SA to mission performance. This definition, written from the pilot's perspective, defines SA as "a pilot's continuous perception of self and aircraft in relation to the dynamic environment of flight, threats, and mission, and the ability to forecast, then execute tasks based on that perception" (Carroll, 1992, p. 5). Although there are a number of other definitions of SA available (e.g., Endsley, 1995a; Sarter & Woods, 1991; Tenney, Adams, Pew, Huggins, & Rogers, 1992), the Air Staff definition was used as the basis for our research efforts. This definition reflects the importance of SA in mission accomplishment, thus capturing the richness and complexity of the concept. It emphasizes perceiving what is important and then using that perception to guide the selection and execution of appropriate behaviors. Unfortunately, it is also very complex because it combines processes, tasks, and the linkages between them into a single construct. Consequently, it is very difficult to separate SA from the other aspects of skilled performance that determine combat proficiency.

## MEASURING SA IN OPERATIONAL FIGHTER SQUADRONS

With the assistance of instructor pilots and other subject-matter experts (SMEs), Waag and Houck (1994) identified 31 behavioral elements of SA. The SMEs felt these elements reflected SA and were important to mission success. Table 6.1 lists these 31 elements and the 8 categories of mission performance the elements represent.

### SA Instruments

The Air Force Research Laboratory developed 4 different instruments to measure SA in operational F-15C squadrons based on these 31 elements. The first instrument required respondents to provide their personal definition of SA. Using their personal definition of SA, each respondent then rated the importance of the 31 elements using a 6-point Likert scale.

The other three instruments, or SA Rating Scales (SARS), measured SA from three different perspectives: self, supervisory, and peer. The self-re-

TABLE 6.1
Categories and Elements of SA

| General Traits | Information Interpretation |
|---|---|
| Discipline | Interpreting air-to-air radar |
| Decisiveness | Interpreting radar warning receiver |
| Tactical knowledge | Ability to use air weapons controller |
| Time-sharing ability | Integrating overall information |
| Reasoning ability | Radar sorting |
| Spatial ability | Analyzing engagement geometry |
| Flight management | Threat prioritization |

| Tactical Game Plan | System Operation |
|---|---|
| Developing plan | Radar |
| Executing plan | Tactical electronic warning system |
| Adjusting plan on-the-fly | Overall weapons system proficiency |

| Communication | Tactical Employment—Beyond Visual Range |
|---|---|
| Quality (brevity, accuracy, timeliness) | Targeting decisions |
| Ability to effectively use information | Fire-point selection |

| Tactical Employment—General | Tactical Employment—Within Visual Range |
|---|---|
| Assessing offensiveness/defensiveness | Maintain track of bogeys/friendlies |
| Lookout (radar, electronic, visual) | Threat evaluation |
| Defensive reaction (chaff, flares, maneuvering) | Weapons employment |
| Mutual support | |

port SARS and supervisory SARS required the respondents to rate either themselves or their subordinates on each of the 31 items. Both SARS used a 6-point scale and the ratings were made relative to other F-15C pilots with whom the respondents had flown. The scale anchors were *acceptable* and *outstanding* because all the pilots were on flying status and were considered mission-ready by Air Force standards. The squadron commander, operations officer, assistant operations officer, weapons officer, and standardization-evaluation flight examiner completed the supervisory SARS on the pilots within their squadron. In addition, squadron flight commanders completed supervisory SARS on the pilots within their flight. The peer SARS required respondents to rate the other mission-ready pilots in the squadron on general fighter pilot ability and SA ability and then to rank order them on their SA ability. Both the peer and supervisory SARS allowed respondents to omit rating and ranking a particular pilot if they felt they did not have enough information to accurately judge that individual. All respondents completed the self-report and peer SARS.

## Results

SARS data was collected on 238 mission-ready F-15C pilots from 11 squadrons stationed at 4 different Air Force bases. Two hundred and six of these pilots provided written definitions of SA. The first column in Table 6.2 lists the seven phrases most frequently used in defining SA. The second column shows the seven most highly rated elements of SA. There is considerable agreement between the phrases used to define SA and the element ratings. In addition, both the phrases and the element ratings indi-

TABLE 6.2
Seven Phases Most Commonly Used by Pilots to Define
SA and the Seven Most Highly Rated Elements of
SA Listed in Decreasing Frequency of Occurrence

| *Most Commonly Used Phases to Define SA* | *Most Highly Rated Elements for SA* |
| --- | --- |
| Composite 3-D image of entire situation | Use of communication information |
| Assimilation of information from multiple sources | Information integration from multiple sources |
| Knowledge of spatial position or geometric relationships among tactical entities | Time-sharing ability |
| Periodic mental update of dynamic situation | Maintaining track of bogies and friendlies |
| Prioritization of information and actions | Adjusting plan on-the-fly |
| Decision making quality | Spatial ability to mentally picture engagement |
| Projection of situation in time | Lookout for threats from visual, radar warning receiver, radar |

TABLE 6.3
SARS Intercorrelations

| | 1 | 2 | 3 | 4 | 5 |
|---|---|---|---|---|---|
| 1. Supervisor SARS | | | | | |
| 2. Peer—Fighter pilot ability | 0.89 | | | | |
| 3. Peer—SA ability | 0.91 | 0.98 | | | |
| 4. Peer—Rank order | 0.92 | 0.91 | 0.92 | | |
| 5. Self-report SARS | 0.45 | 0.56 | 0.57 | 0.49 | |

cate that a significant component of SA involves assimilating and using information to guide action.

Analyses of the peer and supervisory SARS indicated that the pilots reliably classified their fellow pilots in terms of SA. Internal consistency was computed for all 31 items on the supervisory SARS. The resulting measure, Cronbach's coefficient alpha, was 0.99. Interrater reliability was also estimated for the supervisor and peer SARS using an analysis of variance procedure (Guilford, 1954). For the supervisor SARS, these analyses indicated that the average reliability of each supervisor's ratings was 0.50 and the average reliability of the pooled supervisor ratings was 0.88. Similarly, the peer SARS showed an individual reliability of 0.60 and a combined reliability of 0.97. Additional detail concerning the analyses of the SARS data is found in Waag and Houck (1994).

As shown in Table 6.3, there was substantial agreement between supervisor and peer SARS. Table 6.3 also indicates that there is noticeably less agreement between the self-report SARS and the other SARS.

## MEASURING SA IN SIMULATED
## AIR COMBAT MISSIONS

Although the SARS data indicate high reliability and consistency between raters, they do not empirically relate judged SA to pilot performance in air combat missions. In an attempt to determine the relation between SA and mission performance, a composite SA score was developed for each of the 238 pilots. These scores, based on the peer and supervisory SARS, were scaled with a mean of 100 and a standard deviation of 20. Based on this composite score, a sample of 40 mission-ready flight leads was selected to fly a series of multiship air-to-air combat simulations. Mission qualification level was held constant for the simulation portion of this effort because current flight qualification was highly correlated ($r = 0.82$) with the composite SA score (Waag & Houck, 1994). The mean SA scores for the 40 flight leads were 106.3 with a standard deviation of 17.4. An additional 23 mission-ready pilots flew as the flight lead's wing during the course of the simulation.

The combat simulations were flown in the Air Force Research Laboratory's multiship training research facility in Mesa, Arizona. This facility provided the flight simulators and associated simulations necessary to allow the participants to fly realistic combat missions in multibogey, high threat scenarios. The two F-15C pilots flew high fidelity F-15C simulators operating on a secure simulation network. The simulation network also included other manned and computer-controlled aircraft, computer-controlled surface-to-air threats, and a manned air weapons controlled simulator. Additional details concerning the simulation are found in Waag, Houck, Greschke, and Raspotnik (1995).

Each flight lead flew nine simulator sorties over 5 consecutive days. During each sortie four separate engagements were flown. Each engagement consisted of a different scenario representing the same basic mission. After each engagement, the simulation was reinitialized to the appropriate starting conditions and the new scenario began. Scenarios increased in complexity throughout the week.

### Scenario Design

Figure 6.1 illustrates a moderately difficult scenario. In this defensive counterair mission, the two F-15s defended an airfield. The attackers consisted of two bombers escorted by two fighters. The scenario began with the enemy aircraft 80 nautical miles (nm) away from the airfield. The enemy fighters were flying at 20,000 ft and the bombers at 10,000 ft. There was a lateral separation of 10 nm between the fighters and the bombers. At 35 nm, the fighters maneuvered rapidly and descended to 3,500 ft. At 15

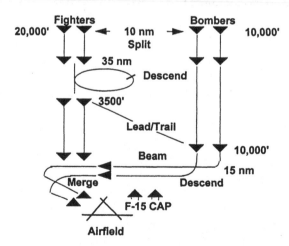

FIG. 6.1. Typical defensive counterair scenario used as part of the combat mission simulations.

nm, the bombers performed a hard right turn and descended to 2,500 ft. The purpose of these maneuvers was to momentarily break the F-15s' radar contact and disrupt the F-15 pilots' ability to identify, target, or engage the enemy aircraft.

**Rating Mission Performance**

The basic approach taken toward SA measurement was through scenario manipulation and performance observation as suggested by Tenney et. al., (1992). Other approaches, such as explicit probes and the Situation Awareness Global Assessment Technique (SAGAT) (Endsley, 1995b), were considered. These other approaches were rejected because of a need for measures that could be used during operational training either in simulators or actual aircraft. All of the scenarios contained enemy maneuvers designed to "trigger" information seeking and decision making by the flight lead. In essence, these trigger events serve as SA probes in a naturalistic environment.

As Kelly (1988) pointed out, measuring air combat skills presents a number of challenges. The fluid, dynamic nature of air combat, combined with the number of alternative tactics and techniques available to the pilot, make objective performance measurement extremely difficult. Even when objective data is available, it is often difficult to interpret the importance of that data. Because of the difficulties involved in establishing the relation between objective measures such as radar locks or engagement parameters and SA, it was decided to rely on behavioral observation by SMEs who were unaware of the SA scores of the pilots they were observing. Two SMEs, retired fighter pilots with extensive experience in air combat and training, watched each engagement in real time and independently completed an observational checklist. To assist them in evaluating pilot performance, cockpit instruments, intraflight communications, and a plan view display of the engagement were available throughout the engagement. After each simulator session, the two SMEs discussed each engagement and completed a consensus performance rating scale containing 24 behavioral indicators based on the SARS. These 24 behavioral indicators were the elements of SA listed in Table 6.1, less the 7 general traits of SA. In addition, the SMEs also wrote a critical event analysis for each mission that identified events that were critical to the outcome of the mission and indicative of the pilot's SA.

**Results**

Figure 6.2 shows the composite SA scores obtained from the SARS and the mean SA score assigned by the SMEs based on their observation of each pilot's performance during the simulated air combat missions. The Pear-

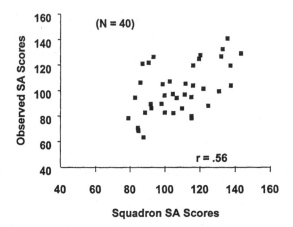

FIG. 6.2.   Mean SA score for combat mission simulation as a function of squadron SA score.

son product moment correlation between these scores is 0.56 ($p < .001$, $df$ = 38) and indicated a significant relation between squadron ratings of SA and performance in these simulated air combat missions.

### SA Ratings and Specific Behaviors

Having established that squadron SA ratings and simulator performance ratings are related, the next question is whether specific behaviors associated with rated SA could be identified. Two aspects of pilot behavior were examined: errors (as identified by SMEs) and communication patterns.

*Identification of Errors.*   Two SMEs reviewed tapes of selected missions and identified errors made by pilots while flying complex scenarios. Four engagements for each of 8 pilots were chosen for this analysis. The pilots selected were the top four and bottom four individuals based on squadron SA rating. The SMEs independently reviewed each of the resulting 32 engagements. They were instructed to identify and record any pilot action or inaction that they considered an error. No strict definition of "error" was provided. The idea was to allow the SMEs to identify without constraint any action or inaction they felt was inappropriate given the situation and the available data. After identifying errors separately, the SMEs reviewed the scenarios together and agreed upon a final set of errors. They also agreed on a list of mutually exclusive error categories and placed each identified error in a category. These categories were later grouped into two more general classes: "decision errors" and "information acquisition errors."

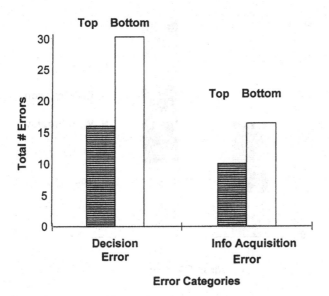

FIG. 6.3. Total number of decision and information errors made during four simulated combat engagements as a function of squadron SA score.

Figure 6.3 shows the number of decision errors and information acquisition errors made by top- and bottom-ranked pilots for the squadron SA ratings. Although the number of errors in each category is small, a relatively large number of decision errors were observed among pilots who were ranked lowest in SA by their supervisors and peers at the squadron. Figure 6.4 shows the breakout of decision errors into more specific categories. Inspection of Figure 6.4 shows that the difference between low-SA and high-SA pilots in the number of decision errors is largely accounted for by errors in tactics selection and flight leadership. Thus, subjective ratings of SA by peers and supervisors are related to more than just performance ratings made by other SMEs; they are also related to observable behaviors in complex combat scenarios.

*Communication Patterns.* Communication is an important source of information to support SA in air combat pilots. Therefore, some aspects of communication behavior may be related to squadron-rated SA. The nature of such a relation is, however, hard to predict. For example, consider the frequency of calls to a wingman or air weapons controller (AWC) to request information. One view is that high-SA pilots will request information less frequently than low-SA pilots because their SA is already good. The opposite view is that high-SA pilots maintain their SA advantage, in part, because they request information more frequently. To address such issues, communications from the top four and bottom four pilots in squadron SA rating

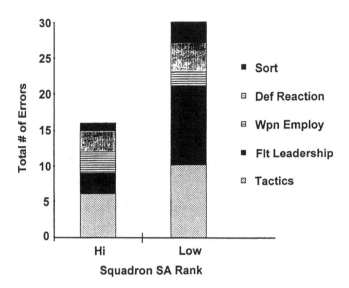

FIG. 6.4.  Types of decisional errors made as a function of squadron SA score.

were transcribed. Three engagements were analyzed for each pilot. Each call made during each of the engagements was transcribed and categorized according to the initiator of the call (lead, wing, AWC); the intended recipient (lead, wing, AWC, entire team); and the purpose (providing information, acknowledging, directing, requesting, informing, or uncodable).

Because engagements differed in length, comparisons of communication frequency were generally made in terms of the number of calls per minute, or call rate. As shown in Fig. 6.5, call rates were higher for high-SA leads than for low-SA leads for every kind of call except acknowledgments. High-SA leads both provided and requested information more frequently than did low-SA leads. High-SA leads also directed other team members more frequently. Like the error data, the communication data analyzed so far suggested that subjective ratings of SA are related to behavioral differences during simulated combat scenarios.

## DISCUSSION

These initial results in developing measures of SA that can be used in a squadron's operational training environment are encouraging. These results indicated that SA is a construct that has meaning and can be used by both peers and supervisors to classify mission-ready pilots. They also indicated that squadron ratings of SA are related to relevant behaviors (e.g.,

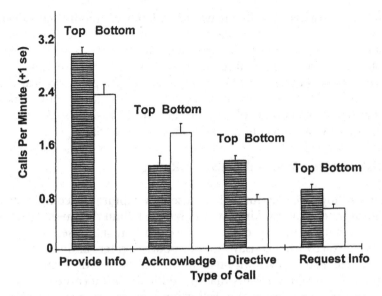

FIG. 6.5. Number of radio calls per minute as a function of squadron SA score and type of call.

errors, communications) and are correlated with mission success in simulated air combat missions.

Although this approach to SA measurement may be classified as subjective rather than objective, this is an oversimplification. All measurement approaches ultimately involve assigning numbers to events according to an explicit set of rules (Stevens, 1951). The distinction between objective and subjective measures simply indicates whether a human observer is an integral component of the measurement instrument. Objective measurement involves data that is generated independently of the human observer. Ideally, this data is generated, recorded, and scored without the intervention of a human observer. Subjective measurement, on the other hand, requires human observers to generate the data itself. Although Muckler (1977) argued that there is no such thing as objective measurement in the strict sense, the distinction continues to be made and objective measures are often preferred to subjective measures. The reason for this preference is that subjective measures are frequently seen as being contaminated by the human observers during the act of measurement. Because objective meas-0ures, on the other hand, are relatively independent of human observers, they are seen as "truer" measures of the construct under study.

Unfortunately, objective measures often fail to capture the richness and complexity of human performance (Kelly, 1988; Meister, 1989; Vreuls & Obermayer, 1985). One reason for this is that objective measures are essentially reductionistic and are therefore best suited for recording the fundamental dimensions of performance (e.g., latency, amount, and deviation). Although these fundamental measures provide us with data that is less subject to error, they also frequently fail to provide us with information concerning the contextual nature of skilled performance. Subjective measures seem more closely related to higher order psychological constructs. The data they produce appears to reflect a synthesis of the more molecular behaviors and to reflect more global dimensions such as interpreting, judging, and deciding—the very essence of SA.

### Potential Problems With Subjective SA Ratings

Subjective ratings are a useful tool because they capture the complexity of SA. Unfortunately, various kinds of problems can limit their usefulness. At the heart of many of these problems is the definition of SA itself. Disagreement about exactly what constitutes SA—between raters and other raters, raters and scientists, or even within an individual rater at different times—will tend to yield rating results that are difficult to interpret.

A key definitional issue is the distinction between momentary SA and SA ability. Momentary SA is demonstrated when one interrupts a mission at a particular point and tests current-situation knowledge (cf., Endsley, 1995a). SA ability is the tendency to maintain good SA in a variety of situations. The Air Staff definition of SA used throughout this effort includes elements of both.

There are serious interpretational problems associated with the assessment of both SA ability and momentary SA. Subjective-ability-measurement problems are illustrated by the task of obtaining ratings of SA ability from squadron supervisors and peers. A primary problem is to communicate to the raters the difference between SA ability and overall piloting skill because several important aspects of pilot skill do not fall within the definition of SA. This problem was resolved by requiring separate ratings of SA and "fighter pilot ability."

Another inherent problem is that raters may be unduly influenced by a pilot's credentials. Rank, hours of experience, qualification status, and participation in special exercises could influence rated SA. Even if the rater is explicitly instructed to disregard credentials in producing a rating, difficulty can be expected because the rater's interpretation of past experiences with the pilot, upon which the SA rating should be based, is colored by the pilot's achievements. The obvious approach to mitigate this problem is to collect data on credentials and experiential factors so that the differential utility of the SA ratings over and above these factors can be assessed. In this study, hours of experience in the F-15 was correlated 0.62 with composite SA rating. When four other experience factors (qualification, Fighter Weapons School graduation, and participation in two exercises) were used to predict composite SA rating, the multiple correlation was 0.85 (Waag & Houck, 1994).

These problems—differentiating SA from performance and credentials—also appear in attempts to assess momentary SA. The SMEs' ratings of SA during each combat scenario can be viewed as a time-averaged momentary-SA measure. SMEs were instructed to rate the pilot's SA during a particular scenario rather than try to assess the pilot's overall SA ability. These instructions were intended to induce the raters to attend to the pilot's SA at various moments during the engagement and arrive at a composite of these momentary SA assessments, as opposed to a general prediction of the pilot's SA ability in other situations. Interviews with the raters, during which they were asked to explain their ratings, suggested that they in fact evaluated SA as it was demonstrated during the engagement rather than predicting overall SA ability. This conclusion is supported by rater comments such as "I can tell that this pilot is very good, but this time he was doing X and didn't notice threat Y, so we had to give him a low rating."

As noted earlier, SMEs were not informed of the pilot's level of experience, most of the pilots were of the same rank, and all of them were at the same qualification level (4-ship flight lead). Although this may have mitigated any tendency for raters to base ratings on pilots credentials, it did not solve the problem of differentiating SA from performance in a scenario. Even though the distinctions between SA, performance, and mission outcome were stressed in instructions to the SMEs, statements by the SMEs in the mission summaries they provided after each scenario suggested that some confounding occurred. In particular, mission success or failure undoubtedly had some influence on SA ratings.

The difficulty of rating SA independently of pilot performance is exacerbated by the inclusion of performance-like elements in the definition of SA. For example, the Air Staff definition includes "... the ability to ... execute tasks ..." (Carroll, 1992, p. 5). This is more than an afterthought—it is a critical feature of the definition for some investigators. For example, Vidulich (1992) wrote "good SA implies a capability to respond appropriately, not just the possession of an accurate assessment" (p. 17). Wickens (1995) also argued that definitions of SA are incomplete without some reference to "the capacity to respond appropriately to events" (p. K2–2). On the other hand, Endsley (1995a) stressed the importance of separating a full assessment and comprehension of the situation from the decisions and responses based on that comprehension. At an academic level, these may not be substantive disagreements. Most investigators seem to agree that while the knowledge of which responses are appropriate to a situation is part of SA, the responses themselves are not. However, to the rater the fact that a pilot responded appropriately is usually good evidence that knowledge of appropriate action was present. Thus, to the extent that knowledge of appropriate courses of action is stressed in the definition of SA, it will be difficult to rate SA independently of performance.

There is another way of defining SA that may be more useful for guiding the process of obtaining accurate ratings. This approach recognizes that all aspects of momentary SA are eventually reducible to some form of knowledge or information in working memory. At any given moment, the information in a pilot's working memory can come from immediate perception (Endsley's Level 1 SA); associations from long-term memory, including retrieved interpretations of perceived information (Endsley's Level 2 SA); and inferences generated by combining and manipulating information in working memory, including predictions about future events (Endsley's Level 3 SA). However, not all working memory information is SA-related (e.g., daydreams are probably not). Therefore, SA could be defined as knowledge (in working memory) about elements of the environment. For a pilot "knowledge about elements of the environment" must be interpreted broadly to include not only information about the location, capabilities, and intentions of threats, but also such things as offensive and defensive status and likely future events. A potential advantage of explicitly focusing on the information that makes up SA is that raters can be instructed to look for evidence that a pilot has various pieces of knowledge.

Of course, not all knowledge is equal in its impact on mission success. Even if the information in a pilot's working memory could be measured objectively, there remains the problem of assigning a weight to each piece of information according to its relevance. These weights would no doubt be constantly changing as the mission evolves. Any objective measurement technique based on assessing specific pieces of knowledge must somehow cope with this daunting weight-assignment problem.

In contrast, a human rater "automatically" takes into account the importance of various components of SA in generating a rating. This should be one of the biggest advantages of using subjective measures; however, this advantage only applies if the rater focuses on SA as knowledge and does not confound SA with performance or mission outcome.

## Generalizability to Other Domains

This research obtained SA ratings of operational pilots from peers and supervisors with minimal disruption of their squadron operations and demonstrated that these ratings were predictive of SA ratings independently obtained during simulated combat missions. It represented a relatively efficient method of obtaining SA ratings in both field and laboratory environments and demonstrated that SMEs are capable of moderate to good agreement regarding an individual's SA.

Although this research focused on the SA of mission-ready F-15C pi-lots, global SA ratings obtained from peers or supervisors, momentary SA ratings based on direct observation of job performance, or both can be ap-plied to measuring SA in other domains such as commercial aviation, medicine, and command and control. Because the specific details of the research methodology will vary depending upon the domain being stud-ied, we believe anyone attempting to apply this approach should adhere to the following guidelines:

1. Regardless of whether global or momentary SA ratings are used, it is essential to develop a set of behavioral indicators that will provide raters a common framework for judging SA.
2. Field raters providing global SA ratings must have had multiple op-portunities to observe the ratees perform in their natural environ-ment and these raters must be given the opportunity to omit rating individuals if they feel they have not had sufficient opportunities to observe that individual.
3. Interrater reliability of global SA ratings improves as the number of raters increases.
4. Job samples used to provide the opportunity for momentary SA rat-ings must provide the opportunity for the individual to exhibit a va-riety of behaviors that are indicative of SA.

### Utility of the SA Concept

The potential for confounding SA with performance strikes at the heart of the utility of the concept. SA is a useful concept to the extent that it sug-gests measures that add to our ability to predict which training techniques, pilots, and displays will lead to the best mission outcomes. A subjective SA rating can be used as an independent, dependent, or intervening variable. As an independent variable, it could help predict which pilots will achieve success in combat. As a dependent variable, it could provide a sensitive method for evaluating the impact of new displays. As an intervening vari-able, it could help explain why different training techniques lead to differ-ent mission outcomes.

Unfortunately, if raters use mission outcomes to infer SA, little is gained from introducing the SA variable; it will have little or no differen-tial utility. Our experience suggests that, to obtain useful SA ratings, raters must focus on what a pilot probably knows about the elements of a situa-tion. As discussed previously, this can include valid inferences, predic-tions, or both about present or future events within the scenario. We used various tools to help raters read the pilot's mind. A video record of the radar display is particularly important. We also used knowledge-based checklists, and, for some scenarios, the pilot's eye movements were recorded and the raters were provided a video showing the pilot's point of gaze. Of course, raters will still make inferences from the actions taken by the pilot. The rater, however, must be constantly aware that executing a maneuver well or making a correct tactical decision is not necessarily proof of good SA. If the rater can maintain this mental set during the rating process, then it is possible to use a pilot's behavior to infer at least some of the knowledge leading to that behavior.

## CONCLUSION

Obviously both subjective and objective measurement approaches are necessary to develop an understanding of SA. Objective measures are important because they provide a necessary check on subjective judgments. Suppose a pilot with impressive credentials misses a critical radar sampling assignment, then pulls the mission out of the fire with superior stick-and-rudder skills. Behavior-based, face-valid indicators of SA would help assess the extent to which ratings in such situations are driven by mission outcome rather than SA. Subjective measures, on the other hand, help to assure that critical aspects of SA are actually being assessed. For example, there are several ways that a pilot can be aware of the location of a particular enemy aircraft (e.g., radar, wing, AWC). An objective measure of the number of enemies sampled on the radar might well be a misleading and incomplete indicator of SA if important information on enemy locations was given by a weapons director and only the critical targets were sampled on radar. Objective measurement technology will doubtless move toward configural assessment of some aspects of SA; however, it will be difficult to mimic the inferences that a human SME makes to assess the cognitive aspects of SA that are not directly measurable.

We believe that the critical SA measurement issues concern how the definition of SA is refined, and which measurements provide the best information for designing and evaluating aircrew training. Regarding the definition of SA, we have argued that the development of both objective and subjective measures will be facilitated by defining momentary SA in terms of specific kinds of knowledge, where knowledge is broadly interpreted to include the pilot's inferences about the enemy's status and intentions, and guesses about likely future events. Viewing SA in terms of constituent knowledge should also facilitate the design and evaluation of aircrew training. It is important, however, not to oversimplify this knowledge-based view of SA. As is always the case in training, simply providing important information is no guarantee that the information will be in the pilot's working memory at the time it is needed. Thus, it is also necessary to focus research and measurement development on the processes by which knowledge is acquired during flight.

## ACKNOWLEDGMENTS

The authors wish to thank Don Vreuls for his helpful comments on an early version of this chapter and Wayne Waag for his role in establishing the Air Force Research Laboratory's situational awareness research program. The authors also wish to thank Bart Raspotnik, Brian Schreiber, and Kelly Lee for their role in analyzing the communication and error data. Portions of this chapter were presented at the International Conference on the Experimental Analysis and Measurement of Situation Awareness, Daytona Beach, FL, 1–3 November 1995.

## REFERENCES

Carroll, L. A. (1992). Desperately seeking SA. *TAC Attack (TAC SP 127–1)*, *32*, 5–6.

Endsley, M. R. (1995a). Toward a theory of situation awareness in dynamic systems. *Human Factors*, *37*, 32–64.

Endsley, M. R. (1995b). Measurement of situation awareness in dynamic systems. *Human Factors*, *37*, 65–84.

Guilford, J. P. (1954). *Psychometric methods*. New York: McGraw-Hill.

Houck, M. R., Whitaker, L. A., & Kendall, R. R. (1993). *An information processing classification of beyond-visual-range air intercepts* (AL/HR–TR–1993–0061, AD A266 927). Williams Air Force Base, AZ: Armstrong Laboratory.

Kelly, M. J. (1988). Performance measurement during simulated air-to-air combat. *Human Factors*, *30*, 495–506.

Meister, D. (1989). *Conceptual aspects of human factors*. Baltimore, MD: Johns Hopkins University Press.

Muckler, F. A. (1977). Selecting performance measures: "Objective" versus "subjective" measurement. In L. T. Pope and D. Meister (Eds.), *Symposium proceedings: Productivity enhancement: Personnel performance assessment in Navy systems* (pp. 169–178). San Diego, CA: Navy Personnel Research and Development Center.

Sarter, N. B., & Woods, D. D. (1991). Situation awareness: A critical but ill-defined phenomenon. *The International Journal of Aviation Psychology*, *1*, 45–57.

Stevens, S. (1951). Mathematics, measurement, and psychophysics. In S. Stevens (Ed.), *Handbook of experimental psychology* (pp. 1–49). New York: Wiley.

Tenney, Y. J., Adams, M. J., Pew, R. W., Huggins, A. W. F., & Rogers, W. H. (1992, July). *A principled approach to the measurement of situation awareness in commercial aviation* (NASA Contractor Report 4451). Langley, VA: National Aeronautics and Space Administration.

Vidulich, M. (1994). Cognitive and performance components of situation awareness: SAINT team task one report. In M. Vidulich, C. Dominguez, E. Vogel, & G. McMillan (Eds.), *Situation Awareness: Papers and Annotated Bibliography* (AL/CF–TR–1994–0085; pp. 17–28). Wright-Patterson Air Force Base, OH: Armstrong Laboratory.

Vreuls, D., & Obermayer, R. W. (1985). Human-system performance measurement in training simulators. *Human Factors, 27*, 241–250.

Waag, W. L., & Houck, M. R. (1994). Tools for assessing situational awareness in an operational fighter environment. *Aviation, Space, and Environmental Medicine, 65*(5, Suppl.), A13–A19.

Waag, W. L., Houck, M. R., Greschke, D. A., & Raspotnik, W. B. (1995). Use of multiship simulation as a tool for measuring and training situation awareness. In *Situation Awareness: Limitations and Enhancement in the Aviation Environment* (AGARD–CP–575; pp. 20-1– 20-8). Neuilly Sur Seine, France: AGARD.

Wickens, C. D. (1995). Situation awareness: Impact of automation and display technology. In *Situation Awareness: Limitations and Enhancement in the Aviation Environment* (AGARD–CP–575; pp. K2-1–K2-13). Neuilly Sur Seine, France: AGARD.

# Chapter 7

# Direct Measurement of Situation Awareness: Validity and Use of SAGAT

Mica R. Endsley
*SA Technologies, Inc.*

The Situation Awareness Global Assessment Technique (SAGAT) is a global tool developed to assess SA across all of its elements based on a comprehensive assessment of operator SA requirements (Endsley, 1987b, 1988b, 1990c). Using SAGAT, a simulation employing a system of interest is frozen at randomly selected times and operators are queried as to their perceptions of the situation at that time. The system displays are blanked and the simulation is suspended while subjects quickly answer questions about their current perceptions of the situation. As a global measure, SAGAT includes queries about all operator SA requirements, including Level 1 SA (perception of data), Level 2 SA (comprehension of meaning) and Level 3 SA (projection of the near future) components. This includes a consideration of system functioning and status as well as relevant features of the external environment.

SAGAT queries allow for detailed information about subject SA to be collected on an element by element basis that can be evaluated against reality, thus providing an objective assessment of operator SA. This type of assessment is a direct measure of SA—it taps into the operator's perceptions rather than infers them from behaviors that may be influenced by many other factors besides SA. Furthermore, it does not require subjects or observers to make judgments about situation knowledge on the basis of incomplete information as subjective assessments do. By collecting samples of SA data in this manner, situation perceptions can be collected immediately (while fresh in the operators' minds), thereby reducing numerous

**147**

problems incurred when collecting data on mental events after the fact, but not incurring intrusiveness problems associated with on-line questioning. Multiple "snapshots" of operators' SA can be acquired that give an index of the SA quality provided by a particular design. By including queries across the full spectrum of an operator's SA requirements, this approach minimizes possible biasing of attention as subjects cannot prepare for the queries in advance because they could be queried over almost every aspect of the situation to which they would normally attend. The method is not without some costs, however, as a detailed analysis of SA requirements is required in order to develop the battery of queries to be administered.

## DEVELOPMENT OF QUERIES

Probably one of the most important issues that must be addressed when using SAGAT is that of determining the queries to use for a particular experimental setting. Asking queries that are relevant to the operator's SA during the freeze is a prime determinant of the utility of the technique. These queries should also be posed in a cognitively compatible manner. That is, they should be phrased to be as similar as possible to how the person thinks about the information and should not require extra transformations or decisions by the operator.

### SA Requirements Analysis

The problem of determining what aspects of the situation are important for a particular operator's SA has frequently been approached using a form of cognitive task analysis called a goal-directed task analysis as illustrated in Table 7.1. In such analysis, the major goals of a particular job class are identified along with the major subgoals necessary for meeting each of these goals. Associated with each subgoal, the major decisions that need to be made are then identified. The SA needed for making these decisions and carrying out each subgoal are identified. These SA requirements focus not only what data the operator needs, but also on how that information is integrated or combined to address each decision. In this

TABLE 7.1
Format of Goal-Directed Task Analysis

| |
|---|
| Goal |
|   Subgoal |
|     Decision |
|       Projection (Level 3 SA) |
|       Comprehension (Level 2 SA) |
|       Data (Level 1 SA) |

analysis process, SA requirements are defined as those dynamic information needs associated with the major goals or subgoals of the operator in performing his or her job (as opposed to more static knowledge such as rules, procedures, and general system knowledge).

Conducting such an analysis is usually carried out using a combination of cognitive engineering procedures. Expert elicitation, observation of operator performance of tasks, verbal protocols, analysis of written materials and documentation, and formal questionnaires have formed the basis for the analyses. In general, the analysis has been conducted with a number of operators who are interviewed, observed, and recorded individually. The resultant analyses are pooled and then validated overall by a larger number of operators.

An example of the output of this process is shown in Table 7.2. This example shows the SA requirements analysis for the subgoal "maintain aircraft conformance" of the major goal "avoid conflictions" for an air traffic controller (Endsley & Rodgers, 1994). In this example, the subgoal is even further divided into lower level subgoals prior to the decisions and SA requirements being listed. In some cases, addressing a particular subgoal occurs through reference to another subgoal in other parts of the analysis, such as the need to readdress aircraft separation. This shows the degree to which a particular operator's goals and resultant SA needs may be very interrelated. The example in Table 7.2 shows one of four major subgoals that are relevant to the major goal of "avoid conflictions." The major goal, "avoid conflictions" is one of three for an air traffic controller.

This analysis systematically defines the SA requirements (at all three levels) that are needed to effectively make the decisions required by the operator's goals. Many of the same SA requirements appear throughout the analysis. In this manner, it can be determined the way in which pieces of data are used together and combined to form what the operator really wants to know.

Although the analysis will typically include many goals and subgoals, they may not all be active at once. In practice, at any given time more than one goal or subgoal may be operational, although they will not always have the same prioritization. The analysis does not indicate any prioritization among the goals (which can vary over time) or that each subgoal within a goal will always be active. For example, unless particular events are triggered, the subgoal of assuring aircraft conformance may not be active for a given controller.

This type of analysis is based on goals or objectives, not tasks. This is because goals form the basis for decision making in many complex environments. Furthermore, tasks tend to be technology dependent. Completely different tasks may be carried out to perform the same goal in two different systems. For example, navigation may be done very differently in an

TABLE 7.2
Example of Goal-Directed Task Analysis
for En Route Air Traffic Control

1.3  Maintain aircraft conformance
    1.3.1  Assess aircraft conformance to assigned parameters
- aircraft at/proceeding to assigned altitude?
- aircraft proceeding to assigned altitude fast enough?
  - time until aircraft reaches assigned altitude
  - amount of altitude deviation
  - climb/descent
    - altitude (current)
    - altitude (assigned)
    - altitude rate of change (ascending/descending)
- aircraft at/proceeding to assigned airspeed?
- aircraft proceeding to assigned airspeed fast enough?
  - time until aircraft reaches assigned airspeed
  - amount of airspeed deviation
    - airspeed (indicated)
    - airspeed (assigned)
    - groundspeed
- aircraft on/proceeding to assigned route?
- aircraft proceeding to assigned route fast enough?
- aircraft turning?
  - time until aircraft reaches assigned route/heading
  - amount of route deviation
    - aircraft position (current)
    - aircraft heading (current)
    - route/heading (assigned)
  - aircraft turn rate (current)
  - aircraft heading (current)
  - aircraft heading (past)
  - aircraft turn capabilities
    - aircraft type
    - altitude
  - aircraft groundspeed
  - weather
  - winds (direction, magnitude)

    1.3.2  Resolve nonconformance
- Reason for nonconformance?
  - Verify data
    - Is presented altitude correct?
      - Aircraft altimeter setting
      - Aircraft altitude (indicated)
    - Is presented airspeed correct?
      - Aircraft airspeed (indicated)
      - groundspeed
      - winds (magnitude, direction)
    - Is presented position/heading correct?
      - Fix distance to Nav aid
      - range/bearing to Fix
      - track code

*(Continued)*

TABLE 7.2
*(Continued)*

---

- Will current behavior cause a problem?
  - Assess aircraft separation (1.1.1)
  - Assess aircraft/airspace separation (1.2.1)
  - Assure minimum altitude requirements (1.4)
- Action to bring into conformance?
  - Provide clearance (2.2)

---

*Note.* From *Situation awareness information requirements for en route air traffic control* (DOT/FAA/AM–94/27) (pp. B6–B7), by M. R. Endsley & M. D. Rodgers, 1994, Washington, DC: Federal Aviation Administration Office of Aviation Medicine. Copyright 1994 by M. R. Endsley and M. D. Rodgers. Reprinted with permission.

automated cockpit as compared to a nonautomated cockpit. Yet, the SA needs associated with navigation are essentially the same (e.g., location or deviation from desired course).

The analysis strives to be as technology free as possible. How the information is acquired is not addressed as this can vary considerably from person to person, from system to system, and from time to time. In different cases the information may be obtained through system displays, verbal communications, other operators, or internally generated within the operator. Many of the higher level SA requirements fall into this category. The way in which information is acquired can vary widely between individuals, over time, and between system designs.

The analysis seeks to determine what operators would ideally like to know to meet each goal. It is recognized that they often must operate on the basis of incomplete information and that some desired information may not be available at all with today's systems. However, for purposes of the design and evaluation of systems, it is necessary to set the yardstick to measure against what they ideally need to know so that artificial ceiling effects based on today's technology are not induced in the process. Finally, it should be noted that static knowledge, such as procedures or rules for performing tasks, is outside the bounds of an SA requirements analysis. The analysis focuses only on the dynamic situational information that affects what the operators do.

These analyses tend to be very long and complex and can take as much as a person-year of effort to complete for a given domain. On the positive side, once completed for a major class of operators, it need not be repeated unless the goals and objectives of the operator change radically. To date, these analyses have been completed for many domains of common concern including en route air traffic control (Endsley & Rodgers, 1994), TRACON air traffic control (Endsley & Jones, 1995), fighter pilots (Endsley, 1993), bomber pilots (Endsley, 1989a), commercial transport pilots (Endsley, Farley, Jones, Midkiff, & Hansman, 1998), aircraft mechanics

(Endsley & Robertson, 1996), and airway facilities maintenance (Endsley, 1994). A similar process was employed by Hogg, Torralba, and Volden (1993) to determine appropriate queries for a nuclear reactor domain.

In each of these domains, SA requirements analyses have formed the basis for the development of appropriate SA queries that can be used and adapted for a wide range of experimental testing in these areas. For the reader who needs to develop queries for a new domain, the methodology described can be readily applied to derive the SA requirements for operators in that area. Although this requires a considerable investment of energy, the resultant analysis also provides a highly useful foundation for directing design efforts and is well worth the effort expended.

### Query Format

Based on the goal-directed task analysis, a list of the SA requirements for a domain can be constructed either as a whole or for particular operator goals and subgoals. These SA requirements form the basis for determining the queries to be used in that domain. Table 7.3 lists the SA queries that we developed for air traffic control (Endsley & Rodgers, 1994; Endsley & Jones, 1995). In each case the query is presented together with categorical response options so that the operator's job in responding to the queries is minimized.

An example of a SAGAT query is shown in Fig. 7.1. The operator serving as a subject in the experiment need only click with a cursor on the desired answer and then click on the next button to continue to the next question. This makes it fairly fast and easy to complete the battery of questions at each freeze point. Another example for a military pilot is shown in Fig. 7.2. Note that in each of these examples an appropriate reference map is provided showing the location of relevant aircraft. These maps are derived from the first question provided to the operator in each case. An appropriate blank map is provided (marked with relevant boundaries and reference points) and the operator is asked to enter the aircraft that he or she knows about at the time in the appropriate location on the map. This completed map then serves as an organizing point for the remainder of the questions. This removes the problem of aircraft callsign confusion: Controllers are very poor at remembering callsigns and pilots may not know them at all. The questions are therefore self-referenced in a way that is compatible with how the operator pictures the situation at the time.

Not all of the queries may be required for a given testing situation. For instance, if the simulation does not employ aircraft emergencies, Query 16 in Table 7.3 may be omitted. Other queries should only be asked at times when they are appropriate. For example, Query 12 and Query 13 deal with proper reception and conformance to clearances. These queries

TABLE 7.3
SAGAT Queries for Air Traffic Control

1. Enter the location of all aircraft (on the provided sector map): aircraft in track control, other aircraft in sector, aircraft will be in track control in next 2 minutes
2. Enter aircraft callsign (for aircraft highlighted of those entered in query 1)
3. Enter aircraft altitude (for aircraft highlighted of those entered in query 1)
4. Enter aircraft groundspeed (for aircraft highlighted of those entered in query 1)
5. Enter aircraft heading (for aircraft highlighted of those entered in query 1)
6. Enter aircraft's next sector (for aircraft highlighted of those entered in query 1)
7. Enter aircraft's current direction of change in each column (for aircraft highlighted of those entered in query 1)

| Altitude change | Turn |
| --- | --- |
| climbing | right turn |
| descending | left turn |
| level | straight |

8. Enter the aircraft type (for aircraft highlighted of those entered in query 1)
9. Enter aircraft's activity in this sector (for aircraft highlighted of those entered in query 1) enroute, inbound to airport, outbound from airport
10. Which pairs of aircraft have lost or will lose separation if they stay on their current (assigned) courses?
11. Which aircraft have been issued assignments (clearances) that have not been completed?
12. Did the aircraft receive its assignment correctly?
13. Which aircraft are currently conforming to their assignments?
14. Which aircraft must be handed off to another sector/facility within the next 2 minutes?
15. Enter the aircraft that are experiencing a malfunction or emergency that is effecting operations.
16. Enter the aircraft which are not in communication with you.
17. Enter the aircraft that will violate special airspace separation standards if they stay on their current (assigned) path.
18. Which aircraft are weather currently an impact on or will be an impact on in the next 5 minutes along their current course?
19. Which aircraft will need a new clearance to achieve landing requirements?
20. Enter all aircraft that will violate minimum altitude requirements in the next two minutes if they stay on their current (assigned) paths?
21. Enter the aircraft that are not conforming to their flight plan.
22. Enter the airport and runway for this aircraft.

*Note.* From *Situation awareness global assessment technique (SAGAT) TRACON air traffic control version user guide* (pp. 19–40), by M. R. Endsley and E. O. Kiris, 1995, Lubbock: Texas Tech University. Copyright 1995 by M. R. Endsley and E. O. Kiris. Reprinted with permission.

should only be asked if the controller indicates in Query 11 that there are aircraft that have been issued a new clearance that is not yet completed. Query 21 should only be provided if the controller indicates in Query 9 that the aircraft will be landing in the sector. Query 22 is only relevant for landing aircraft and is generally not provided in a test involving en route air traffic control.

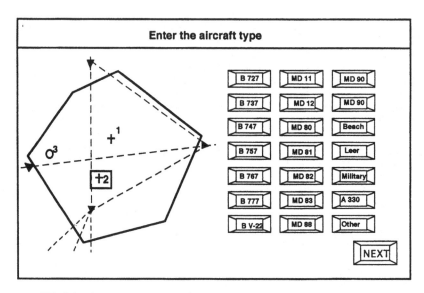

FIG. 7.1. SAGAT query example for ATC: aircraft type. *Note.* From *Situation awareness global assessment technique (SAGAT) TRACON air traffic control version user guide* (p. 26), by M. R. Endsley and E. O. Kiris, 1995, Lubbock: Texas Tech University. Copyright 1995 by M. R. Endsley and E. O. Kiris. Reprinted with permission.

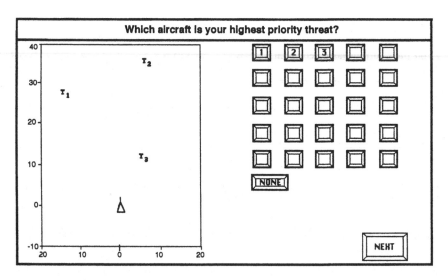

FIG. 7.2. SAGAT query example for military pilot: aircraft priority.

A determination of which queries should be provided for a given test can be made based on the SA requirements analysis, the capabilities of the simulation, the simulation scenarios, and the objectives of the test. The queries should not be too narrowly focused on just one or two items of interest, such as only measuring altitude when evaluating a new head-up display. This is because, if only a few probes are used, subjects are much more likely to shift their attention to the factors they know they will be questioned on, thereby artificially affecting SA. It has also been found that changes to one part of a system may inadvertently affect SA on other issues (Endsley, 1995a). A wide range of queries is needed to detect these changes. Vidulich (chap. 13, this volume) finds that narrowly focused queries are less sensitive for detecting changes in SA associated with a design than are a broad range of queries. For this reason, it is recommended that a broad range of queries be used except those that are clearly not relevant to SA in a particular simulation or situation.

To ease the administration of the queries and analysis of the data, SAGAT queries are frequently computerized (e.g., Endsley & Kiris, 1995b). The computerized battery ensures that certain queries are not asked when they are not appropriate. Although it is not required that administration be computerized (queries can be administered verbally or in written form), it has been found that administration in this manner greatly reduces the burden of analysis. The computer can collect the subject's answers to each query into a database that can be compared to the correct answers from the simulation's computer database. Relational database programs for correlating and scoring this data are a great benefit in the analysis process.

## EXAMPLES OF SAGAT RESULTS

The rich data provided by SAGAT provides some justification for the effort involved in using it. It has been used to examine changes in SA induced by changes in display formats, display hardware, and automation concepts; individual differences in SA among operators and factors related to SA; and major changes in the rules of operation.

### Evaluation of Sensor Hardware

Figure 7.3 shows an example of SAGAT results from a study that examined two different avionics concepts for a fighter aircraft system (Endsley, 1988b). In that study, a new avionics system that pilots subjectively felt was superior to the old system was evaluated. Simulation testing did not show an improvement in mission performance, however. The SAGAT testing revealed that the new system did indeed provide better SA. This was evidenced by a better knowledge of enemy aircraft location as well as im-

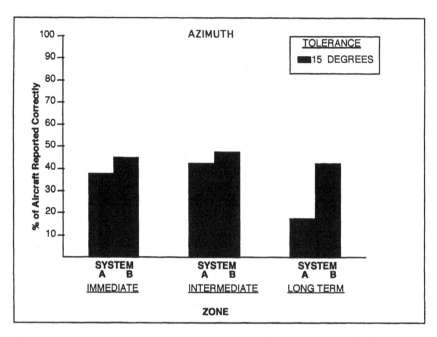

FIG. 7.3.  Example of SAGAT results: knowledge of aircraft location (by azimuth) for two avionics systems. *Note.* From "Situation awareness global assessment technique (SAGAT)," by M. R. Endsley, 1988, *Proceedings of the National Aerospace and Electronics Conference* (NAECON; p. 795), New York: IEEE. Copyright 1988 by M. R. Endsley. Reprinted with permission.

proved knowledge on several other factors important for SA in this domain. As shown in Fig. 7.3, pilots had significant improvement in SA with the new system for aircraft at various ranges with the greatest improvement for aircraft at the farthest range (> 70 miles). From these results, it was concluded that the new system was providing the desired benefit to SA. The lack of improvement in mission performance was most likely due to the fact that the pilots had not yet discovered how to modify their tactics to take advantage of the improved SA, and that insufficient testing had been conducted to see improvements in the less sensitive performance metrics available for fighter combat scenarios. As an aid for making design decisions, this sort of discrimination is highly useful.

### Evaluation of Free Flight

In another example, a totally new form of distributing roles and responsibilities between pilots and air traffic controllers was examined. Termed "free flight," this concept was originally described to incorporate major changes in the operation of the national airspace. It could include aircraft filing direct routes to destinations rather than along predefined fixed airways, and the authority for the pilot to deviate from that route with the air traffic controller's permission or perhaps even fully autonomously (RTCA, 1995). As it was felt that such changes could have a marked effect on the ability of the controller to keep up as the monitor in such a new system, a study was conducted to examine this possibility (Endsley, Mogford, Allendoerfer, Snyder, & Stein, 1997).

Results showed a trend towards poorer controller performance in detecting and intervening in aircraft separation errors with these changes in the operational concept and poorer subjective ratings of performance. Finding statistically significant changes in separation errors during ATC simulation testing is quite rare however. More detailed analysis of the SAGAT results provided more diagnostic detail as well as support for this finding. As shown in Fig. 7.4, controllers were aware of significantly fewer aircraft in the simulation under free flight conditions. Attending to fewer aircraft under higher workload has also been found in other studies (Endsley & Rodgers, 1998). In addition to reduced Level 1 SA, however, controllers also had a significantly reduced understanding (Level 2 SA) of what was happening in the traffic situation. This was evidenced by lower SA regarding those aircraft that weather would impact and a reduced awareness of the aircraft in a transitionary state. They were also less aware of which aircraft had not yet completed a clearance, and for those aircraft whether it was received correctly and whether they were conforming. Controllers also demonstrated lower Level 3 SA with free flight. Their knowledge of where the aircraft was going (next sector) was significantly lower under free flight conditions.

These findings were useful in pinpointing whether concerns over this new and very different concept were justified or whether they merely represented resistance to change. The SAGAT results showed that the new concept did indeed induce problems for controller SA that would prevent them from performing effectively as monitors to provide pilots with separation assistance and how these problems were manifested. This information is very useful diagnostically because it allows a determination of what sorts of aids might be needed to assist operators in overcoming these deficiencies. For instance, in this example, a display that provided enhanced information on flight paths for aircraft in transitionary states might be recommended as a way of compensating for the lower SA observed. Far from merely providing the approval or disapproval of a concept under evaluation, this rich source of data is very useful in developing iterative design modifications and making trade-off decisions.

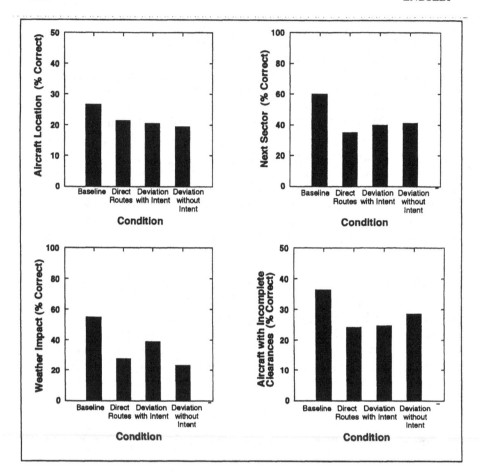

FIG. 7.4.   Example of SAGAT results: free flight implementations. *Note.*
From *Effect of free flight conditions on controller performance, workload and situa-
tion awareness: A preliminary investigation of changes in locus of control using ex-
isting technology* (DOT/FAA/CT–TN 97/12) (pp. 20–23), by M. R. Endsley, R.
Mogford, K. Allendoerfer, M. D. Snyder, and E. S. Stein, 1997, Atlantic
City, NJ: Federal Aviation Administration William J. Hughes Technical
Center. Copyright 1997 by M. R. Endsley, R. Mogford, K. Allendoerfer,
M. D. Snyder, and E. S. Stein. Reprinted with permission.

## Evaluation of Automation Concepts

SAGAT has also been used to provide information on questions of more
scientific interest. For instance, it has long been observed that operators
who act as monitors of automated systems may become "out-of-the-loop"
and less able to take over manual control when needed. This was hypothesized to be due partially to a lessening of manual skills, but primarily due

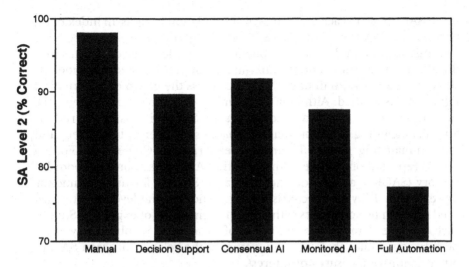

**Level of Automation Prior to Automation Failure**

FIG. 7.5. Example of SAGAT results: level of automation. *Note.* From "The out-of-the-loop performance problem and level of control in automation," by M. R. Endsley and E. O. Kiris, 1995, *Human Factors*, *37*(2), p. 390. Copyright 1995 by Human Factors and Ergonomics Society. Reprinted with permission.

to a loss of SA through both a decreased vigilance and a lower understanding that may be associated with passive monitoring (Endsley, 1987a, 1995a). Applying SAGAT at the end of a short simulation involving an expert system for aiding in automobile navigation (Endsley & Kiris, 1995a), we found that Level 2 SA was significantly lower when subjects were operating under full automation as compared to several intermediate levels of automation and purely manual control (see Fig. 7.5). This lower level of SA corresponded to the predicted increase in the time required to take over manual performance following an automation failure. Porter (1996) found a similar result; lower SA under high automation in a high-fidelity simulation involving an air traffic control tool called Center TRACON Automation System (CTAS). These studies demonstrate that SAGAT can help address basic issues involving questions about observed changes in human performance with certain types of system implementations. They may also direct us towards solutions to problems like the level of automation manipulation.

## Investigations of Individual Differences in SA

In some cases, SA itself is the subject of the investigation rather than merely an index of merit for some other factor being investigated. In these studies a direct measure of SA is highly desirable. For instance, we used mean SAGAT scores across a number of simulation trials as an index of individual pilot SA ability (Endsley & Bolstad, 1994). This measure provided a useful index of SA that was independent of simple experience measures, the effect of teammates and the aircraft system they had been provided. It showed a 10 to 1 ratio in mean SA score across the group of experienced fighter pilots studied. Although the number of subjects was limited (21), this study found significant correlates to SA ability as measured by SAGAT, including attention sharing, spatial skills, perceptual speed, and pattern-matching ability. In another experiment, Collier and Folleso (1995) reported on the use of the Situation Awareness Control Room Inventory (SACRI), a technique adapted after SAGAT for use in a nuclear power plant. They found consistent differences in the level of SA measured between two operators with very different levels of expertise. Studies such as these demonstrate the utility of measuring SA objectively when performing studies that seek to investigate the relation between SA and other cognitive measures of interest.

## ISSUES OF VALIDITY AND RELIABILITY OF SAGAT

The SAGAT technique has thus far been shown to have a high degree of validity, sensitivity, and reliability for measuring SA.

### Sensitivity

The examples referenced here show that SAGAT has good sensitivity to system manipulations, automation manipulations, expertise differences, and operational concepts in a variety of system domains. Vidulich (chap. 13, this volume) found good sensitivity for the technique across a wide range of studies when a broad range of queries is used.

### Criterion Validity

SAGAT also has predictive validity with SAGAT scores indicative of pilot performance in a combat simulation (Endsley, 1990b). This study found that fighter pilots who were able to report on an enemy aircraft's existence via SAGAT were three times more likely to later kill that target in the simulation.

Responses to SA queries also appear to be sensitive to features of the task situation. Fracker reported that SA measured using a freeze methodology was sensitive to the importance of the aircraft (degree of threat) and the number of enemy aircraft present in a low-fidelity task. Gronlund, Ohrt, Dougherty, Perry, and Manning (1998) reported that memory probes are sensitive to the importance of the aircraft in a high-fidelity air traffic control task. Another study (Endsley & Smith, 1996) similarly found that recall accuracy for aircraft location on a tactical situation display by experienced fighter pilots was significantly related to the aircraft's importance to their decision tasks. Gugerty (1997) found that the percentage of cars correctly reported decreased with increases in load (number of cars) in a driving task with attention focused on a subset based on the car's importance to the driver. In Endsley and Rodgers (1998), we found that the percentage of aircraft reported by experienced controllers via SAGAT significantly decreased with increases in the number of aircraft in an air traffic control simulation. As to those aircraft that the controllers knew were there, we found that controllers were significantly less accurate on most other factors related to the aircraft as the number of aircraft present increased; however, controllers interestingly preserved their knowledge of aircraft separation under load. These findings indicated that this type of measure is sensitive to changes in taskload and to factors that effect operator attention.

## Reliability

Measurement reliability has been demonstrated in a study that found high reliability ( test-retest scores of 0.98, 0.99, 0.99, and 0.92) of mean SAGAT scores for four fighter pilots who participated in two sets of simulation trials (Endsley & Bolstad, 1994). Fracker (1991) reported low reliability for his measure of aircraft location, however, this may be reflective of the absolute error score used in his test, as well as the low-fidelity of the simulation and use of inexperienced subjects. Collier and Folleso (1995) reported good reliability for their measure involving two experienced nuclear power plant operators. Gugerty (1997) reported good reliability for the percentage of cars recalled, recall error, and composite recall error (even-odd reliabilities of 0.93, 0.92, and 0.96) in his study involving a driving task. In general, these results support the reliability of the measure.

## Construct Validity

Probably the greatest two areas of concern about SAGAT have centered around the perceived intrusiveness of freezes in a simulation to collect SAGAT data and the degree to which it reflects memory and as such is limited. Each of these issues will be addressed separately.

### Intrusiveness

Several studies have shown that a temporary freeze in a simulation to collect SAGAT data does not impact performance. In previous work (Endsley, 1990a, 1995a), I reported on a study in which one, two, or three freezes were introduced for periods of 30 seconds, 1 minute, or 2 minutes to collect SAGAT data. This study found no impact by the stops on subject performance measures. Other studies have used SAGAT in some trials and not others in order to examine whether SAGAT interfered with subject performance (Bolstad & Endsley, 1990; Endsley, 1989b; Northrop Corporation, 1988). No effect was found in any of these studies. Subjectively, the pilots in the studies appeared to adjust to the technique quite well and were able to freeze and return to the action fairly readily. Hogg, Torralba, and Volden (1993) reported that the power plant operators in their study subjectively reported no effect from the freezes and considered it similar to their training exercises.

Despite the conflicting evidence of these studies, a concern over the possibility of intrusiveness has been voiced (Sarter & Woods, 1991) and continues to be repeated. In an effort to dispel these concerns, another study was conducted on the issue of intrusiveness. This study investigated whether operator performance could be effected by the mere threat of a stop to collect SAGAT data. That is, are operators somehow altering their behavior during simulation trials when they feel they may be stopped and tested on their SA? To answer this question, a study was conducted to compare trials where subjects were told that only performance would be measured with trials where subjects were told that a stop to collect SAGAT data might occur. In the latter case, SAGAT stops occurred only half of the time. Any effect of the actual SAGAT stop could therefore be differentiated from the mere threat of the stop and compared to trials with subjects who knew they would not be stopped.

A set of trials was conducted for an air-to-air fighter sweep mission. The subject, flying as the pilot of single aircraft, was to penetrate enemy territory and maximize enemy fighter kills while maintaining a high degree of survivability. Four computer controlled aircraft were the adversaries in these engagements. Subject instructions were manipulated during the test. In one-third of the trials, the subjects were told that only performance would be measured and no SAGAT stops would be made. In the other two-thirds of the trials, the subjects were told that there might be a stop to collect SAGAT data in addition to performance measurement. Half of these trials were stopped once for 2 minutes at a random point to collect SAGAT data. Half were not stopped. Each of six subjects completed five trials under each of the three conditions: no stop/none expected, no stop/stop expected, stop/stop expected. The conditions were presented in a random order. A total of 90 trials were completed. Pilot performance in terms of kills and losses was collected as the dependent measure.

The test was conducted using a medium-fidelity mission simulation on a Silicon Graphics 4D-220 computer. The system had a high-resolution 19″ color display monitor and realistic stick and throttle controls. A simulated head-up display, tactical situation display, vertical situation display, and fuel and thrust gages were provided. Six subjects participated in the test. The subjects were all experienced former military fighter pilots. The mean subject age was 43.6 years (range of 33–57). They had an average of 2803 hours (range of 1500–3850) and an average of 15.2 years (range of 7–25) of military flight experience. Two of the six subjects had combat experience.

Analysis of variance was used to evaluate the effect of the test condition (no stop/not expected, no stop/stop expected, and stop/stop expected) on each of the two performance measures: aircraft kills and losses. The test condition had no significant impact on either performance measure, $F(2, 87) = 0.15, p = 0.861, F(2, 87) = 1.53, p = 0.223$, as shown in Fig. 7.6. In viewing the data, it can be seen that the number of kills was almost identical and independent of whether subjects expected a stop or not or whether they actually experienced a stop. Although the subject "died" slightly more often in those trials where a stop was expected but not received, this difference was not significant. This data supports the null hypothesis; a stop or even the threat of a stop to collect SAGAT data does not have a significant impact on performance.

The results of this study confirmed the previous findings that have not found a demonstrable effect on performance of freezes in a simulation to collect SAGAT data. In addition, it expanded on these studies to reveal that even the threat of a stop does not significantly impact performance. Subjective comments by the subjects after the study confirm this. They reported that the information about whether to expect a SAGAT stop was irrelevant

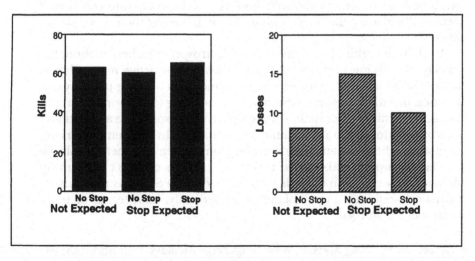

FIG. 7.6.   Pilot performance in simulations as a function of SAGAT freezes and expectations.

to them. At least on a conscious level, they were not preparing in any way for the SAGAT test. The results of this study indicated that they were not doing so unconsciously either. These results are also meaningful because the "opponents" in the trial were digital aircraft (computer controlled). This eliminated the possibility that effects on performance may have been masked by freezes of both the red team and the blue team equally.

Overall, the results of this study indicated that using SAGAT to collect data on SA is not intrusive on subject performance and provided an additional indication of the method's validity for directly measuring subject SA during simulations. Although it is never possible to "prove" the null hypothesis (SAGAT does not influence performance) all of the studies collected so far indicate that it does not appear to significantly influence performance as long as the stops are unpredictable to the subject (Endsley, 1988b). Based on the utility that has been found with the technique, it is believed that any such risk is well worth the information collected.

### Memory

***Retrospective Versus Concurrent Memory.*** The second issue concerning the use of SAGAT is whether it provides a good representation of the operator's SA or whether it is hindered by being dependent on memory. Addressing this issue depends on carefully examining the construct of memory. First, it has been widely reported that retrospective memory for past cognitive events is rather poor (Nisbett & Wilson, 1977). A technique that is dependent on retrospective memory would clearly represent a problem for SA measurement. SAGAT, however, seeks to tap into concurrent memory by placing the queries immediately following the event's occurrence.

In addition to thinking of memory as a retrospective phenomenon, the concept of working memory is also pertinent to the active manipulation and use of information. Baddeley's (1986) model of working memory, for instance, included a central executive for processing (comparing, manipulating, combining) information along with a visual-spatial scratchpad and a verbal-auditory loop for retaining information. In this sense, working memory is where dynamic information from the environment would be resident. It would make sense therefore that the contents of working memory are exactly what needs to be tapped by an SA measure. This information would need to be obtained immediately, however, as memory decay can be an issue.

***Access to Working Memory and Long-Term Memory.*** In addition, experiments have shown that SA is not solely dependent on working memory. Endsley (1988a, 1995b) and Fracker (1988) have both emphasized the importance of long-term memory stores for SA, particularly for experienced operators. Hartman and Secrist (1991) discussed the concept of skilled memory, positing access to long-term memory stores relevant to SA. In previous work (Endsley, 1990a, 1995a), I found that pilots' ability to report their SA via SAGAT was unaffected by how long after the freeze the question was asked (testing intervals were from around 20 seconds to up to 6 minutes). This manipulation was obtained by providing the queries in a random order. Thus, pilots were answering SAGAT questions during the entire interval, however, the timing of a particular query could be altered and compared to accuracy on the same query provided at different intervals across subjects and freezes. This study showed that SA information was available for quite some time after a freeze (when other conflicting tasks or information are not introduced as interference), indicating that long-term memory has a role in SA as well as working memory.

Two explanations are offered for these findings. First, the study used expert subjects performing a realistic task. Most laboratory studies of working memory employ stimuli with little or no inherent meaning to the subjects. The storage and utilization of task relevant information has been found to be very different from irrelevant information (Chase & Simon, 1973). Information has been found to be more retrievable when it is processed effortfully and with awareness (Cowan, 1988) as would be the case with a meaningful task. It is also likely that experienced operators have long-term memory stores (such as schemata and mental models) that serve to organize information and have an effect on its availability for a measure such as SAGAT. By contrast, Fracker (1991) found very poor performance on SA queries in his tests that probed student subjects (nonexperts) performing a very artificial computer task.

These results may be most parsimoniously explained by a more integrated model of working memory and long-term memory. Cowan (1988) proposed a model in which working memory was depicted as an activated subset of long-term memory. The focus of attention was represented within that activated memory and controlled by the central executive. Information can become activated effortfully through that focus of attention or habituated information may bypass that focus. Spontaneous activation of long-term memory based on associations may also bring information into activation. Durso and Gronlund (in press) concluded that an integrated model of working memory and long-term memory might best explain research on SA and memory. They drew on a model developed by Ericsson and Kintsch (1995) that showed that working memory contains pointers to information in long-term memory. Durso and Gronlund concluded that working memory may be a constraint on the processing of information to form SA (e.g., via the central executive), but not to the storage of SA.

*Explicit Versus Implicit Memory.* In terms of SAGAT and the measurement of SA via queries, these findings indicate that SA information should be retrievable for a short period of time during a simulation freeze via some combination of working and long-term memory stores. The situational representation of the operator should be reportable. This forms an explicit measure of memory. Other research has posited the idea of implicit memory—information that is available to affect performance, but which operators cannot report (Lewandowsky, Dunn, & Kirsner, 1989). Gugerty (1997) specifically addressed this issue in a driving task. He compared subject responses to recall probes regarding other cars (a measure of explicit SA) to subject performance on crash avoidance and detection of blocking cars (measures of implicit SA). He found that both the recall measures and the performance-based measures were reliable and significantly correlated with each other. Gugerty concluded that the two measures tapped the same knowledge base, thereby supporting their validity and agreeing with researchers who have found the distinction between implicit and explicit memory to be artificial and based on insensitive measurement.

A final concern revolves around whether information even needs to be in the subject's memory (or mental representation) in the first place. Perhaps it is sufficient that subjects merely reference the information on their displays when needed (Durso, Hackworth, Truitt, Crutchfield, Nikolic, & Manning, 1998). If this were the case, SA, as a dynamic mental representation of the environment, would not be needed. To address this question, I will leave an experiment to the reader. The next time you are pulled over by a policeman for speeding (you may have to break the law to carry out this experiment), and the officer asks, "Do you know how fast you were going," your reply should be, "No, but it doesn't matter because I can look at my speedometer any time I want." My hypothesis is that the policeman will not be amused by this argument. Getting information from a display is not tantamount to providing an operator with SA (or with effective performance). One of the most frequent causal factors for SA errors involves situations where all of the information is available to the operator but it is not attended to for various reasons (Jones & Endsley, 1996). The real challenge in today's systems is the ability of the operator to dynamically locate and integrate needed information from the plenitude that is available in the system. For this reason, SA, as a measure of the operator's ability to form the needed understanding of the status of a dynamic system under the constraints of many opposing demands, is the measure of interest.

In summary, these studies indicated that SA measures such as SAGAT can reliably tap into memory stores (either working memory or long-term memory) as an index of SA. They do not appear to be hampered by problems of retrospective memory recall or implicit memory constraints.

## IMPLEMENTATION RECOMMENDATIONS

Several recommendations for SAGAT administration have been made based on previous experience in using the procedure (Endsley, 1995a).

### Training

An explanation of SAGAT procedures and detailed instructions for answering each query should be provided to subjects before testing. The idea of using a "surprise" to collect this data will probably only work once and may be hampered by misunderstandings regarding the questions themselves. For this reason, it is usually better to thoroughly brief participants. Several training trials should be conducted in which the simulator is halted frequently to allow subjects ample opportunity to practice responding to the SAGAT queries. Usually three to five samplings are adequate for a subject to become comfortable with the procedure and to clear up any uncertainties about how to answer the queries.

### Test Design

SAGAT requires no special test considerations. The same principles of experimental design and administration apply to SAGAT as to any other dependent measure. Measures of subject performance and workload may be collected concurrently with SAGAT as no ill effect from the insertion of breaks has been shown. To be cautious, however, half of the trials should be conducted without any breaks for SAGAT so that a check is provided for this contingency.

### Procedures

Subjects should be instructed to attend to their tasks as they normally would with the SAGAT queries considered as secondary. No displays or other visual aids should be visible while subjects are answering the queries. If subjects do not know or are uncertain about the answer to a given query, they should be encouraged to make their best guess. There is no penalty for guessing as this allows for consideration of the default values and other wisdom gained from experience that subjects normally use in decision making (e.g., embedded schemata information). By giving subjects "credit" for this type of knowledge, the true benefits of systems that seek to supplant that knowledge can be more fairly assessed. If subjects do not feel comfortable enough to guess, they may go on to the next question. Talking or sharing of information between subjects should not be permitted. If multiple subjects are involved in the same simulation, all subjects should be queried simultaneously and the simulation resumed for all subjects at the same time. (Later comparisons of team members' SAGAT data can also be used to assess the degree to which they possessed a common picture of the situation—shared SA.)

## Which Queries to Use

### Random Selection

As it may be impossible to query subjects about all of their SA requirements during a stop because of time constraints, a portion of the SA queries may be randomly selected and asked each time. A random sampling provides consistency and statistical validity, thus allowing SA scores to be easily compared across trials, subjects, systems, and scenarios.

Due to attentional narrowing or a lack of information, certain questions may seem unimportant to a subject at the time of a given stop. It is important to stress that they should attempt to answer all queries anyway. This is because even though they think it unimportant, the information may have at least a secondary importance, they may not be aware of the information that makes a question very important (e.g., the presence of a pop-up aircraft), and if only questions of the highest priority were asked, subjects might be inadvertently provided with artificial cues about the situation that will direct their attention when the simulation is resumed. Therefore, a random selection from a constant set of queries is recommended at each stop.

### Experimenter Controlled

In certain tests it may be desirable to have some queries omitted due to simulation limitations or the scenarios' characteristics. For instance, if the simulation does not incorporate aircraft malfunctions, the query related to this issue may be omitted. Additionally, with particular test designs it may be desirable to ensure that certain queries are presented every time. When this occurs, it is important that subjects also be queried on a random sampling from all SA requirements and not just on those related to a specific area of interest to the evaluation being conducted because of the ability of subjects to shift attention to the information they know they will be tested on. What may appear to be an improvement in SA in one area may just be a shift of attention from one aspect of the situation to another. When the SAGAT queries cover all of the SA requirements no such artificial cueing can occur and attention shifts can be detected.

## When to Collect SAGAT Data

It is recommended that the timing of each freeze for SAGAT administration be randomly determined and unpredictable enough so the subjects cannot prepare for them in advance. If the freeze occurrence is associated with the occurrence of specific events or specific times across trials, prior studies showed that the subjects will be able to figure this out (Endsley, 1988b). This allows the subjects to prepare for them or actually improve SA through the artificiality of the freeze cues. An informal rule has been to ensure that no freezes occur earlier than 3 to 5 minutes into a trial so that subjects can build up a picture of the situation and that no two freezes occur within 1 minute of each other.

The result of this approach is that the activities occurring at the time of the stops will be randomly selected. Some stops may occur during very important activities that are of interest to the experimenter and others when no critical activities are occurring. This gives a good sampling of the subjects' SA in a variety of situations. During analysis the experimenter may want to stratify the data to take these variations into account.

### How Much SAGAT Data to Collect

The number of trials necessary will depend upon the variability present in the dependent variables being collected and the number of data samples taken during a trial. This will vary with different subjects and designs, but between 30 and 60 samplings per SA query (across subjects and trials) with each design option are adequate in a within-subjects test design.

Multiple SAGAT stops may be taken within each trial. There is no known limit to the number of times the simulator can be frozen during a given trial. No ill effects have been found for as many as three stops during a 15 minute trial (Endsley, 1995a). In general, it is recommended that a freeze last until a certain amount of time has elapsed, regardless of how may questions have been answered. Freezes as long as 2 minutes in duration were used with no undue difficulty or effect on subsequent performance. Freezes as long as 5 to 6 minutes allow subjects access to SA information without memory decay (Endsley, 1995a).

### Data Collection

The simulator computer should be programmed to collect objective data corresponding to the queries at the time of each freeze. Because some queries will pertain to higher level SA requirements that are unavailable from the computer, an expert judgment of the correct answer may be made by an experienced observer who is privy to all information and reflects the SA of a person with perfect knowledge. A comparison of the subjects' perceptions of the situation (as input into SAGAT) to the actual status of each variable (as collected per the simulator computer and expert judgment) results in an objective measure of subject SA. Questions asked of the subject but not answered should be considered incorrect. No evaluation should be made of questions not asked during a given stop.

It is recommended that the answer to each query be scored as correct or incorrect based upon whether it falls into an acceptable tolerance band around the actual value. For example, it may be acceptable for a subject to be 10 mph off the actual groundspeed. This method of scoring poses less difficulty than dealing with absolute error (see Marshak, Kuperman, Ramsey, & Wilson, 1987). A tabulation of the frequency of correctness can then be made within each test condition for each SA element. As data scored as correct or incorrect are binomial, the conditions for analysis of variance are violated. A correction factor [$Y' = \arcsin(Y)$] can be applied to binomial data that allows analysis of variance to be used. In addition, a chi-square, Cochran's Q, or binomial $t$ test (depending on the test design) can be used to evaluate the statistical significance of differences in SA between test conditions.

Finally, it is recommended that each query be evaluated separately rather than in some combined form. In general, analysis has not supported the combination of SAGAT queries into a single measure or into groups by SA level. SA on each query can be independent of the others as reflections of shifts in operator attention in association with various display manipulations and as dictated by the demands of their tasks. Combinations of queries are likely to reduce sensitivity of the metric by losing these important distinctions.

## LIMITATIONS AND APPLICABILITY FOR USE

This technique has primarily been used within the confines of high-fidelity and medium-fidelity part-task simulations. This provides experimenter control over freezes and data collection without any danger to the subject or processes involved in the domain. It may be possible to use the technique during actual task performance if multiple operators are present to ensure safety. For example, it might be possible to verbally query one pilot in flight while another assumes flight control. Such an endeavor should be undertaken with extreme caution, however, and may not be appropriate for certain domains.

Sheehy, Davey, Fiegel, and Guo (1993) employed an adaptation of this technique by making videotapes of an ongoing situation in a nuclear power plant control room. These tapes were then replayed to naive subjects with freezes for SAGAT queries employed. It is not known how different the SA of subjects passively viewing a situation may be from subjects actually engaged in task performance, however, this approach may yield some useful data. For instance, in Endsley and Rodgers (1998), we used computer generated replays of air traffic control errors to study controller SA. Similar to Sheehy et al., this study involved passive viewing of the displays. Nonetheless, we were able to examine shifts in controller SA that corresponded to workload changes and that were indicative of SA problems as experienced in real-time operations.

## CONCLUSION

To date, SAGAT has been used for a wide variety of system evaluations and construct investigations in such diverse applications as aircraft, ATC, driving, nuclear power operations, and teleoperations. Perhaps the most widely tested measure of SA, it has been shown to have good levels of sensitivity, reliability, and predictive validity.

## REFERENCES

Baddeley, A. D. (1986). *Human memory*. Oxford, England: Clarendon.

Bolstad, C. A., & Endsley, M. R. (1990). *Single versus dual scale range display investigation* (NOR DOC 90–90). Hawthorne, CA: Northrop.

Chase, W. G., & Simon, H. A. (1973). Perceptions in chess. *Cognitive Psychology, 4*, 55–81.

Collier, S. G., & Folleso, K. (1995). SACRI: A measure of situation awareness for nuclear power plant control rooms. In D. J. Garland & M. R. Endsley (Eds.), *Experimental analysis and measurement of situation awareness* (pp. 115–122). Daytona Beach, FL: Embry-Riddle University Press.

Cowan, N. (1988). Evolving conceptions of memory storage, selective attention, and their mutual constraints within the human information processing system. *Psychological Bulletin, 104*(2), 163–191.

Durso, F. T., & Gronlund, S. D. (in press). Situation awareness. In F. T. Durso, R. Nickerson, R. Schvaneveldt, S. Dumais, M. Chi, & S. Lindsay (Eds.), *Handbook of applied cognition*. New York: Wiley.

Durso, F. T., Hackworth, C. A., Truitt, T. R., Crutchfield, J., Nikolic, D., & Manning, C. A. (1998). Situation awareness as a predictor of performance for en route air traffic controllers. *Air Traffic Control Quarterly, 6*(1), 1–20.

Endsley, M. R. (1987a). The application of human factors to the development of expert systems for advanced cockpits. In *Proceedings of the Human Factors Society 31st Annual Meeting* (Vol. 2, pp. 1388–1392). Santa Monica, CA: Human Factors Society.

Endsley, M. R. (1987b). *SAGAT: A methodology for the measurement of situation awareness* (NOR DOC 87–83). Hawthorne, CA: Northrop.

Endsley, M. R. (1988a). Design and evaluation for situation awareness enhancement. In *Proceedings of the Human Factors Society 32nd Annual Meeting* (Vol. 1, pp. 97–101). Santa Monica, CA: Human Factors Society.

Endsley, M. R. (1988b). Situation awareness global assessment technique (SAGAT). In *Proceedings of the National Aerospace and Electronics Conference* (NAECON) (pp. 789–795). New York: IEEE.

Endsley, M. R. (1989a). *Final report: Situation awareness in an advanced strategic mission* (NOR DOC 89–32). Hawthorne, CA: Northrop.

Endsley, M. R. (1989b). *Tactical simulation 3 test report: Addendum 1 situation awareness evaluations* (81203033R). Hawthorne, CA: Northrop.

Endsley, M. R. (1990a). A methodology for the objective measurement of situation awareness. In *Situation Awareness in Aerospace Operations* (AGARD–CP–478; pp. 1/1–1/9). Neuilly Sur Seine, France: NATO–AGARD.

Endsley, M. R. (1990b). Predictive utility of an objective measure of situation awareness. In *Proceedings of the Human Factors Society 34th Annual Meeting* (Vol. 1, pp. 41–45). Santa Monica, CA: Human Factors Society.

Endsley, M. R. (1990c). *Situation awareness in dynamic human decision making: Theory and measurement.* Unpublished doctoral dissertation, University of Southern California, Los Angeles.

Endsley, M. R. (1993). A survey of situation awareness requirements in air-to-air combat fighters. *International Journal of Aviation Psychology, 3*(2), 157–168.

Endsley, M. R. (1994). *Situation awareness in FAA Airway Facilities Maintenance Control Centers (MCC): Final Report.* Lubbock: Texas Tech University.

Endsley, M. R. (1995a). Measurement of situation awareness in dynamic systems. *Human Factors, 37*(1), 65–84.

Endsley, M. R., (1995b). Toward a theory of situation awareness in dynamic systems. *Human Factors, 37*(1), 32–64.

Endsley, M. R., & Bolstad, C. A. (1994). Individual differences in pilot situation awareness. *International Journal of Aviation Psychology, 4*(3), 241–264.

Endsley, M. R., Farley, T. C., Jones, W. M., Midkiff, A. H., & Hansman, R. J. (1998). *Situation awareness information requirements for commercial airline pilots* (ICAT–98–1). Cambridge: Massachusetts Institute of Technology International Center for Air Transportation.

Endsley, M. R., & Jones, D. G. (1995). *Situation awareness requirement analysis for TRACON air traffic control* (TTU–IE–95–01). Lubbock: Texas Tech University.

Endsley, M. R., & Kiris, E. O. (1995a). The out-of-the-loop performance problem and level of control in automation. *Human Factors, 37*(2), 381–394.

Endsley, M. R., & Kiris, E. O. (1995b). *Situation awareness global assessment technique (SAGAT) TRACON air traffic control version use guide.* Lubbock: Texas Tech University.

Endsley, M. R., Mogford, R., Allendoerfer, K., Snyder, M. D., & Stein, E. S. (1997). *Effect of free flight conditions on controller performance, workload and situation awareness: A preliminary investigation of changes in locus of control using existing technology* (DOT/FAA/CT–TN 97/12). Atlantic City, NJ: Federal Aviation Administration William J. Hughes Technical Center.

Endsley, M. R., & Robertson, M. M. (1996). *Team situation awareness in aircraft maintenance.* Lubbock: Texas Tech University.

Endsley, M. R., & Rodgers, M. D. (1994). *Situation awareness information requirements for en route air traffic control* (DOT/FAA/AM–94/27). Washington, DC: Federal Aviation Administration Office of Aviation Medicine.

Endsley, M. R., & Rodgers, M. D. (1998). Distribution of attention, situation awareness, and workload in a passive air traffic control task: Implications for operational errors and automation. *Air Traffic Control Quarterly, 6*(1), 21–44.

Endsley, M. R., & Smith, R. P. (1996). Attention distribution and decision making in tactical air combat. *Human Factors, 38*(2), 232–249.

Ericsson, K. A., & Kintsch, W. (1995). Long term working memory. *Psychological Review, 102*, 211–245.

Fracker, M. L. (1988). A theory of situation assessment: Implications for measuring situation awareness. In *Proceedings of the Human Factors Society 32nd Annual Meeting* (Vol. 1, pp. 102–106). Santa Monica, CA: Human Factors Society.

Fracker, M. L. (1990). Attention gradients in situation awareness. In *Situational Awareness in Aerospace Operations* (AGARD–CP–478; pp. 6/1–6/10). Neuilly Sur Seine, France: NATO–AGARD.

Fracker, M. L. (1991). *Measures of situation awareness: An experimental evaluation* (AL–TR–1991–0127). Wright-Patterson AFB, OH: Armstrong Laboratory.

Gronlund, S. D., Ohrt, D. D., Dougherty, M. R. P., Perry, J. L., & Manning, C. A. (1998). Role of memory in air traffic control. *Journal of Experimental Psychology: Applied, 4*, 263–280.

Gugerty, L. J. (1997). Situation awareness during driving: Explicit and implicit knowledge in dynamic spatial memory. *Journal of Experimental Psychology: Applied, 3*, 42–66.

Hartman, B. O., & Secrist, G. E. (1991). Situational awareness is more than exceptional vision. *Aviation, Space and Environment Medicine, 62*, 1084–9.

Hogg, D. N., Torralba, B., & Volden, F. S. (1993). *A situation awareness methodology for the evaluation of process control systems: Studies of feasibility and the implication of use* (1993–03–05). Storefjell, Norway: OECD Halden Reactor Project.

Jones, D. G., & Endsley, M. R. (1996). Sources of situation awareness errors in aviation. *Aviation, Space and Environmental Medicine, 67*(6), 507–512.

Lewandowsky, S., Dunn, J. C., & Kirsner, K. (Eds.). (1989). *Implicit memory: theoretical issues.* Hillsdale, NJ: Lawrence Erlbaum Associates.

Marshak, W. P., Kuperman, G., Ramsey, E. G., & Wilson, D. (1987). Situational awareness in map displays. In *Proceedings of the Human Factors Society 31st Annual Meeting* (Vol. 1, pp. 533–535). Santa Monica, CA: Human Factors Society.

Nisbett, R. E., & Wilson, T. D. (1977). Telling more than we can know: Verbal reports on mental processes. *Psychological Review, 84*(3), 231–259.

Northrop Corporation. (1998). *Tactical simulation 2 test report: Addendum 1 situation awareness test results.* Hawthorne, CA: Author.

Porter, A. W. (1996). *Investigating the effects of automation on situation awareness* (DRA/LS(LSC4)/CHCI/CD299/1.0). Malvern, England: Defense Research Agency.

RTCA (1995). *Report of the RTCA Board of Directors select committee on free flight.* Washington, D.C.: Author.

Sarter, N. B., & Woods, D. D. (1991). Situation awareness: A critical but ill-defined phenomenon. *The International Journal of Aviation Psychology, 1*(1), 45–57.

Sheehy, E. J., Davey, E. C., Fiegel, T. T., & Guo, K. Q. (1993, April). *Usability benchmark for CANDU annunciation—lessons learned.* Paper presented at the ANS Topical Meeting on Nuclear Plant Instrumentation, Control, and Man–Machine Interface Technology, Oak Ridge, TN.

# Chapter 8

# Strategies for Psychophysiological Assessment of Situation Awareness

Glenn F. Wilson
*Air Force Research Laboratory*

Very few experiments have been conducted using psychophysiological measures to study *situation awareness* (SA). In fact, at this time only one such study is known. The goal of this chapter is to examine where psychophysiological measures could be used to increase our understanding of operator SA. The approach is to use knowledge generated from other areas of operator performance and to extrapolate those findings to SA research. Psychophysiological measures have been used to study complex cognitive domains such as mental workload and fatigue. From these examples it is possible to speculate about how these metrics might prove useful in furthering an understanding of SA. If this expectation is realized, then additional work can lead to the development of batteries of SA measures that would include psychophysiological data. Because SA involves complex cognitive activity, psychophysiological measures would no doubt be used in conjunction with subjective and performance techniques.

Psychophysiology is at the interface of cognition and physiology. It strives to understand the relationship between a person's cognitive activity and the changes in their physiology that produce the cognitive activity or result from it. Both central and peripheral nervous system activity is examined in order to provide a better understanding of the relation between cognition and physiology and also to understand the interdependence among the various physiological systems during mental activity. Electrical and hormonal measures are typically included with newer techniques (e.g., Positron Emission Tomography and functional Magnetic Resonance Imaging) that examine the metabolic activity of the brain during cognitive activity. Much of the work done in this area has been performed using simple, single task paradigms. However, complex tasks are being more widely used in order to approach real-world actuality. These complex tasks often involve the simultaneous performance of multiple tasks in order to approximate real-world situations. Further, there are an increasing number of projects in which physiological data are collected while operators perform their jobs in their actual work environments. This type of data is required to achieve the validity needed for application of psychophysiological measures in the work setting. If the results of these studies demonstrate that the psychophysiological measures have utility, then they can be applied to day-to-day situations.

Psychophysiological measures have unique properties that should make them attractive to investigators in the SA field. Some SA measures, such as the query technique, require that ongoing task performance be stopped while the operator is interrogated about their level of SA. This interference with the primary task performance is problematic and its use is limited to situations where it is possible to stop the flow of the task. For example, these procedures are difficult to apply during flight. On the other hand, psychophysiological measures can be unobtrusively obtained without interfering with performance. A second beneficial feature to the SA researcher is that, in contrast to the discrete nature of some SA measures such as subjective reports, psychophysiological measures are continuously available. The continuous nature of psychophysiological measures can be especially useful in situations where the timing of critical events cannot be precisely controlled or when events of interest are unplanned or uncontrolled. Because the events are unpredictable, the SA associated with these novel or unexpected events would be missed by the more standard SA measurement techniques. Additionally, because the physiological data are continuously recorded, it is possible to go back and assess the situation as it existed when the events occurred and it is also possible to examine the antecedent conditions. A relevant example is cited in Wilson (1995): a pilot's heart rate responses were recorded during a bird strike. Immediately following the unexpected event the pilot's heart rate increased by 50% within about 25 seconds; however, his heart rate returned to his pre-bird strike level within about 60 seconds. This can be taken as evidence that the pilot responded to the emergency event and within approximately 90 seconds evaluated the situation and attained the appropriate SA. He evaluated the unique situation, assessed the potential damage to his aircraft, and initiated actions to gain altitude and return safely to base.

As previously stated, psychophysiological measures have not been used in SA research to any great extent; however, they have been used in numerous mental workload studies. Because of the postulated overlap of mental workload with SA in some situations, it is worth examining these data (Endsley, 1995a; Taylor, 1989). Several examples will be presented from workload studies that bear upon the study of SA. Hopefully, these examples will be of sufficient interest to SA investigators to inspire them to consider using psychophysiological measures in their future work.

Additional information about the various psychophysiological measures and where they have been applied is found in the following sources: A current overview of psychophysiology methods and theory in Cacioppo and Tassinary (1990); a shorter review of the application of psychophysiological methods to the aerospace environment in Caldwell, et al. (1994); and a review of psychophysiological methodology applied to mental workload in Kramer (1991) and Wilson and Eggemeier (1991).

In the only known SA study to use psychophysiological measures it was reported that both electroencephalographic activity (EEG) and eyeblinks showed effects that could be related to the subject's SA maintenance (Vidulich, Stratton, Crabtree, & Wilson, 1994). The task was a simulated air-to-ground combat flight mission in which the type of displayed information available to the subjects was manipulated. Some of the display types provided information in a format that should have been conducive to good SA whereas other formats made the acquisition of good SA more difficult. The results showed that the EEG data evidenced higher levels of theta band activity and lower levels of alpha band activity in the condition that was associated with the lowest levels of SA. Further, the shortest blink durations and the highest rate of blinking were found in the condition that produced the lowest levels of SA as determined by the subjective and performance measures. This was interpreted as being consistent with previous literature that associated these changes with situations having higher levels of mental workload. In effect, the subjects were working harder to maintain the best possible level of SA with the information presented to them. In the difficult condition they were required to integrate information from different sources as compared to the conditions associated with better SA in which the information was much easier to perceive. It is not clear if these psychophysiological data are providing measures of SA or workload or are indicative of the interaction between them. SA and workload are interdependent at some levels and this may be an example of such a case. Further research is needed in a wide range of situations in order to determine the utility of psychophysiologcial data to the assessment of SA. However, these results demonstrate that psychophysiological measures have some utility, when collected as part of a larger battery of measures, in the investigation of SA. The ability of the psychophysiological measures to provide continuous monitoring of operate state is especially useful in determining SA in field settings.

It is not clear that psychophysiological measures can directly tap the high-level cognitive processes involved in SA. For example, it is not currently possible to use psychophysiological data to infer whether information is correctly assessed as it is processed by an operator who is trying to maintain proper SA. However, it is possible to use psychophysiological activity to determine whether a relevant stimulus was detected and ascertained to be important to the task at hand. Whether the information was *correctly* processed in the current context to maintain SA is not clear; however, by using a classification analysis, it may be possible to determine whether the operator is aware of the crucial aspects of their environment or the overall situation. Another potential application is to use psychophysiological measures to determine whether the operator's overall cognitive state is conducive to good SA.

## APPLICATION OF PSYCHOPHYSIOLOGICAL MEASURES TO SA QUESTIONS

Because few examples of research exist in which psychophysiological measures were applied to SA questions, this chapter considers examples of how psychophysiological measures have been used in other areas and speculates as to how they might be applied to SA questions. There seem to be at least five approaches that can be used when applying psychophysiological measures to SA. The first is to use them to determine whether the overall functional state of the operator is conducive to good SA. This approach includes using EEG, eyeblinks, and cardiac activity to monitor an operator's functional state. The functional states of interest are found at the extremes of the consciousness continuum. That is, psychophysiological measures can be used to determine if the person is asleep, fatigued at one end of the continuum, or mentally overloaded at the other end.

The second approach is to use psychophysiological measures to determine whether an operator perceives critical environmental cues. These measures include event related potentials (ERP), event related desynchronization (ERD), transient heart rate (HR), or electrodermal activity (EDA). This is an expansion of the approach suggest by Byrne (1995).

The third potential use of psychophysiological measures is to determine whether the operator has the appropriate level of SA by examining their expectancies regarding environmental stimuli. An example is found in a study by Stern, Wang, and Schroeder (1995). They suggested that good SA means that a person showed expectancy to upcoming events. If this is true, the strategy then would be to monitor the physiological responses to those events to provide an estimate of an operator's current level of SA.

Another approach is to use psychophysiological data to identify times when it is advantageous to insert subjective, query, or performance measures of SA. Because the physiological data are continually available, it is possible to monitor operator responses throughout the work session. This approach utilizes the observation of significant changes in the physiological responses recorded during job performance to insert subjective and performance probes. This is based upon the idea that when an operator recognizes salient features in the environment or mentally perceives significant situations, there will be changes in the operator's physiology. For example, if a potentially serious condition is recognized there will be a concomitant alteration in the heart rate and EEG patterns. These changes could be monitored to signal the appropriate times to interject probes.

The last approach is to apply classification techniques to SA data to determine whether it is possible to correctly detect different levels of SA based upon psychophysiological data. This involves using psychophysiological data to train computerized classifiers to see if the classifiers can be used to distinguish between good and bad SA. This technique has been successfully applied in mental workload research and because of the similarities with SA, this approach may prove successful.

**Operator Functional State Assessment**

SA has been defined as a person's "perception of the elements of the environment within a volume of time and space, the comprehension of their meaning and the projection of their status in the near future" (Endsley, 1988, p. 792). Because this requires high-level cognitive functioning, the operator must possess the skills necessary to perform the required steps to attain SA and possess a functional state such that this process can occur. For example, the operator must be sufficiently alert to intake and process sensory information, to access memory and decide if intervention is needed, and then to perform the required actions. Operators must be aware and alert and not so cognitively compromised that they cannot function. Although a large range of operator states permit this level of processing, there are also a number that would interfere. Sleep, high levels of fatigue, inattention, and cognitive overload are some operator states that are common and interfere with an operator's ability to maintain proper SA. Also, levels of consciousness surrounding these states, such as sleep inertia—the state often found between awakening and full functional alertness, may not be conducive to maintaining proper SA. The time prior to cognitive overload is also a time of decreased functionality. Psychophysiological methods can be used to determine the functional state of the operator; this information identifies periods when proper SA is not possible.

Sleep and fatigue can be detected with various psychophysiological measures. In fact, sleep stages are defined entirely upon electrophysiological criteria (Rechtshaffen & Kales, 1968). Performance and subjective data are not available with regard to sleep architecture. Horne (1978) reviewed the psychophysiological effects of sleep deprivation. Heart rate, respiration, oral temperature, and saccadic eye movements all decrease as a function of sleep loss. Palmer, Wilson, Reis, and Gravelle (1995) reported that one night's loss of sleep was associated with significant increases in heart rate variability. Their subjects performed several single tasks and one dual task before one night's sleep loss, after the sleep loss, and after a day of recovery. Of their psychophysiological measures, heart rate variability showed the most consistent sleep loss changes. Drowsiness and fatigue also present characteristic psychophysiological markers. Typically, the dominant EEG frequency slows as the person progresses from alert to drowsy to asleep. This is usually accompanied by a slowing of the heart rate and an increased number of eyeblinks with longer eye closure duration. EEG alpha abundance typically decreases as the period of sleep loss increases. Makeig and Inlow (1993) reported that fatigue caused lapses in performance that were accompanied by changes in the EEG spectra. Although not an absolute sign of lowered SA, these states are certainly conducive to lowered SA.

On the other end of operator functionality is cognitive overload. This state is not as well defined psychophysiologically but has been determined in some studies. Brookings, Wilson, and Swain (1996) studied the effects of mental workload on air traffic controllers while they performed a simulated terminal radar approach control task. As the task demands progressed from low to an overload condition, they found increasing levels of frontal EEG theta band activity with corresponding decreases in alpha band activity. Eyeblink rate also decreased and respiration rate increased as the task became more difficult with the lowest blink rates and highest breath rates occurring during the overload condition.

By continuously monitoring an operator's functional state, it is possible to determine the operator's status and to take appropriate action to alert both the operator and the system so that the operator's functional state could be changed to be more conducive to proper SA maintenance. Psychophysiological measures can be used to nonintrusively provide continuous estimates of operator state by monitoring for sleep, fatigue, and mental overload. By avoiding the extremes, it could be ensured that the operator's condition supports optimal SA.

## Determining the Perceptual Aspects of SA

Once it is resolved that the operator's functional state is in the appropriate range there still remains the problem of determining if the awake operator is maintaining a good level of SA. Because essentially no research has been conducted to evaluate the utility of psychophysiological measures to determine an operator's level of SA, it is not clear what advantage these measures may have. However, it is possible to assess which aspects of SA would be appropriate for investigations using these methods by reviewing the psychophysiological literature in related areas. Mental workload research shares several dimensions with SA (Endsley, 1995a). The level of cognitive activity is high in most real-world environments in which SA is of interest. It is possible to find results in the mental workload literature that suggest strategies that can be used to assess the merits of psychophysiological measures to determine SA.

Brain ERPs have been used to assess whether environmental stimuli were detected. There exists a large body of literature detailing the relation between ERPs and cognition. ERPs are the small brain electrical potentials that are related to information processing. Because of their small size relative to the background EEG, several responses are generally averaged together so that the ERPs can be viewed. There are two broad classes of ERPs, exogenous and endogenous. The characteristics of exogenous ERPs are primarily determined by the nature of the environmental stimuli. The sensory modality and stimulus intensity and other factors determine the nature of the ERP. Endogenous ERPs are primarily determined by cognitive aspects of the stimulus processing. The presence of ERPs (exogenous) to relevant stimuli can be used to determine whether the operator detected the stimuli and recognized its importance to the task (endogenous). For example, task relevant stimuli could be presented during different phases of a task. If the stimuli are relevant to good SA and are presented to an operator who is cognizant of the task, then a person would expect to observe ERPs. However, if the operator does not possess good SA, then they might miss the importance of the stimuli and that would result in small or absent ERPs.

**Expectancy Evaluation**

Stern, Wang, and Schroeder (1995) suggested that psychophysiological measures could be used to determine whether operators expected certain classes of stimuli by using heart rate and eye movement activity. Cardiac deceleration is found when subjects expect to acquire certain information and anticipate an event that requires them to make a response. Expectancy of a stimulus also inhibits blinking and is associated with shorter eye closure durations. If an operator shows these responses to appropriate stimuli, it can be ascertained whether the operator has the proper SA that leads them to expect those stimuli. Wilson (1995) cited data collected during a low altitude parachute extraction. A C-130 transport aircraft flew about 10 feet above the ground to deliver cargo. As several tons of cargo was extracted from the aircraft by a parachute the flight dynamics changed very dramatically. In order to avoid collision with the ground, the pilot must maintain a high level of SA. The maneuver caused drastic changes in the pilot's eyeblinks. Prior to the drop, the eyeblinks became uncharacteristically slow and showed a very regular pattern with short closure durations. During the drop, in anticipation of the requirement to maintain SA, there is a period of blink inhibition. These data demonstrated that the pilot had good SA and the eyeblink activity showed that the pilot correctly anticipated the situation.

Brain-evoked activity, as found in the P300 amplitude, is widely used as a measure of expectancy. The P300 component of the ERP is a positive deflection in the ERP that is large and appears 250 to 600 milliseconds following stimulus presentation. Its cognitive concomitants are well established and the amplitude of this ERP component changes with the level of expectancy. Lower levels of expectancy are associated with larger amplitude P300s. Wilson, Swain, and Ullsperger (1998) reported that the magnitude of the P300 amplitude is modulated by a warning stimulus that alerted the subject to the difficulty of the succeeding task. However, as pointed out by Endsley (1995b), the P300 did not indicate whether the information was correctly registered relative to the situation. With clever experimental design, it may be possible to elicit some of this information from the P300, however, the problem remains that a number of eliciting stimuli must be presented in order to acquire the P300.

By inserting critical stimuli at appropriate times, other measures such as ERDs and evoked cardiac changes could be used to determine the saliency of particular stimuli. ERDs represent changes in EEG activity that are related to stimulus processing (Pfurtsheller, Stancák, & Neuper, 1996). Typically, the energy in the alpha band of the EEG is monitored following the presentation of a stimulus. Reductions in alpha band amplitude are associated with information processing, selective attention, and the preparation for motor responses. Increases in task complexity are also accompanied by increased ERDs. Potentially, the ERD could be used to determine whether an operator has detected an SA relevant stimulus; however, the utility of ERDs in complex task situations has been questioned (Fournier, Wilson, & Swain, 1999).

## SA Probe Assessment Technique

Another potential application of psychophysiological measures in SA research is to indicate specific times when other SA measures should be applied. This is similar to the strategy used by Rokicki (1987) in a workload context. In that study, heart rate activity data collected during flight testing were used as a debriefing tool where increases in the operator's heart rate were used to focus discussion. Transitory changes in heart rate were often useful to jog the operator's memory or to highlight significant episodes during the mission. The same strategy can be used to determine when best to query operators about their SA during a mission. This can be useful for understanding SA in a particular situation and for identifying commonalties in the mechanisms underlying SA in general. This approach may also be applied online. For example, sudden changes in physiological measures (e.g., increased heart rate or changes in EEG activity) can be used as markers to indicate that something of importance had happened to the operator. This activity may be used to initiate an inquiry to determine whether anything noteworthy had been detected. The cause for the change in the physiology may be external or internal. For example, important information may have been detected from the system displays or the operator may have realized that the situation was critical because they synthesized the relevant information held in memory.

## SA Level Classification

A problem related to SA assessment is that of determining an operator's mental workload level. Several studies (Wilson & Fisher, 1991, 1995; Russell, Wilson, & Monett, 1996; Green et al., 1996), used various classification tools to determine how well operator's mental workload can be predicted using psychophysiological data collected from laboratory, simulator, and flight data. Wilson and Fisher (1995) showed that it was possible to determine which of seven task categories subjects were engaged in and which of two levels of difficulty they were experiencing with an average accuracy of 86%. In another study (Wilson & Fisher, 1991), they demonstrated that data recorded during flight could be used to correctly identify the flight segment with accuracies of better than 90%. Current work indicated that at least an 85% overall accuracy can be achieved in determining the level of mental workload during a simulated air traffic control task using psychophysiological data (Wilson, Monett, & Russell, 1997).

The approach taken thus far to classify operator functional state has been based on dividing the data set into training and test subsets. If future work can demonstrate that data from different sessions can be used to achieve comparable accuracy, then it will be possible to apply these procedures to the work environment. Several classifiers have been used in this context, including linear discriminant analysis, Bayes classifiers, and a variety of artificial neural network classifiers. These classifiers are trained on data of known categories first and then unknown data are submitted to the developed classifier to test its adequacy. Based upon the percentage of correct classifications of test data the acceptability of the classifier can be judged. Further, the weights assigned to the data features given by the classifier can be examined to determine the important ones for each operator. These selected features provide information regarding the underlying mechanisms of the task under investigation.

The above studies suggested that these procedures could be applied to determine if SA features can also be extracted. Using subjective and performance data to determine operator's SA levels, simultaneously collected physiological data may be used to develop classification algorithms. The method requires selecting data recorded during periods of good and bad SA based upon subjective and performance criteria. These data are then randomly divided into two sets with one set used to train the classifier. During this training one data set is provided to the classifier and each example is identified as either good or bad SA. Then the test data set is given to the classifier without the good or bad SA labels. The classifier then assigns each example to the good or bad SA categories. The accuracy of the classifier is evaluated by determining the percentage of correct classifications of the good and bad SA segments. An advantage of this approach is that the features selected by the classifier to provide information about the nature of the changes in the physiological data associated with good and bad SA can be examined. From these results it is possible to design parametric studies to yield further insights into the relation between SA and psychophysiology.

Because these data can be acquired noninvasively and are continuously available, they make possible online monitoring of operator state. Current technology permits the recording of multiple channels of physiological data without interfering with the operator's work routine. Small electrodes and battery-powered recording systems can be easily worn by operators at their work stations. These data can be used to determine the operator's functional state and estimates can be made about the nature of the task in which they are currently engaged.

### Reformulation in SA Context

Another framework in which the applicability of psychophysiological measures to SA can be discussed is found in Endsley (1995b). In her theory, she proposes a hierarchy of three SA phases where each succeeding phase depends on the successful completion of the lower ones. Level 1 SA requires that the operator properly perceive "the status, attributes, and dynamics of relevant elements in the environment" (p. 36). The operator must be able to acquire all of the information relevant to the system being controlled. Level 2 SA is the "comprehension of the situation . . . based on a synthesis of disjointed Level 1 elements" (p. 37). This step requires the operator to correctly combine information from various elements and arrive at an appropriate overall view. The individual environmental elements must be integrated with the current operator goals. Level 3 SA incorporates "the ability to project the future actions of the elements in the environment" (p. 37). This is achieved on the successful acquisition of Levels 1 and 2 SA. The operator must understand the meaning of the environmental information in the context of the current goals and be able to project into the future.

From the previous discussion of psychophysiological measures, it is clear that they can be applied to an analysis of Level 1 SA components. A variety of measures have been used to determine whether environmental stimuli are detected, including ERPs, ERDs, and changes in HR and blink activity. In Level 2 SA, the operator builds a model of the environment and therefore develops certain expectancies about events. According to Endsley (1995b), the ability to recognize "critical cues" from the environment that fit the operator's model of the situation is important to achieving good SA. Several psychophysiological measures have been successfully employed to determine a person's expectations. For example, Stern, Wang, and Schroeder (1995) suggested that cardiac, eyeblink, and eye movements can be used to determine an operator's model of the environment by examining their responses to critical stimuli that show whether the stimuli were expected. Brain activity, such as ERPs and ERDs, may be used to monitor expectancy. The P300 component of the ERP has been used for a number of years to examine subject's expectancies in a given context. It seems plausible to apply these same procedures to questions regarding SA. Responses should differ depending on whether the operator had the proper situational model and was or was not expecting certain critical cues.

The Level 3 SA phase of Endsley's theory concerns the ability to predict future events. Although there is no known psychophysiological metric for knowing if an operator is correctly predicting future events (assuming that the operator is not asleep or cognitively overloaded), it is not clear how psychophysiology can be used here. However, post hoc examinations can determine the saliency of the events the operator should have correctly predicted if they had good SA. This is in effect the same strategy as the evaluation proposed for Level 2 SA, only waiting until the future events occur. This may determine whether the preceding SA evaluation was correct or whether the operator's appraisal was incorrect. It may be possible to assess an operator's evaluation of a situation and its consequences in the future by examining situations in which the future contains serious problems. For example, if the current environment is benevolent but contains critical cues that the future is very different and contains problems that must be solved, then the operator's psychophysiological responses should show this concern for the future. If the psychophysiology showed that the operator was tranquil, then it can be deduced that the operator does not have proper SA.

## CONCLUSION

The purpose of this chapter is to review existing knowledge concerning the application of psychophysiological measures in applied settings and to extrapolate from this data to provide suggestions about how they might be successfully applied to SA research. Five areas are suggested as possibilities for testing the utility of psychophysiological measures. They include: determining whether the functional state of the operator was conducive to maintenance of proper SA, using these measures to determine whether environmentally critical SA cues were detected, monitoring the operator's expectancy for task cues, using the operators' physiology to identify periods when significant incidents occurred and inserting probes to further query operators about their SA, and applying classification procedures to see if levels of SA could be identified on the basis of the psychophysiological data. If it is shown by further research that these methods do provide valuable information a next step will be to implement them online to provide continuous SA assessment.

Future advances in brain functional imaging may provide technologies that will permit the investigation of complex brain functioning relevant to SA. High-resolution EEG and functional Magnetic Resonance Imaging may provide techniques that show high-resolution brain activity in near real time while operators are engaged in complex task performance. The high spatial resolution could reveal areas of the brain that are necessary for the accomplishment of good SA. It is then possible to monitor these areas to assess the level of SA shown by an operator during task performance.

## REFERENCES

Brookings, J. B., Wilson, G. F., & Swain, C. R. (1996). Psychophysiological responses to changes in workload during simulated air traffic control. *Biological Psychology, 42,* 361–377.

Byrne, E. A. (1995). Role of volitional effort in the application of psychophysiological measures to situation awareness. In D. J. Garland & M. R. Endsley (Eds.), *Experimental analysis and measurement of situation awareness* (pp. 147–153). Daytona Beach, FL: Embry-Riddle Aeronautical University Press.

Cacioppo, J. T., & Tassinary, L. G. (1990). *Principles of psychophysiology: Physical, social, and inferential elements.* New York: Cambridge University Press.

Caldwell, J. A., Wilson, G. F., Centiguc, M., Gaillard, A. W. K., Gundel, A., Lagarde, D., Makeig, S., Myhre, G., & Wright, N. A. (1994). *Psychophysiological assessment methods* (AGARD–AR–324). Neuilly Sur Seine, France: NATO–AGARD.

Endsley, M. R. (1988). Situation awareness global assessment yechnique (SAGAT). In *Proceedings of the National Aerospace and Electronics Conference* (pp. 789–795). New York: IEEE.

Endsley, M. R. (1995a). Measurement of situation awareness in dynamic systems. *Human Factors, 37*(1), 65–84.

Endsley, M. R. (1995b). Toward a theory of situation awareness in dynamic systems. *Human Factors, 37*(1), 32–64.

Fournier, L. R., Wilson, G. F., & Swain, C. R. (1999). Electrophysiological, behavioral, and subjective indexes of workload when performing multiple tasks: Manipulations of task difficulty and training. *Biological Psychology, 31*, 129–145.

Greene, K. A., Bauer, K. W., Kabrisky, B., Rogers, S. K., Russell, C. A., & Wilson G. F. (1996). A preliminary investigation of selection of EEG and psychophysiological features for classifying pilot workload. In C. H. Dagli, M. Akay, C. L. P. Chen, B. R. Fernandez, & J. Ghosh (Eds.), *Proceedings of the Artificial Neural Networks in Engineering (ANNIE '96) Conference* (Vol. 6, pp. 691–697). New York: ASME Press.

Horne, J. A. (1978). A review of the biological effects of total sleep deprivation in man. *Biological Psychology, 7*, 55–102.

Kramer, A. F. (1991). Physiological metrics of mental workload: A review of recent progress. In D. L. Damos (Ed.), *Multiple task performance* (pp. 279–328). London: Taylor & Francis.

Makeig, S.,& Inlow, M. (1993). Lapses in alertness: Coherence of fluctuations in performance and the EEG spectrum. *Electroencephalography and Clinical Neurophysiology, 86*, 23–35.

Palmer, B., Wilson, G. F., Reis, G., & Gravelle, M. (1995). *The effects of one night's loss of sleep and recovery on physiological, performance, and subjective indices* (Report No. AL/CF–TR–1995–0128). Wright-Patterson AFB, OH: Armstrong Laboratory.

Pfurtscheller, G., Stancák, A., & Neuper, C. (1996). Event-related synchronization (ERS) in the alpha band—an electrophysiological correlate of cortical idling: A review. *International Journal of Psychophysiology, 24*, 39–46.

Rechtshaffen, A., & Kales, A. (1968). *A manual of standardized techniques, terminology and scoring system for sleep stages of human subjects*. Bethesda, MD: U.S. Department of Health, Education, and Welfare.

Rokicki, S. M. (1987). Heart rate averages as workload/fatigue indicators during OTE. In *Proceedings of the Human Factors Society 31st Annual Meeting* (Vol. 2, pp. 784–785). Santa Monica, CA: Human Factors Society.

Russell, C. A., Wilson, G. F., & Monett, C. T. (1996). Mental workload classification using a backpropagation neural network. In C. H. Dagli, M. Akay, C. L. P. Chen, B. R. Fernandez, & J. Ghosh (Eds.), *Proceedings of the Artificial Neural Networks in Engineering (ANNIE '96) Conference* (Vol. 6, pp. 685–690). New York: ASME Press.

Stern, J. A., Wang, L., & Schroeder, D. (1995). Physiological measurement techniques: What the heart and eye can tell us about aspects of situational awareness. In D. J. Garland & M. R. Endsley (Eds.), *Experimental analysis and measurement of situation awareness* (pp. 155–162). Daytona Beach, FL: Embry-Riddle Aeronautical University Press.

Taylor, R. M. (1990). Situational awareness rating technique (SART): The development of a tool for aircrew systems design. In *Situational Awareness in Aerospace Operations* (AGARD–CP–478, pp. 3/1–3/17). Neuilly Sur Seine, France: NATO–AGARD.

Vidulich, M. A., Stratton, M., Crabtree, M., & Wilson, G. (1994). Performance-based and physiological measures of situational awareness. *Aviation, Space and Environmental Medicine, 65*(5 Suppl.), A7–A12.

Wilson, G. F. (1995). Psychophysiological Assessment of SA? In D. J. Garland & M. R. Endsley (Eds.), *Experimental analysis and measurement of situation awareness* (pp. 141–145). Daytona Beach, FL: Embry-Riddle Aeronautical University Press.

Wilson, G. F., & Eggemeier, F. T. (1991). Physiological measures of workload in multi-task environments. In D. Damos (Ed.), *Multiple-task performance* (pp. 329–360). London: Taylor & Francis.

Wilson, G. F., & Fisher, F. (1991). The use of cardiac and eyeblink measures to determine flight segment in F4 crews. *Aviation, Space and Environmental Medicine, 62*, 959–961.

Wilson, G. F., & Fisher, F. (1995) Cognitive task classification based upon topographic EEG data. *Biological Psychology, 40*, 239–250.

Wilson, G. F., Monett, C. T., & Russell, C. A. (1997). Operator functional state classification during a simulated ATC task using EEG. In *Proceedings of the Human Factors and Ergonomics Society 41st Annual Meeting*, 2, pp. 1382. Santa Monica, CA: Human Factors Society.

Wilson, G. F., Swain C. R., & Ullsperger, P. (1998). ERP components elicited in response to warning stimuli: The influence of task difficulty. *Biological Psychology, 47*, 137–158.

# Chapter 9

# Use of Testable Responses for Performance-Based Measurement of Situation Awareness

Amy R. Pritchett
*Georgia Institute of Technology*

R. John Hansman
*Massachusetts Institute of Technology*

Performance-based measurement of situation awareness can be a valuable measurement technique for testing a wide range of systems. Unlike measurement techniques that attempt to ascertain the subject's mental model or knowledge at different times throughout an experiment, performance based testing focuses solely on the subject's outputs. This quality makes it ideal for comparing the desired and achieved performance of a human–machine system, and for ascertaining weak points of the subject's situation awareness. When combined with other measures of situation awareness, performance-based measures can also identify broader concerns about the ultimate actions taken by the operator. For example, the root cause of incorrect operator actions can be identified as a problem with situation awareness or as a situation where the operator has correct situation awareness but then has problems with making and executing a satisfactory decision.

This chapter first describes several measures of situation awareness by discussing the points in the decision process they examine. Using this framework, the comparative benefits and weaknesses of knowledge-based measurements, subject verbalization, and performance-based measures are compared. Methods for combining different types situation awareness measures are discussed.

This chapter then focuses on performance based measures that use situations with testable responses. During the simulation runs, the subjects are presented with situations for which, if the subject has sufficient situation awareness, an action is required. This technique provides an unambiguous accounting of the types of tasks for which the pilots had sufficient situation awareness. The use of situations with testable responses in representative flight simulator studies is detailed as examples. Finally, because the subject's responses depend heavily on the precision with which the situations are generated, considerations in using situations with testable responses in experiments are discussed.

## COMPARISON OF SITUATION AWARENESS MEASURES

### Steps in the Decision Making Process

Different measures of situation awareness can be defined by the points in the decision making process at which measurements are taken. Performance-based measures of situation awareness evaluate only the actions taken by the subject and their impact on the system. Other measures attempt to examine more directly the characteristics internal to the subject.

In order to describe the difference between performance-based measures and other measures of situation awareness, it is useful to identify the intermediate steps in the decision process. One such breakdown is shown in Fig. 9.1, and is used for the remainder of the discussion as an illustrative example. This breakdown was used in studies of pilot interaction with alerting systems (Pritchett & Hansman, 1997). However, it bears many similarities to models of decision-making tasks in more general studies. (For example, Cacciabue, 1993; Endsley, 1995a; Knaeuper & Rouse, 1985; Rasmussen, 1986).

This model follows the different stages of processing of a control signal during monitoring for, and reacting to, hazardous situations; the different stages correspond to different levels of automation possible with sophisticated alerting systems. For example, in a collision avoidance task, the pilot, with or without the assistance of alerting systems or automation, must process the available information about the other aircraft, assess whether a collision hazard exists and action must be taken, and, if action is required, decide on and execute a resolution maneuver that provides safe separation from the other aircraft.

The structure of this model assumes a purely serial nature of the task. Each subfunction receives an input, processes the input, and outputs a new set of information to act as input for the next subfunction. These new sets of information delineate boundaries between each subfunction. A weakness at any subtask, therefore, has implications for subsequent subtasks that rely on its outputs.

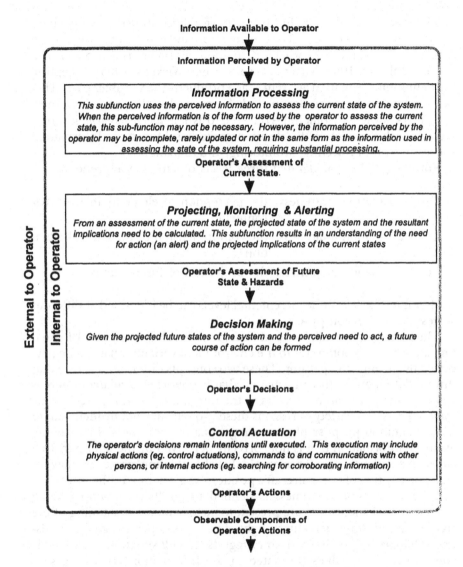

FIG. 9.1.  Four-stage model of decision-making tasks.

The model also delineates the information that is available for measurement externally to the operator, and the information that is internal to the operator. The external information consists of the information made available to the operator and the discernible actions they take; these values can be measured unambiguously. On the other hand, as is discussed later, the values internal to the operator have limited use for predicting performance.

This model provides a useful abstraction for illustrating the points where situation awareness is created by the operator, and can be measured. Some debate has occurred over whether situation awareness is a product or a process (Billings, 1995; Sarter & Woods, 1995; Wickens, 1995). Accepting the definition of situation awareness as a product—the variables passed between subfunctions in the alerting task, for example—the process of generating this knowledge, sometimes referred to as *Situation Assessment*, is represented by this model as the decision maker's subfunctions Information Processing, and Projecting, Monitoring, and Alerting.

Endsley (1995a) has defined three levels of situation awareness:

Level 1 situation awareness: the perception of elements in the environment

Level 2 situation awareness: the comprehension of the current situation

Level 3 situation awareness: the projection of future status.

As shown in Fig. 9.2, these incremental levels can be identified at different stages in the decision process.

The information available to the operator correlates with Level 1 situation awareness, as both refer to the environment variables that can be perceived by the human operator. These two concepts differ, however, in that this model identifies the information that is physically available, whereas Level 1 situation awareness refers to the information that is perceived by the human. For a variety of reasons these two sets may not be identical; in fact, a common purpose of measuring situation awareness is to identify available information that is not correctly perceived by the operator. In a collision avoidance task, for example, Level 1 situation awareness may include perception of other aircraft's positions as given by a traffic display.

Level 2 situation awareness is defined to go "beyond simply being aware of the elements that are present to include an understanding of the significance of those elements in light of pertinent operator goals" (Endsley, 1995b, p. 37). This definition suggests Level 2 situation awareness is comprised of the values calculated in the Information Processing subfunction for use in the Projecting, Monitoring, and Alerting. In the collision avoidance task, for example, Level 2 situation awareness would include knowledge of the other aircraft's position relative to the own aircraft. This knowledge may require processing of display information to convert the other aircraft's absolute positions to an understanding of their motion relative to the own aircraft.

Finally, Level 3 situation awareness is defined as "the ability to project the future actions of the elements in the environment . . . [Situation awareness] includes comprehending the meaning of that information in an inte-

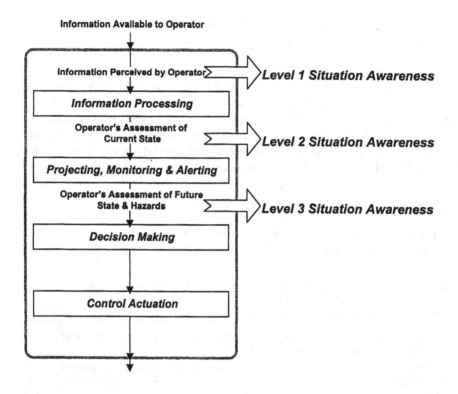

FIG. 9.2.   Levels of situation awareness the decision process.

grated form, comparing it with operator goals, and providing projected future states of the environment that are valuable for decision making" (Endsley, 1995b, p. 37). These values are analogous with the projections of the hazard and need for an alert. For example, in the collision avoidance task, Level 3 situation awareness would include the projected collision hazard and the need to initiate an avoidance maneuver in the near future.

A "disconnect" has been noted between situation awareness and decision making. Although many poor decisions have been attributed to poor situation awareness, the causality between situation awareness and decision making, in some situations, is neither absolutely necessary nor sufficient. This disconnect can be illustrated by the serial nature of the model illustrated in Figs. 9.1 and 9.2. In general, for decision making to be performed successfully, proper inputs to the decision making process must be available, as created by the operator's situation awareness. However, in some situations, complete situation awareness may not be necessary for good decision making. For example, a complete assessment of the system state may not be required for a simple, effective strategy. Procedures or luck may negate any detrimental effects of poor situation awareness. Likewise, procedures and training may remove the need for detailed, comprehensive, situation awareness.

In other situations, complete situation awareness may not be sufficient to guarantee good decision making. For example, the decision making process may be sufficiently complex or so little understood that the operator, even with complete situation awareness, is unable to consistently make good decisions. For example, in a flight simulator evaluation of collision avoidance during closely spaced parallel approaches, pilots frequently indicated awareness of a potential collision but then did not select avoidance maneuvers that provided adequate aircraft separation (Pritchett & Hansman, 1997).

## MEASUREMENT POINTS IN THE DECISION MAKING PROCESS

The model of the decision process and situation awareness given in Figs. 9.1 and 9.2 provides a useful abstraction for also describing various measures of situation awareness. Figure 9.3 displays the measurement points for knowledge-based, performance-based, and verbalization methods of measuring situation awareness.

Performance based measures examine the external data points. With the information available to the operator available as a reference point, performance-based measures consider only the physical actions taken by the operator and the resulting outputs of the system they act on.

Knowledge-based measures attempt to directly ascertain the subject's internal awareness. One way this can be accomplished is by using structured, directed probes. For example, the Situation Awareness Global Assessment Technique (SAGAT) freezes the operator's displays at random times during experiment runs, and queries the subjects about their knowledge of the environment (Endsley, 1995a). These queries can probe for knowledge corresponding to any of the levels of situation awareness, from the most basic of facts to complicated predictions of future states.

Verbalization methods attempt to gather insight into both the operator's knowledge and the operator's strategies at each subtask throughout the experiment runs by asking the operator to describe (verbalize) his or her current thought process.

The comparative strengths of these measures are summarized in Table 9.1. As pointed out by Billings (1995), situation awareness involves both external considerations (the situation), and internal considerations (awareness). The different types of situation awareness measures each have a different emphasis on external or internal measures. Measurement of the external considerations, as given by performance based measures, has the

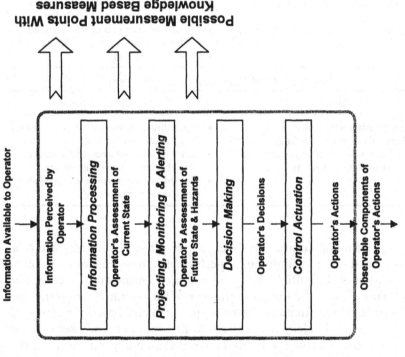

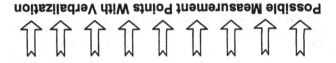

FIG. 9.3. Measurement points with knowledge-based measures.

195

TABLE 9.1
Strengths and Limitations of Different
Situation Awareness Measures

| *Knowledge-Based Measures of Situation Awareness* | |
|---|---|
| *Strengths* | *Potential Limitations* |
| • Isolates components of current situation awareness<br>• Provides insight into development and maintenance of situation awareness | • Can not necessarily be used to predict final performance of operator<br>• Easier to measure declarative knowledge than procedural knowledge<br>• Possibly intrusive into operator's task |

| *Verbalization Measures of Situation Awareness* | |
|---|---|
| *Strengths* | *Potential Limitations* |
| • Provides insight into both situation awareness and thought processes<br>• Provides insight into perceived importance of information | • Can not necessarily be used to predict final performance of operator<br>• Easier to measure declarative knowledge than procedural knowledge<br>• Limited by user's ability to relate all considerations during experiment runs<br>• Possibly distracting |

| *Performance-Based Measures of Situation Awareness* | |
|---|---|
| *Strengths* | *Potential Limitations* |
| • Assesses final performance of system and records operator's actions<br>• Sufficiency of situation awareness can be inferred in some situations | • Not a direct measurement of operator's situation awareness<br>• Easier to measure procedural knowledge than declarative knowledge<br>• Limited by descriptiveness of available performance measures |

advantage of being comparatively objective: physical variables are typically recorded directly. In doing so, performance-based measures provide an indication of the sufficiency of the operator's situation awareness and additionally records the final performance of the operator within the specific content of the situation. However, performance based measures examine directly only elements external to the operator; the user's awareness of the situation can only be inferred. Depending on the task, this inference may or may not be strongly supported. Additionally, performance measures are generally suited for demonstrations of procedural knowledge, rather than declarative knowledge.

Knowledge-based and verbalization measures, on the other hand, examine the information and processes internal to the operator. These measures have the advantage of providing direct insight into the operator's awareness and the mechanisms used to maintain it. Verbalization measures also provide insight into the strategies used by the operator. However, these measures have limitations. Because they may interrupt or change the operator's task, they may be intrusive and therefore be biased. These measures are also limited by the operator's ability to communicate. For example, these methods may be better suited to declarative knowledge that can be easily expressed by the operator, compared to procedural knowledge. With verbalization measures, subjects also may not describe all of their knowledge or strategies, making complete identification of their knowledge difficult. Finally, because of the disconnect between situation awareness and decision making, the predictive ability of these measures for estimating final performance is limited.

These measurements serve somewhat different but complementary roles. For providing a detailed, theoretical assessment of the subject's situation awareness, the knowledge-based techniques are more accurate, as they measure these variables directly. Performance-based measurement can only make inferences based on the particular information the subject acted on, and how it was interpreted.

However, performance-based measurements can satisfy several goals that knowledge-based techniques can not. As shown in Fig. 9.4, performance is affected by several factors, including insufficient situation awareness. By measuring the final performance of the operator, several factors can be identified that are difficult to predict based solely on knowledge-based or verbalization measures: the final performance of the subject; the related considerations in decision making and control actuation that require situation awareness; and the sufficiency of the operator's situation awareness, especially in the presence of factors such as time pressure and uncertainty.

First, the most apparent strength of performance-based measures is their ability to ascertain the timing and substance of a user's reaction to realistic situations. For testing of systems, final decisions must be based on whether the user will be provided with sufficient situation awareness to perform the correct actions, which performance-based techniques measure directly. Knowledge-based measurement techniques, on the other hand, can only make reasonable guesses about the likely user's real-time actions given their knowledge state, and have limited predictive ability (Adams, Tenney, & Pew, 1995). Verbalization methods may be biased by operator perceptions about their performance and opinions about other factors, such as a preference for one system. For example, in a flight simulator study examining autopilot mode-awareness, the pilot's actual, real-time reactions often varied significantly from those they named as "what they would do" during non-time critical questioning afterward (Johnson & Pritchett, 1995).

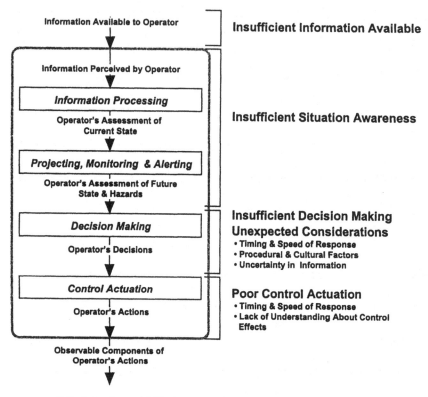

FIG. 9.4.   Potential blocks to satisfactory actions by the operator.

Second, performance based measures can assess the requirements placed on situation awareness by the decision and control actuation strategies used by the operator. Because the actual, exact strategies the operator will use are typically unknown, the elements of the situation that must be held in awareness can not be exactly specified. Pew (1995) contrasted "ideal, attainable and actual" situation awareness. Whereas knowledge-based and verbalization measures can focus on the differences between these three levels of situation awareness, performance-based measures focus solely on "required" or "sufficient" situation awareness. For example, in initial studies of pilot collision avoidance during closely spaced parallel approaches, the strategies pilots would use for alerting and selecting an avoidance maneuver were unknown; therefore, the knowledge of the traffic situation they would require to maintain adequate aircraft separation was unknown and required testing (Pritchett & Hansman, 1997).

Finally, performance-based measures can examine, in the context of the test situation, the relative impact of situation awareness and other potential blocks to satisfactory performance, as shown in Fig. 9.4. For example, in a flight simulator study by Midkiff and Hansman (1993), ATC neglected to turn the pilots toward the landing runway although the pilots could overhear the aircraft before and after them being giving the proper instructions. Although the pilots' discussions indicated they were aware of the situation, they did not take a strong reaction because of their reticence to assume the air traffic controller had made an error, and because of the perceived uncertainty in information overheard from other aircraft. A knowledge-based measurement of the pilots situation awareness also would have provided a measurement, in this case, of the pilot's awareness of the problem. Only performance-based measurement, however, could ascertain how the pilots would act on this information within an established set of air traffic control procedures.

In the development of a human-in-the-loop system, performance-based measurement may be seen as complementary to knowledge-based measurement. Each is useful at different times, and for different purposes, throughout the design process. For final testing of a system, performance-based measurement is appropriate for its ability to ascertain the resulting performance of the entire system, and to point to areas of situation awareness that are deficient. Although performance-based measurement does not provide as pure a measurement of an operator's internalized state of awareness as other techniques, it is able to illustrate the interrelationship between the operator's awareness and the operator's resulting actions.

## COMBINING DIFFERENT SITUATION AWARENESS MEASURES

The previous section discussed the relative merits of different measures of situation awareness when applied individually. In developing experiments involving decision making and situation awareness, combining different measures of situation awareness can provide a measure greater in scope than the sum of their individual parts. The different measures examine different points in the decision process. By comparing their results, the specific blocks to performance illustrated in Fig. 9.4 can be identified by comparing verbalized situation awareness and measured performance.

The most relevant factor that can be identified is whether the primary block to satisfactory performance stems from poor situation awareness or poor decision making. This distinction provides valuable guidance to the system designer on the requirements for the system's displays.

Another factor that can be identified is a difference between the declarative knowledge stated by the operator in knowledge-based measurements, and the procedural knowledge displayed by the operator through measurable actions—a comparison between what the operators say and what they do.

Finally, a comparison of the different measures is required to ascertain unexpected obstacles to actions, such as the procedural blocks to action discussed in the previous section. Although this ability was primarily listed as a feature of performance-based measures, such results are the most conclusive when a disparity between awareness and action can be explicitly identified.

As a practical concern, different measures of situation awareness are generally easy to combine. Because the measures examine different data at different times during the experiment, their timing does not conflict, and their measurements can be kept independent of each other. In addition, performance measures and subject verbalization are generally easy to implement, and their inclusion does not necessarily change the experimental protocol significantly.

However, there may be concern that some types of measures may be intrusive, such as knowledge-based measures that query the subject during the runs. For example, it has been debated whether techniques that periodically ask subjects to name pieces of information may change the subject's strategies for situation assessment and decision making to include that information more than normal (Endsley, 1995a; Vidulich, 1995). If any of the measures in the experiment are intrusive, the changes in subject behavior may affect the other measures.

## APPLYING PERFORMANCE BASED MEASUREMENTS

### Types of Performance Based Measures of Situation Awareness

Using its most general definition, performance-based measurement of situation awareness is any measurement that infers subjects' situation awareness from their observable actions or the effects these actions ultimately have on system performance.

Endsley (1995b) described several different types of performance based measures. *Global measures* infer situation awareness from final performance criteria, such as whether a mission or task was completed. *Imbedded task measures* consider measurements more specific to the system being tested; for example "when evaluating an altitude display, deviations from prescribed altitude levels or time to reach a certain altitude can be measured." *External task measures* examine subject reactions to changes to, or removal of, information relevant to the task at hand.

The most general concern with the use of performance-based measures is the uncertainty in inferring situation awareness from observable performance. Good performance may result despite poor situation awareness through the use of proceduralized responses or luck. Poor performance may result even with good situation awareness because of poor decision making or other blocks to satisfactory action.

To resolve this ambiguity, a fourth type of performance-based measurement of situation awareness has been suggested: the use of *testable responses* (Pritchett, Hansman, & Johnson, 1995; also called *implicit measures* by Vidulich, 1995). The remainder of this chapter focuses on the experimental use of testable responses.

## THE USE OF TESTABLE RESPONSES FOR PERFORMANCE BASED MEASUREMENT OF SITUATION AWARENESS

Testable responses provide a mechanism to unambiguously ascertain subjects' situation awareness from their performance. As shown in Fig. 9.5, events with testable responses are isolated, experimentally controlled events that can not be anticipated through any means other than good situation awareness, and that require a discernible, identifiable action (or set of actions) from the operator.

This section details considerations in the use of testable responses, using several flight simulator studies as examples. At the end of this section, the benefits and limitations of testable responses are discussed.

### Ensuring Unambiguous Measurement of Situation Awareness

Situations with testable responses must have the ability to mandate a clear and unambiguous response. When expert users, such as airline pilots, are used as subjects, situations can be chosen for which training or procedures demand a certain response. For example, one situation in a flight simulator study allowed pilots to overhear communications that suggested that another aircraft had not departed the runway the subjects were very close to landing on. In this case, action was required to avert a collision; a lack of

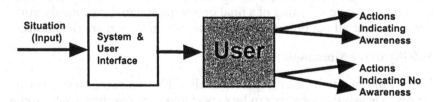

FIG. 9.5.   Use of testable responses for measuring situation awareness.

action by the pilots could be considered to represent a lack of pilot situation awareness (Midkiff & Hansman, 1993).

These types of situations, which require practiced responses, are not likely to have blocks to final performance due to poor decision making. Therefore, strong reactions can be considered an indication of good situation awareness; correspondingly, the lack of response can be considered an indication of insufficient situation awareness.

### Experiment Measurements for Discerning Awareness

The testable responses should be capable of examining the range of all probable actions and in-actions by the subject throughout the experiment. Care must be taken to ensure actions will be recorded that are different, less severe, or incorrect in addition to just looking for the expected or desired result. For example, the response to the situation "Aircraft on Landing Runway" might be expected to be an immediate go-around. However, in a flight simulator study, the subjects' actions were often less severe, with pilots instead attempting to query ATC or each other to verify the knowledge they had gained from party line information (Midkiff & Hansman, 1993).

Performance-based measurement does not preclude other concurrent methods of assessing situation awareness. For example, Midkiff and Hansman also debriefed their subjects in an attempt to get pilot opinions on their situation awareness during the experiment. Other flight simulator studies have asked the pilots to verbalize their actions (Johnson & Pritchett, 1995) or have incorporated simple queries (Mykityshyn, Kuchar, & Hansman, 1994).

### Scope of Events with Testable Responses

The technique of using events with testable responses tests subjects' situation awareness at occasional, discrete events. In order to conduct a comprehensive study covering the full scope of potential conditions, situations should be chosen to cover the domain in which the system is expected to perform. For example, in the Midkiff and Hansman simulator study (1993), the nine situations tested were the testable situations that had received the highest importance ratings in a pilot survey of party line information importance. Testing of a final prototype system may include situations that test all conditions given in the system design specifications.

### Fidelity of the Situations

Finally, the situations must represent believable and recognizable occurrences to which the subjects can be expected to react as they would in the real environment. For example, in a flight simulator study, the subjects were flying an air transport simulator and believed they were over-hearing other air transport aircraft. Therefore, the "Potential Collision" situations were staged to happen at a rate that was physically reasonable and were carefully scripted to portray to the subject a believable scenario of pilot confusion and/or mechanical failure on the part of the intruding aircraft (Midkiff & Hansman, 1993).

### Benefits and Limitations on the Use of Testable Responses

This section discussed general considerations in implementing situations with testable responses. In summary, this form of situation-awareness measurement requires that the subjects be exposed to believable events that require measurable responses. In doing so, the ambiguity in inferring situation awareness from performance is removed, and the sufficiency of the subject's situation awareness can be measured directly.

However, the use of testable responses is limited to tasks and situations that require unambiguous, proceduralized responses. Some tasks may involve untrained operators, who may not react in foreseeable, measurable ways, but instead be variable and ambiguous. Likewise, some tasks may be sufficiently complex, or the required cues are so subtle, that a large number of interpretations may be possible, both with and without awareness, creating the potential for ambiguities.

The use of testable responses may also be impractical in situations that are too complex, which include cues too subtle to be artificially manipulated, or include too many operators to reliably control throughout the experimental runs. For example, the difficulty in controlling and recreating events during surgery may limit the use of testable responses (e.g., Small, 1995).

Finally, testable responses are best for examining the sufficiency of an operator's situation awareness to off-nominal or emergency conditions. Therefore, these events may not provide more general insight into the operator's situation awareness required for more routine tasks that do not demand a singular, distinct responses.

## GENERATING TESTABLE RESPONSES

The use of situations with testable responses requires practical considerations during experiment design. Several general mechanisms can be used to control the dynamics of the environment during the experiment runs in order to cause the situations of interest. These different mechanisms each have implications for how exactly a scenario can be replicated in different experiment runs and the type of environment each may be practical for. The extent to which the subject can observe the situations developing also impacts the method of generating the situation. This section will discuss the different mechanisms available for generating situations with testable responses in order to measure subject situation awareness.

## Flexibility & Robustness in Event Generation

In order to make consistent conclusions and comparisons between different experimental conditions, situations with testable responses may be required to occur reliably with the same set of core characteristics in different experiment runs and between different subjects. This robustness requirement may be confounded by fidelity requirements that mandate that the subjects control the system in a realistic manner, even if variance in their actions prevent situations from occurring, has situations occur at inappropriate times, or changes their characteristics.

For example, in a preliminary flight simulator evaluation of the Traffic alert and Collision Avoidance System (TCAS), a pair of pilots during one experimental flight did not experience any of the potential aircraft collision situations that were hard-wired into the simulation. The pilots were tired from the 10-hour day of flying and had heard beer was going to be imbibed after their experiment run was complete. To speed things up, they decided to fly at Mach 0.82, faster than their cleared speed of Mach 0.80. By the time the experimenters recognized the deviation from the expected course of events, the subject aircraft was ahead of schedule. As shown schematically in Fig. 9.6, the pseudo-aircraft flew their predetermined courses, passing well behind the subject aircraft at the time of each expected traffic situation (Billings, 1997, personal communication).

If the subjects' actions before the desired situations can not unwittingly prevent them from occurring, then these situations can be hard-wired into

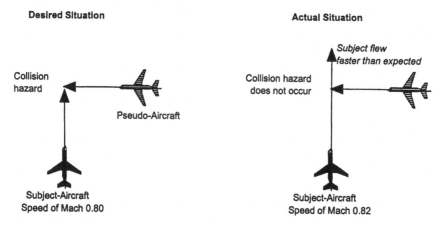

FIG. 9.6.   Effect of variations in subject actions on situation generation.

the experiment setup. However, in many experimental simulations, to maintain fidelity and allow the subject a range of normal actions, the subject is free to control the vehicle as he or she sees fit. In these cases, some measure of flexibility may be required in the method used for generating the desired situations in order to prevent a lack of robustness. During the experiment design, the amount of flexibility required can be estimated by comparing the extent of potential subject deviations with the actions required to still generate the desired situations.

## Effects of Subject Ability to Observe Situation Development

Another consideration in using situations with testable responses is the extent to which the subjects can observe the dynamics that are used during the experiment to create the desired situations. Some sudden situations, such as a mechanical failure, do not require a great deal of development and therefore are not observable until the situation has occurred. However, other situations such as traffic collisions evolve over time and may present a problem. As the subjects can observe more of the underlying dynamics of the situations, more care must be taken during the experiment design in order to make the situations appear believable while maintaining situation robustness to variations in the subjects' actions. This section discusses several general considerations for creating situations with testable responses in conditions observable by subjects. The next section illustrates specific techniques used in flight simulator experiments to meet these considerations.

If the controllable system dynamics are observable by the subject, these dynamics must be constrained to follow a commensurate level of fidelity in the development of the situation. For example, when testing cockpit displays of traffic information with very crude information, surrounding aircraft may be given erratic, extreme steering commands in order to create the desired traffic situation, without the subject recognizing any anomaly; however, once the subject can view additional information about nearby aircraft such as heading, speed and intended flight path, the other aircraft must follow smooth, believable trajectories.

A similar consideration in observable fidelity requires, when elements of the developing situation are observable, all information available to the subject, such as background or peripheral cues, must corroborate with the primary source of information. In addition to providing a base requirement on the fidelity of the experiment, such secondary cues may be used as an opportunity to temporarily explain unusual dynamics in the system that will eventually lead to an event. For example, in flight simulator tests of traffic displays, the background air traffic control communications can be used to explain any erratic behavior by other aircraft that results from those other aircraft jockeying into position from which to cause a collision.

In many cases, high observability of the system dynamics by the subject requires more frequent feedback to the mechanisms controlling the scenario, such that the resulting update in system dynamics at any one time is sufficiently small to not be questionable. This may require predetermination of an experiment script with enough detail to guide frequent feedback and a succession of small changes in system dynamics.

Finally, care should be taken to prevent any relationship between subject actions and system dynamics from being observable by the subject. For example, during flight simulator testing of cockpit traffic displays, when a pseudo-aircraft is homing on the subject, a slight delay or lag may be added to the pseudo-aircraft's actions to prevent a noticeable, immediate correlation between subject maneuvers and the pseudo-aircraft's change in trajectory.

Observability of the system by the subject can limit the accuracy and robustness of the situations. For example, in flight simulator experiments, if the subject pilot is free to make deviations from his or her flight path, and the behavior of the surrounding aircraft is highly observable, then the changes in pseudo-aircraft trajectory that are small and slow enough to not be observable may not have sufficient effect to cause a desired traffic situation.

### Mechanisms for Generating Testable Responses

Several mechanisms exist for generating situations with testable responses. First, events can be hard-wired or programmed into the experimental run. For example, at a specific time, an event is automatically triggered without any outside influence. Hard-wiring situations remain the simplest mechanisms. However, as previously mentioned, this mechanism is not flexible in the face of variations in the subject's actions leading up to the situation, and therefore may not be sufficiently robust for the situation to occur, or for the situation to occur in the same way, in different experiment runs.

Several other mechanisms have more flexibility and robustness. One common technique is the use of experimenters as confederate agents in the scenarios. Confederate agents are visible to the subject and play the part of other people the subject would normally interact with in real operations. The confederate agents act within the capabilities of their simulated role in such a way to cause the situation requiring a testable response. For example, in a flight simulator study of autopilot mode awareness, a confederate, acting as the subject-pilot's first officer, mistakenly set the commanded autopilot descent rate to a different setting (Johnson & Pritchett, 1995). Confederate agents have been used successfully in many instances. However, the use of confederate agents is limited in cases where subject has significantly more control over the dynamics of the situation than the confederate, or when the subject is in a simulated environment that does not include other people.

Another technique is the use of pseudo-agents—system variables and external agents controlled by experimenters—to cause situations with testable responses. For example, in flight simulator experiments, the trajectories of many other aircraft may be controlled by high-level steering commands generated by a few experimenters following a predefined script; experimenters may read from the script to create a believable set of air traffic control transmissions on the commonly available air traffic control frequencies; and experimenters may control key aspects of the own aircraft's status, such as a mechanical failure.

Situations requiring testable responses can also be created using automatic control of the scenario. This technique requires the experimenter to define, at a high level, the situations in predetermined scripts. During experiment runs feedback of any variations of the subject's actions from that expected are used to modify the behavior of the pseudo-agents in order to ensure the desired situations occur with the prescripted characteristics. This feedback process is shown in Fig. 9.7 (Johnson & Hansman, 1995).

Each of these techniques has comparative benefits. Automatic control can enable more consistent robustness in the situations, and may be able to cause situations to occur within tighter parameters than experimenters, acting as confederates or pseudo-agents, may be able to provide in conditions with complex dynamics or intricate scenarios. For example, the exact miss distance in a simulated near mid-air collision between the subject and a pseudo-aircraft can be specified and attained with greater accuracy than the use of pseudo-pilots alone can achieve.

The use of experimenters, on the other hand, can provide more flexibility in the situation generation in cases where possible subject variations may not be anticipated in the predetermined script. The use of experimenters as confederate agents may also be preferable in environments involving substantial interaction between the subject and other operators.

## SUMMARY AND CONCLUSIONS

Performance-based measurement is a valuable technique for examining the sufficiency of an operator's situation awareness to deal with real-life situations. By measuring actions and performance metrics external to the operator, performance-based measures are comparatively unbiased, nonintrusive and objective. Performance-based measures are also a central part of system design and testing, where the ultimate performance of the system is of primary concern and limitations to performance must be identified.

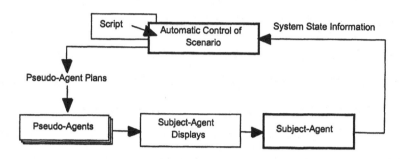

FIG. 9.7. Use of real time feedback to continuously update pseudo-agent actions.

Performance-based measures are complimentary to other techniques for measuring situation awareness, such as knowledge based measures and subject verbalization. Used independently of each other, the different measures serve different purposes throughout the design of a system; performance-based measures evaluate the real-time actions of the operator, while other measures evaluate operator situation awareness more directly but have limited ability to predict final operator behavior and system performance. When performance based measures are used in conjunction with other types of measures, the combination may additionally be able to discern where in the decision process obstacles to satisfactory performance are occurring.

The difficulty in using performance-based measures lies in the potential ambiguity between the measured performance and the inferred situation awareness. To resolve this issue, the use of events with testable responses has been suggested. Events with testable responses are isolated, experimentally controlled events that can not be anticipated through any means other than good situation awareness, and which require a discernible, identifiable action from the operator. Typically, events with testable responses involve trained, proceduralized or obvious requirements for action. In other words, the decision-making required by these events should be minimal, making situation awareness the dominant factor in subject performance. For many types of environments, events with testable responses provide a practical, nonintrusive, objective experimental measure of situation awareness.

## REFERENCES

Adams, M. J., Tenney, Y. J., & Pew, R. W. (1995). Situation awareness and the cognitive management of complex systems. *Human Factors, 37*(1), 85–104.

Billings, C. (1995). Situation awareness measurements and analysis: A commentary. *Proceedings of the International Conference on Experimental Analysis and Measurement of Situation Awareness*, Daytona Beach, FL, 1–6.

Cacciabue, P. C. (1993, October). A methodology of human factors analysis for systems engineering. *Proceedings of the 1993 International Conference on Systems, Man and Cybernetics*, Le Touquet, France, 689–694.

Endsley, M. (1995). Measurement of situation awareness in dynamic systems. *Human Factors, 37*(1), 65–84.

Endsley, M. (1995). Towards a theory of situation awareness in dynamic systems. *Human Factors, 37*(1), 32–64.

Johnson, E. N., & Hansman, R. J. (1995, January). Multi-agent flight simulation with robust simulation generation. *MIT Aeronautical Systems Laboratory Report, ASL-95*(2).

Johnson, E. N., & Pritchett, A. R. (1995, June). Experimental Study of Vertical Flight Path Mode Awareness. Paper presented at *6th IFAC/IFIP/IFOR/IEA Symposium on Analysis, Design and Evaluation of Man-Machine Systems*, Cambridge, MA.

Knaeuper, A., & Rouse, W. B. (1985). A rule-based model of human problem-solving behavior in dynamic environments. *IEEE Transactions on Systems, Man and Cybernetics, SMC-15*(6), 708–719.

Midkiff, A. H., & Hansman, R. J. (1993). Identification of important 'party line' information elements and implications for situational awareness in the datalink environment. *Air Traffic Control Quarterly, 1*(1), 5–30.

Mykityshyn, M., Kuchar, J., & Hansman, R. J. (1994). Experimental study of advanced electronic cockpit displays for instrument approach information. *International Journal of Aviation Psychology, 4*(2), 141–166.

Pew, R. W. (1995). The state of situation awareness measurement: Circa 1995. *Proceedings of the International Conference on Experimental Analysis and Measurement of Situation Awareness*, Daytona Beach, FL, 7–16.

Pritchett, A. R., Hansman, R. J., & Johnson, E. N. (1995). Use of testable responses for performance-based measurement of situation awareness. *Proceedings of the International Conference on Experimental Analysis and Measurement of Situation Awareness*, Daytona Beach, FL, 75–82.

Pritchett, A. R., & Hansman, R. J. (1997). Pilot non-conformance to alerting system commands during closely-spaced parallel approaches. *MIT Aeronautical Systems Lab Report ASL-97*(2).

Rasmussen, J. (1986). *Information processing and human-machine interaction*. New York: North-Holland.

Sarter, N., & Woods, D. (1995). How in the world did we ever get into that mode? Mode error and awareness in supervisory control. *Human Factors, 37*(1), 5–19.

Small, S. (1995). Measurement and analysis of situation awareness in anesthesiology. *Proceedings of the International Conference on Experimental Analysis and Measurement of Situation Awareness*, Daytona Beach, FL, 123–126.

Vidulich, M. A. (1995). The role of scope as a feature of situation awareness metrics. *Proceedings of the International Conference on Experimental Analysis and Measurement of Situation Awareness*, Daytona Beach, FL, 69–74.

Wickens, C. (1995). The tradeoff of design for routine and unexpected performance: Implications of situation awareness. *Proceedings of the International Conference on Experimental Analysis and Measurement of Situation Awareness*, Daytona Beach, FL, 57–64.

# Chapter 10

# The Trade-off of Design for Routine and Unexpected Performance: Implications of Situation Awareness

Christopher D. Wickens
*University of Illinois at Urbana-Champaign,*
*Institute of Aviation, Aviation Research Laboratory*

Recent efforts to define *situation awareness* (SA) by investigators such as Endsley (1995), Dominguez (1994), and Adams, Tenney, and Pew (1995) have allowed some degree of consensus to emerge regarding this important concept. The definition I use is closely related to that used by Endsley and is as follows: SA is the continuous extraction of information about a dynamic system or environment, the integration of this information with previously acquired knowledge to form a coherent mental picture, and the use of that picture in directing further perception of, anticipation of, and attention to future events.

The utility of any such definition, however, requires that the user carefully define the properties of the environment within which the "situation" in question evolves. In the aerospace community, three such environments are prominent and each has different implications for objective measurement:

1. The 3D geographical space around the aircraft as occupied by hazards, such as other air traffic (friend and foe), weather, and terrain (Wickens, 1995, 1996).
2. Internal systems of the aircraft, in particular the automation systems (Sarter & Woods, 1995b, 1997).
3. Responsibility for the array of tasks confronting the pilot, his or her crew, and various automated agents (Chou, Madhavan, & Funk, 1996; Funk, 1991).

For each of these domains—geographical hazard awareness, system awareness, and task awareness—the specific metrics of "the situation" are expressed in different qualitative languages, and hence, the objective measures are superficially quite different between them. Geometry plays a key role in geographical hazard awareness, Boolean logic plays an important role in system awareness, and checklists and queues are important to task awareness. Yet in all cases, there are certain common themes underlying the preservation of SA—themes that impact its objective measurement and the design decisions that support its maintenance.

Four examples of such commonalities are:

1. The dissociation between the contents of SA—a cognitive phenomenon sometimes associated with long-term working memory (Kintch & Ericsson, 1995)—and the responses or actions taken on the basis of those contents.
2. The dissociation between the relatively rapidly evolving state of SA, dictated by the **dynamic** attributes in my definition, and the more enduring characteristics of long-term memory.
3. The distinction between the **product** of SA (e.g., the contents of working memory and long-term working memory relevant to the situation) and the **processes** (attention, mental models) necessary to attain it (Adams, Tenney, & Pew, 1995).
4. The trade-offs that exist between design for the routine performance and design for global SA and the relation of this trade-off to the psychological construct of **expectancy**.

It is this fourth theme that is the focus of this chapter. I initially illustrate the specific nature of this trade-off with regard to hazard awareness, then illustrate its parallels in systems and task awareness, and finally discuss some of the implications of the trade-off for both design and performance evaluations.

## DESIGN TRADE-OFFS

### Hazard Awareness

In the laboratory my colleagues and I completed a series of studies examining the optimal format for design of aviation control, navigation, and hazard awareness displays (cf. Olmos, Liang, & Wickens, 1997; Wickens, in press; Wickens, Liang, Prevett, & Olmos, 1996; Wickens & Prevett,

1995). In this research the pilot was characterized as confronting two generic types of navigational tasks: local guidance and global awareness.

Local guidance is the process of maintaining precision along the flight path and characterizes the routine flight that occupies perhaps 95 to 99% of a pilot's flight time. The information needs for local guidance are of depictions of actual and predicted deviations from the flight path; that is, information that is ego-referenced, presenting a view directly ahead of the aircraft (i.e., forward-looking or three-dimensional) and relatively close in (i.e., the flight path within a few thousand feet ahead of the aircraft as defined by the time constant of the aircraft and its speed). Hence, the ideal display for local guidance is a forward-looking 3D ego-referenced "highway (or tunnel) in the sky" (Haskell & Wickens, 1993; Reising, Liggett, Solz, & Hartsock, 1995; Theunissen, 1994, 1995; Wickens & Prevett, 1995). Such a view is represented schematically in Fig. 10.1a.

On relatively infrequent occasions, the pilot may be suddenly called on to utilize far more global hazard information within a much greater volume of space (e.g., 360 degrees around the aircraft). Such is the case if the pilot unexpectedly finds him or herself lost or disoriented or if unexpected weather or traffic calls for a sudden radical departure from the intended flight path. Here the highly ego-referenced view of the tunnel in the sky is ill-suited. It either provides a very narrow "keyhole" view of the world or, if the geometric field of view is expanded (Fig. 10.1b; Barfield, Rosenberg, & Furness, 1995; Wickens & Prevett, 1995), it provides a highly distorted picture of where things (e.g., hazards) are located. Instead, ideal displays for hazard awareness tend to be those that are more two-dimensional (look down), zoomed-out, and world-referenced. In a large number of experiments (summarized in Wickens, 1996, 1999), I utilized several different measures of hazard awareness (see Olmos, Liang, & Wickens, 1995) to establish the advantage of more exocentric displays for SA (Figs. 10.1c and 10.1d), and thereby identify the trade-off of display support for the two different kinds of tasks.

For example, the tunnel provides a high gain depiction of motion that is necessary for stable, closed-loop control. Because global hazard awareness is less of a control function, a display to support such awareness can sacrifice this precision in order to attain a greater representation of a wider space. Although there is only a low probability that some part of this wider space may need to be occupied by the pilot's aircraft in the future, such probability is not zero should the unexpected maneuver become necessary.

Hence, the dilemma for designers: are designs made for the routine that occupies most of a pilot's time or for the unusual that occurs only rarely, but has potentially dangerous consequences if not well-supported?

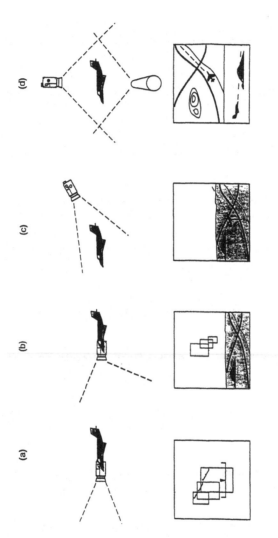

FIG. 10.1. Four frames of reference (viewpoints) by which navigational information can be conveyed to the pilot. The panel depicts the view on the pilot's display: (a) immersed with a conformal view of the world (keyhole view precludes global awareness), (b) expanded geometric field of view to depict a greater range of hazards (distortion of apparent position), (c) tethered exocentric view behind ownship (slight azimuth offset), and (d) two coplanar views.

Before addressing this question, analogous forms of this trade-off in systems and task awareness are described.

### Systems Awareness

Emerging from early applications of cognitive engineering to the design of nuclear power control rooms (Goodstein, 1981; Landeweerd, 1979; Rasmussen, 1986; Rasmussen, Pejtersen, & Goodstein, 1995), has been a realization that system characteristics ideally suited for routine operation may be very poorly suited for dealing with system failures. In routine operations, operators utilize skill- and rule-based behavior to assist them in causal or "forward" reasoning and relatively convergent thinking (e.g., what procedures do I need to follow to attain a well-defined and **routine** goal?). In contrast, under conditions of failure and fault management, they must use knowledge-based behavior to assist them in diagnostic or "backward" reasoning and relatively divergent thinking (e.g., what conditions caused this **unusual** state of the plant to occur?; Vicente & Rasmussen, 1992; Wickens & Hollands, 2000). Displays and procedures designed to best support the routine circumstances are not necessarily well-suited for fault management. Although displays for routine operations need not consider probabilistic information, those that support diagnostic reasoning may well need to consider such information (e.g., base rates of different possible failure types).

Applying similar logic to the design of flight deck automation, a series of investigations by Sarter and Woods (1992, 1994, 1995a, 1995b, 1997) revealed that the automated systems' designs for routine circumstances are poorly suited for the infrequent occasions when the automated system fails or it is asked to perform its duties under relatively improbable (but not impossible) circumstances (e.g., executing a missed approach or in the face of an incorrect "set up"). In the routine case, effective design may merely require an economical display of those automation modes in operation and those "armed" to be activated in the near future (this is analogous to the preview offered by the pathway for local guidance). In the unexpected case, however, there is a need for the pilot to receive considerably more elaborate feedback regarding why the system is responding as it is: what assumptions it is making regarding the control of the flight path given the current state of the aircraft. Such a need imposes a more elaborate display design challenge.

Finally, in both nuclear or process control and in flight deck management, even within the conditions of fault and failure themselves, there is evidence for a tradeoff. The procedures to be followed for relatively "routine" simple faults—that can be addressed by following straightforward checklists—may be quite inappropriate for the less expected, more complex multiple faults. In the latter circumstances, blindly following check-lists may be a very "brittle" tactic (Roth & Woods, 1988) that can lead the fault manager down some dangerous paths and may make a bad situation worse. This is true simply because the designers of checklists or fault management procedures are never able to anticipate all of the possible modes of failure and failure combinations that could occur, particularly as those modes include possible human error components.

**Task Awareness**

A third, less investigated example of the trade-offs that exist is found in task awareness. This describes the operator's awareness of what tasks have been done, need to be done, and, for the latter, in what priority (Adams, Tenney, & Pew, 1991; Chou, Madhavan, & Funk, 1996; Schutte & Trujillo, 1996). In either automated systems or multioperator systems, task awareness includes this knowledge of task queuing as well as task responsibility: who is responsible for the performance of each task. Here again, however, there is a difference in design support for the routine and for the unexpected. During the routine, checklists, whether paper or automated, provide excellent support for task awareness (Degani & Wiener, 1993). During the unexpected or unusual operating circumstances, the checklists are no longer as effective because departures from preplanned sequences and unexpected shifts in responsibility may be required.

A study by Schutte and Trujillo (1996) found that flight crew members who adopted procedural responses to unexpected failures or adhered to the standard priority scheme of "aviate-navigate-communicate-system monitoring" were considerably less successful in coping with an unexpected system failure than those crew members who were flexible in their task management strategies.

**Implications of Trade-offs for Test and Evaluation**

I have outlined three important aerospace performance domains in which the design for optimal support of routine performance may not effectively support the SA necessary to function effectively during unexpected circumstances and vice versa. This generic trade-off poses a very real dilemma for those engaged in system (or operator) test and evaluation. On what criterion should the merits of a system be judged—the routine that characterizes perhaps 90 to 95% of performance or that very small sample of time when the unexpected occurs? One logical argument is that the figure of merit of such a system should be weighted 95% on how well it supports the routine and 5% on how well it supports SA in unexpected circumstances. The problem with such an argument, however, is that it is often during the unexpected events that systems are most vulnerable to the sorts of catastrophic events that characterize incidents like the Three Mile Island nuclear power plant disaster (Reason, 1990, 1997) or the recent crash of a commercial airliner near Cali, Columbia (Aeronautica Civil, 1996; Strauch, 1997). During unanticipated events when the operator must respond under the stress of an emergency the expected cost of inadequate system operation is far greater than it is during the anticipated and routine. Although low-frequency events (i.e., failures) **are** adequately sampled by test pilots in aircraft certification programs, these events are generally anticipated by the pilot as part of the test flight plan and cannot truly be described as "unexpected." Indeed, the failure of a test pilot to cope with a truly unexpected automation failure was tragically illustrated in the crash of a highly automated aircraft in France (Dornheim, 1995).

A second issue related to the ability of operators to handle the unexpected concerns **individual differences** in operator personnel. As noted, it can be argued that the system is most vulnerable in the unexpected circumstance. Furthermore, it is also likely that this system vulnerability is amplified when the unexpected circumstance is encountered by relatively unskilled operators. I believe that such behavior (performance of low-skilled operators in unexpected circumstances for which they are ill-prepared) must be adequately sampled and disproportionately weighted in aerospace system evaluation because the consequences of poor human response to unexpected events have lethal implications. Such circumstances contribute only a small proportion to the total system operation time, but contribute a disproportionately high expected cost of system failure. Indeed, it is often pointed out that incidents in complex systems (and those in aircraft in particular) occur so rarely precisely **because** they result from a very low probability **combination** of many low-frequency events (Diehl, 1991; Reason, 1990). However, as long as the goal remains to further reduce accident rates, it is essential to focus design considerations on those that better support the detection of, response to, and management of the low-frequency event combinations. A combination of low operator skills together with inadequate training and preparation for unexpected events is certainly not incorporated in flight tests using highly qualified test pilots. The combination appears to be sampled in traditional LOFT training, but the lessons learned from such training are rarely applied to system design, as discussed later.

The preceding arguments concerning the importance of measuring responses to very unexpected events, as implicit measures of SA, gives rise to two important methodological issues. The first concerns the issue of **expectancy**. This concept, related to probability, has a long history in psychological laboratory-based research on perception and reaction time (Broadbent, 1972; Fitts, 1966; Posner, 1978; Wickens, 1992). However, we argue that there may be a real qualitative difference between low-ex-pectancy (as in low-frequency events in a laboratory reaction time experi-ment) and truly surprising events. In human factors evaluations, the latter must be constrained to events that typically occur once per experiment, and participants may not be prebriefed that these are even a part of the experimental protocol.

Examples of truly surprising events include simulated runway incur-sions to assessment head–up displays (Fischer, Haines, & Price, 1980; Wickens & Long, 1995), and automation failures in aircraft evaluation (Beringer, 1999). Response times to such events may be an order of mag-nitude longer than to low-frequency events that an experiment participant nevertheless knows could occur as part of the protocol (Summala, 1981). Furthermore, the pattern of response to such events may be **qualitatively** different than the pattern of response to more expected events. As one ex-ample, head–up displays appear to support the processing of both routine and rare events better than head–down displays; however, the processing of the truly surprising event (e.g., unbriefed, once per experiment) ap-pears better supported by the head–down displays (Wickens, 1997).

The second methodological issue concerns the use of performance, rather than user opinion, to evaluate SA directly or to evaluate the ability of a particular display device to support SA. With regard to the first, it is important to bear in mind that because operators are not aware of that which they are unaware, a subjective rating of SA is itself inherently flawed as a measure. As to the second use of performance as a measure, it is im-portant to consider that a user's subjective evaluation of how well a given interface might support response to the truly unexpected event (or how important providing that support might be) can be quite inaccurate or bi-ased if the user has never experienced the particular event (Andre & Wickens, 1995). A user may seriously underestimate how disorienting an event could be for themselves or for a less skilled operator (Kelley, 1998).

**The Trade-off in the Design Process**

The trade-offs previously discussed explicitly dictate that optimal designs be very different for the routine and for the unexpected. Addressing the concerns more specifically to the depiction of the geographical environ-ment (local guidance and global hazard awareness displays), I consider four alternative solutions to the trade-offs.

*The Compromise Display.* As Fig. 10.1 illustrates, there are display schemes at the midpoint of egocentricity; those between the ego-ref-erenced display best suited for local guidance (Fig. 10.1a) and the world-referenced display best suited for global awareness (Fig. 10.1d). Such a compromise display will "satisfice" in the sense of providing adequate support for both and appears to have distinct advantages if display space is at a premium. Research revealed that a rotating "tether" display concept, schematically illustrated in Fig. 10.1c appears to achieve such a compromise (Wickens & Prevett, 1995). It provides an adequate view around and behind ownship, yet preserves some of the ego-referenced aspects of the pilot's forward view within the cockpit. If the viewpoint is tethered behind ownship, then a fairly high gain motion field is evident as the aircraft pitches or rolls, thereby satisfying many of the requirements for guidance. The one item of concern is the ambiguity of perceiving the relative position of specific hazards from ownship.

*The Dual Display.* An alternative to the compromise is to design the optimal display for each task (guidance and awareness) and present the two simultaneously (Olmos, Wickens, & Chudy, 1999). Such a solution encounters three limitations. First, this is not an economical use of limited space in environments such as the flight deck and shrinking the two panels into a smaller size reduces the resolution of both. Second, it imposes added scanning demands on the operator. Third, it hampers the ability of the operator to mentally translate information presented on one display into where the information is located on the other display. For example, the location of a dangerous hazard may be depicted on the global hazard awareness display, but the pilot may need to know where it is likely to be seen on the local guidance display. Research found that this "cognitive linkage" between separate display panels can, to some degree, be supported by adopting techniques of **visual momentum** (Woods, 1984). An example here is to physically depict the field of view of the local guidance display in terms of a "wedge" depicted on the rendering of the global awareness display (Fig. 10.2; Aretz, 1991; Olmos, Liang, & Wickens, 1997; Olmos, Wickens, & Chudy, 1999).

*The Sequential Display.* The space problem created by the dual display can be solved by the use of flexible or multifunction displays; different renderings optimally suited for one task or another are depicted at different times within the same physical viewport. Numerous examples of this strategy are found with the various modes of depicting either electronic maps (the horizontal situation display) in commercial glass cockpits or radar coverage in combat aircraft. Whether such displays are user-chosen or adaptively selected by automation (Parasuraman & Mouloua, 1996), they have the advantage of space economy. There are, however, three drawbacks to implementing a sequential strategy. First, if information on one display must be related to that on the other, working memory limitations sometimes impose on the user's ability to carry over information from one to the other (Seidler & Wickens, 1992; Wickens & Seidler,

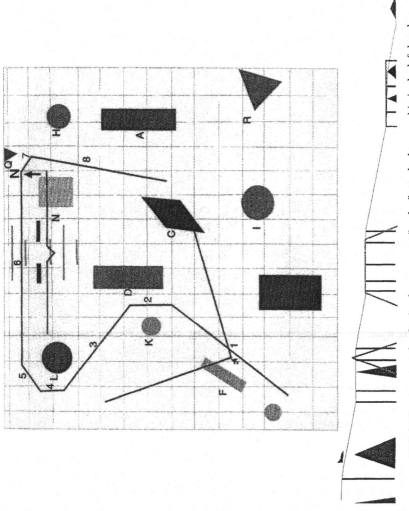

FIG. 10.2.   An example of a visual momentum "wedge" attached to ownship in the left-hand side of the display. This wedge depicts the visual angle of the forward view seen either in a more egocentric display (Fig. 10.1b) or seen looking forward out the cockpit windscreen.

1997). Second, when the number of possible displays exceeds a handful, manually operating whatever device or menu selection tool is required to "navigate" from one screen to another adds another burden (Seidler & Wickens, 1992). Finally, such a situation creates a certain amount of inconsistency of representation. For example, the same physical space may, at one moment, represent a rotating map (stable aircraft) and at the next moment, represent a rotating aircraft (fixed map). Such inconsistency may be confusing, and at times even dangerous, if the operator temporarily forgets what mode he or she is viewing.

*The Automated System.*   The final, and perhaps most radical, solution to the design trade-offs between local guidance and hazard awareness is to design primarily for the support of global SA (support for the unexpected) and assume that local guidance can be, will be, and often *is* easily automated. By definition, local guidance during routine operations is fairly predictable and is the sort of skill-based task that easily lends itself to effective and reliable automation. Hence, display support to the operator for such a routine task can be less than optimal. Such an approach then allows greater efforts to be focused on the design of effective displays for global awareness. Indeed, research indicated that the "tether display," shown schematically in Fig. 10.1c, represents such a choice; it serves global hazard awareness somewhat better than human performance in local guidance (e.g., flight path tracking accuracy). If the guidance function was under autopilot control most of the time, such minor deficiencies in human performance would be less critical to full system performance.

If this approach (to weight design for optimal support of global awareness and assume automation of routine) is adopted, then the decision to assume that routine performance is automated should **only** be made after giving thought to two important considerations. First, the operator must be provided with some level of active choice in the navigational decisions of the system even if the decisions are intermittent and at higher levels (Endsley & Kiris, 1995; Wickens, Mavor, Parasuraman, & McGee, 1998). Second, designers should **never** assume that automation will not fail, despite assurances that may be offered by manufacturers to the contrary. Design should always support the human's ability to respond to unexpected system failure. Painful lessons revealed that automation is rarely, if ever, entirely "failure free."

## CONCLUSION

I have argued that a system-wide analysis of performance differences in routine and unexpected circumstances by the unskilled as well as the highly skilled operators should be undertaken (coupled with understanding of the constraints of display space and human cognition), before designs are implemented and before test programs are conducted to evaluate the implemented designs. I feel that weighting of the optimal design for (and measurement of) global SA should be shifted to a level that greatly exceeds the proportion of time such awareness is actually required in order to handle the unexpected circumstances.

## ACKNOWLEDGMENTS

Much of the research summarized in the report was supported by a grant from the NASA Ames Research Center, Moffett Field, CA (NASA NAG 2–308). Vernol Battiste and Sandra Hart were the technical monitors. Several students at the University of Illinois contributed to the thinking and data underlying this chapter and include Bradley Boyer, Ian Haskell, Chia-Chin Liang, Oscar Olmos, Tyler Prevett, and Karen Seidler.

## REFERENCES

Adams, M. J., Tenney, Y. J., & Pew, R. W. (1991). *Strategic workload and the cognitive management of advanced multi-task systems* (Report No. SOAR/CSERIAC 91–6). Wright-Patterson AFB, OH: Crew System Ergonomics Information Analysis Center.

Adams, M. J., Tenney, Y. J., & Pew, R. W. (1995). Situation awareness and the cognitive management of complex systems. *Human Factors, 37*(1), 66–85.

*Aeronautica Civil of The Republic of Columbia* (1996). Aircraft accident report: Controlled flight into terrain, American Airlines Flight 965, Boeing 757–223, N651AA, near Cali, Colombia, Dec. 20, 1995. Santafe de Bogota, D.C. Colombia.

Andre, A. D., & Wickens, C. D. (1995). When users want what's *not* best for them: A review of performance-preference dissociations. *Ergonomics in Design,* 10–13.

Aretz, A. J. (1991). The design of electronic map displays. *Human Factors, 33*(1), 85–101.

Barfield, W., Rosenberg, C., & Furness, T. A., III (1995). Situation awareness as a function of frame of reference, computer-graphics eyepoint elevation, and geometric field of view. *The International Journal of Aviation Psychology, 5*(3), 233–256.

Beringer, D. B. (1999). Automation in general aviation: Responses of pilots to autopilot and pitch trim malfunctions. In *Proceedings of the 40th Annual Meeting of the Human Factors and Ergonomics Society* (pp. 86–90). Santa Monica, CA: Human Factors & Ergonomics Society.

Broadbent, D. (1972). *Decision and stress.* New York: Academic Press.

Chou, C., Madhavan, D., & Funk, K. (1996). Studies of cockpit task management errors. *International Journal of Aviation Psychology, 6,* 307–320.

Degani, A., & Wiener, E. L. (1993). Cockpit checklists: Concepts, design, and use. *Human Factors, 35*(4), 330–345.

Diehl, A. E. (1991). Human performance and systems safety considerations in aviation mishaps. *International Journal of Aviation Psychology, 1*(2), 97–106.

Dominguez, C. (1994). Can SA be defined? In M. Vidulich, C. Dominguez, E. Vogel, & G. McMillan (Eds.), *Situation awareness: Papers and annotated bibliography* (AL/CF–TR–1994–0085). Wright-Patterson AFB, OH: Armstrong Laboratory.

Dornheim, M. A. (1995, January 30). Dramatic incidents highlight mode problems in cockpits. *Aviation Week and Space Technology, 1,* 57–59.

Endsley, M. R. (1995). Measurement of situation awareness in dynamic systems. *Human Factors, 37*(1), 65–84.

Endsley, M. R., & Kiris, E. O. (1995). The out-of-the-loop performance problem and level of control in automation. *Human Factors, 37*(2), 381–394.

Fischer, E., Haines, R. F., & Price, T. A. (1980). *Cognitive issues in head-up displays* (NASA Technical Paper No. 1711). Moffett Field, CA: NASA Ames Research Center.

Fitts, P. M. (1966). Cognitive aspects of information processing III: Set for speed versus accuracy. *Journal of Experimental Psychology, 71,* 849–857.

Funk, K. (1991). Cockpit task management: Preliminary definitions, normative theory, error taxonomy, and design recommendation. *The International Journal of Aviation Psychology, 1*(4), 271–286.

Goodstein, L. P. (1981). Discriminative display support for process operations. In J. Rasmussen & W. B. Rouse (Eds.), *Human detection and diagnosis of system failures.* New York: Plenum.

Haskell, I. D., & Wickens, C. D. (1993). Two- and three-dimensional displays for aviation: A theoretical and empirical comparison. *The International Journal of Aviation Psychology, 3*(2), 87–109.

Kelley, C. (1998). Subjective experience as a basis for metacognitive judgments. In D. Gopher & A. Koriat (Eds.), *Attention and performance, Vol. 16.* Orlando, FL: Academic Press.

Kintch, W., & Ericsson, K. A. (1995). Long term working memory. *Psychological Review, 102,* 211–245.

Landeweerd, J. A. (1979). Internal representation of a process fault diagnosis and fault correction. *Ergonomics, 22,* 1343–1351.

Olmos, O., Liang, C. C., & Wickens, C. D. (1995). Construct validity of situation awareness measurements related to display design. In D. J. Garland & M. R. Endsley (Eds.), *Proceedings of the International Conference on Experimental Analysis and Measurement of Situation Awareness* (pp. 219–255). Daytona Beach, FL: Embry-Riddle Aeronautical University Press.

Olmos, O., Liang, C. C., & Wickens, C. D. (1997). Electronic map evaluation in simulated visual meteorological conditions. *International Journal of Aviation Psychology, 7*(1), 37–66.

Olmos, O., Wickens, C. D., & Chudy, A. (1999). Tactical displays for combat awareness: An examination of dimensionality and frame of reference concepts, and the application of cognitive engineering. *International Journal of Aviation Psychology.*

Parasuraman, R., & Mouloua, M. (Eds.). (1996). *Automation and human performance.* Mahwah, NJ: Lawrence Erlbaum Associates.

Posner, M. I. (1978). *Chronometric explorations of the mind.* Hillsdale, NJ: Lawrence Erlbaum Associates.

Rasmussen, J. (1986). *Information Processing and Human-Machine Interaction: An Approach to Cognitive Engineering.* New York: North-Holland.

Rasmussen, J., Pejtersen, A., & Goodstein, L., (1995). *Cognitive engineering: Concepts and applications.* New York: Wiley.

Reason, J. (1990). *Human Error.* New York: Cambridge University Press.

Reason, J. (1997). *Managing the Risks of Organizational Accidents.* Brookfield, VT: Ashgate.

Reising, J. M., Liggett, K. K., Solz, T. J., & Hartsock, D. C. (1995). A comparison of two head-up display formats used to fly curved instrument approaches. In *Proceedings of the 39th Annual Meeting of the Human Factors and Ergonomics Society* (pp. 1–5). Santa Monica, CA: Human Factors & Ergonomics Society.

Roth, E. M., & Woods, D. D. (1988). Aiding human performance I: Cognitive analysis. *Le Travail Human, 51,* 39–64.

Sarter, N. B., & Woods, D. D. (1992). Pilot interaction with cockpit automation: Operational experiences with the flight management system. *The International Journal of Aviation Psychology, 2,* 303–322.

Sarter, N. B., & Woods, D. D. (1994). Pilot interaction with cockpit automation: II. An experimental study of pilots' model and awareness of the flight management system. *The International Journal of Aviation Psychology, 4,* 1–28.

Sarter, N. B., & Woods, D. D. (1995a). How in the world did we ever get into that mode? Mode error and awareness in supervisory control. *Human Factors, 37*(1), 5–19.

Sarter, N. B., & Woods, D. D. (1995b). *Strong, silent, and 'out-of-the-loop': Properties of advanced (cockpit) automation and their impact on human-automation interaction* (CSEL Report No. 95–TR–01). Columbus: Cognitive Systems Engineering Lab., Ohio State University.

Sarter, N. B., & Woods, D. D. (1997). Team play with a powerful and independent agent. *Human Factors, 39,* 553–569.

Schutte, P. C., & Trujillo, A. C. (1996). Flight crew task management in non-normal situations. In *Proceedings of the 40th Annual Meeting of the Human Factors and Ergonomics Society* (pp. 244–248). Santa Monica, CA: Human Factors & Ergonomics Society.

Seidler, K. S., & Wickens, C. D. (1992). Distance and organization in multifunction displays. *Human Factors, 34*(5), 555–569.

Strauch, B. (1997). Automation and decision making: Lessons learned from the Cali accident. In *Proceedings of the 41st Annual Meeting of the Human Factors and Ergonomics Society* (pp. 195–199). Santa Monica, CA: Human Factors & Ergonomics Society.

Summala, H. (1981). Driver steering response latencies. *Human Factors, 23*(6), 683–692.

Theunissen, E. (1994). Factors influencing the design of perspective flight path displays for guidance and navigation. *Displays, 15*(4), 241–254.

Theunissen, E. (1995). Influence of error gain and position prediction on tracking performance and control activity with perspective flight path displays. *Air Traffic Control Quarterly, 3*(2), 95–116.

Vicente, K., & Rasmussen, J. (1992). Ecological interface design: Theoretical foundations. *IEEE Transactions on Systems, Man, & Cybernetics, 22*(4), 589–606.

Wickens, C. D. (1995). *Integration of navigational information for flight* (Technical Report No. ARL–95–11/NASA–95–5). Savoy: University of Illinois Institute of Aviation.

Wickens, C. D. (1996). Situation awareness: Impact of automation and display technology. In *AGARD Conference Proceedings 575: Situation Awareness: Limitations and Enhancement in the Aviation Environment* (pp. K2–1/K2–13). Neuilly Sur Seine, France: AGARD.

Wickens, C. D. (1997). Attentional issues in head-up displays. In *Engineering Psychology and Cognitive Ergonomics: Integration of Theory and Application* (pp. 3–22). London: Avebury Technical.

Wickens, C. D. (1999). Frame of reference for navigation. In D. Gopher & A. Koriat (Eds.), *Attention and performance, Vol. 16* (pp. 113–143). Cambridge, MA: MIT Press.

Wickens, C. D., & Hollands, J. (2000). *Engineering psychology and human performance* (3rd ed.). Saddle Brook, NJ: Prentice-Hall.

Wickens, C. D., Liang, C. C., Prevett, T., & Olmos, O. (1996). Electronic maps for terminal area navigation: Effects of frame of reference on dimensionality. *International Journal of Aviation Psychology, 6*(3), 241–271.

Wickens, C. D., & Long, J. (1995). Object- vs. space-based models of visual attention: Implications for the design of head-up displays. *Journal of Experimental Psychology: Applied, 1*(3), 179–194.

Wickens, C. D., Mavor, A. S., Parasuraman, R., & McGee, J. P. (Eds.). (1998). *The future of air traffic control: Human operators and automation.* Washington, DC: National Academy Press.

Wickens, C. D., & Prevett, T. (1995). Exploring the dimensions of egocentricity in aircraft navigation displays: Influences on local guidance and global situation awareness. *Journal of Experimental Psychology: Applied, 1,* 110–135.

Wickens, C. D., & Seidler, K. S. (1997). Information access in a dual-task context: Testing a model of optimal strategy selection. *Journal of Experimental Psychology: Applied, 3*(3), 196–215.

Woods, D. D. (1984). Visual momentum: A concept to improve the cognitive coupling of person and computer. *International Journal of Man-Machine Studies, 21,* 229–244.

# Chapter 11

# Testing the Sensitivity of Situation Awareness Metrics in Interface Evaluations

Michael A. Vidulich
*Air Force Research Laboratory*

Improving *situation awareness* (SA) has become a vital goal for system designers. SA refers to the "continuous extraction of environmental information, integration of this information with previous knowledge to form a coherent mental picture, and the use of that picture in directing future perception and anticipating future events" (Dominguez, 1994, p. 11). SA has probably been most closely associated with fighter pilots engaged in air combat. Before becoming a popular research topic, SA was commonly used in the course of debriefings to explain the outcome of training missions (Waddell, 1979) and has been referred to as the "ace factor" that separates fighter aces from average fighter pilots (Spick, 1988). As SA has become an increasingly popular topic in human factors research it has been linked to performance in other domains, such as air traffic control, commercial aviation, or emergency management. Typically the discussion of SA in any of these domains centers around how to maintain SA in difficult situations or how to improve SA by some sort of human factors intervention.

Any of the standard human factors interventions (such as training, selection, or interface design) might be useful in improving SA in some circumstances (see Vidulich, Dominguez, Vogel, & McMillan, 1994). However, probably the most common approach for trying to improve SA has been to change the human–machine interface. Typically the interface for some demanding task is altered to add more task relevant information or displays are reformatted to display the information more compatibly with human cognition. Unfortunately, the application of new information or

**227**

new display formats to an interface is no guarantee that SA will be improved. It is essential that such interface modifications be tested to demonstrate their efficacy. Naturally, such testing presupposes the availability of appropriate metrics.

The goal of this chapter is to determine whether the appropriate metrics for measuring the impact of interface manipulations upon operator SA are available. In order to do this, the strategy selected was to collect the appropriate tests from the human factors literature to create a database and then to conduct a meta-analysis examining the existing metrics. The goal was to establish the basic sensitivity of existing metric approaches and to begin identifying the guidelines for using SA metrics for interface evaluations.

## ASSEMBLING THE DATABASE

Step one of the meta-analytic process is to assemble a database of the relevant literature. For the present review, any paper that conducted an empirical test of SA was collected for the original database. After assembly of the database, each paper was examined more closely to determine its appropriateness for inclusion in the present meta-analysis. To be included the paper had to discuss an experiment that: involved a manipulation of a human-machine interface, used a measure that was either explicitly identified as a SA metric or from which changes in SA were inferred, and described the experiment in sufficient detail to allow the generation of the necessary meta-analytic descriptors and outcomes. Once a paper had been added to the final database, its independent and possible moderating variables were classified and its dependent measures were assessed.

### Main Independent Variable

The major independent variable was defined by the selection process described previously. A paper was only included in the final database if there was some identifiable interface manipulation that was either intended or inferred to affect SA.

### Dependent Measures

Ideally it would have been simplest if all experimental outcomes could have been considered examples of some unitary "SA" metric. However, there is considerable variability in how researchers go about assessing SA. Although it is possible to aggregate meta-analytic statistics across all of these different approaches, to do so seems to welcome the possibility of the classic "adding apples and oranges" problem. To avoid this, the results from the different major categories of SA metrics were analyzed separately. The different categories were: performance-based, memory probes, subjective ratings, and physiological. To be included in the analysis, each study had to report data on at least one of these metric types. In many cases a study contributed data on more than one metric type.

In addition to the four SA metric categories, any reported mental workload results were also analyzed. It is common for advocates of new interface designs to claim that the new design will simultaneously improve SA and reduce mental workload. Therefore, checking to see if there was any typical relation between SA and mental workload seemed advisable.

**General Moderator Variables**

In addition to assessing the overall effect of the independent variable in a meta-analysis, another question is whether the information in the database suggests variables that might influence the strength of any relation or at least provide context for the research. Potential moderator variables that were considered for each study in the current review were: (a) type of interface manipulation (new format, additional information, or automation), (b) year of publication, (c) type of publication (referred journal, technical report, or proceedings paper), (d) type of task (simple laboratory task vs. simulated or real task), (e) number of subjects, (f) type of subjects (novices or experts), and (g) type of experimental design for the interface variable (between-subject or within-subject).

**Moderator Variables for Specific Dependent Measures**

Just as there are several types of dependent measures present in the literature there are debates among practitioners using each type of measure regarding the best approach for employing it. So, for several of the specific dependent-measure databases, there were specific moderator variables identified:

*Task Performance.* Task performance was unusual because in some studies the authors appeared to infer SA directly from performance, whereas in other studies, performance was assessed as an adjunct to some other SA metric. An "inferred SA" variable was used to record whether the experimenter inferred SA from performance data. In some cases the experimenter might explicitly state that a manipulation was intended to improve SA and that the improvement would be demonstrated through improved performance. In other cases, such an explicit statement might not be present, but the experimenter might claim that SA was improved while the only supporting data offered was performance-based. Either of these cases would be considered positive cases of inferred SA and contrasted to cases where multiple measures were used and some other measure was explicitly identified as the SA measure.

Another variable was used to identify the general category of the performance metric. The possible types of performance metrics were: measures of effectiveness and measures of performance. Measures of effectiveness were considered to be some global measure that would presume performance from integrated performance; such as number of points scored, or number of kills. Measures of performance were more microscopic measures of performance on specific tasks or subtasks; such as tracking error, or reaction time to a specific event.

*Memory Probes.* The two main dimensions for differentiating between types of memory probes concerned the breadth of focus or "scope" for the questions used as queries and the timing of the queries. Two levels were used to describe the breadth of the memory probe questions: wide or narrow. A wide breadth of focus would include many questions about different aspects of the task. For example, in a piloting task a wide breadth of questions would be expected to include questions about the tactical situation, the aircraft's geographical location, the aircraft's attitude, and the status of aircraft subsystems. The popular Situation Awareness Global Assessment Technique (SAGAT, Endsley, 1995) is a good example of a procedure using a wide breadth of focus. A narrow breadth of focus would use a very small set of questions (possibly only one question). There were also two possibilities regarding the timing of the queries. A task trial might be interrupted to introduce the queries (as in SAGAT) or the queries might occur at the end of a trial or session.

*Subjective Ratings.* This included any technique that extracted and quantified the subjects' subjective opinion about the SA experienced in different task conditions. Two studies (Tsang & Vidulich, 1994; Vidulich & Tsang, 1987), found three variables that were useful in classifying subjective workload metrics. The three main variables were dimensionality, immediacy, and evaluation style. These three variables were used in the present meta-analysis to classify subjective SA ratings. Dimensionality referred to whether the subjective technique asked the subjects to rate SA directly along a single dimension or required assessments along multiple dimensions that were later combined to generate the SA rating. The immediacy dimension distinguished between ratings collected immediately after a specific task experience and ratings of generic task conditions collected as part of either a session or experiment debriefing. The evaluation style distinguished between ratings that required an absolute rating of a task along a dimension versus an evaluation style based on comparing two tasks along a dimension.

### "Scoring" Each Study

After each study had been added to the database and classified on every relevant moderating variable the next step was to score each study according to its outcome. In a typical meta-analysis this scoring would consist of using the statistical results reported in the paper to calculate an effect size for the relevant independent variable. Once all of the results from all studies are converted to the same measure of effect size it is possible to combine the results across the studies. This allows the researcher to determine the overall strength of the independent variable's effect on the dependent measure and whether any of the moderating variables tend to influence the size of that effect (if any). Procedures for calculating effect sizes from a wide variety of commonly reported inferential statistics and for using the effect sizes in the meta-analysis have been published (e.g., Rosenthal, 1984).

Unfortunately, the calculation of effect sizes was complicated in the present meta-analysis by the strong tendency for SA researchers to use within-subject experimental designs. It has recently been shown that the calculation of the standard meta-analytic effect size statistics from typical within-subject designs is problematic and that the seriousness of this problem has not been accounted for in standard applications of meta-analysis (Dunlap, Cortina, Vaslow, & Burke, 1996). Due to the heavy reliance of SA researchers on within-subject designs, it was deemed unwise to use the traditional effect size calculations for the present meta-analysis. Consequently, it was decided to use a meta-analytic "vote-counting" procedure (Bushman, 1994). This is a decidedly less powerful and less well quantified approach to meta-analysis, but it does allow the main questions of the present meta-analysis to be addressed.

The details of the scoring of each individual study are presented elsewhere (Vidulich, in press). In general, each study was classified according to the general moderating variables described previously and the results for any of the categories of dependent measures were listed. In many cases there were multiple instances within a general category of dependent measure (e.g., tracking error, reaction time, and error frequency present as performance measures). The results of each specific measure was scored into one of three categories: *positive*, a statistically significant result in the appropriate direction for the expected SA enhancement; *no effect*, no statistically significant result; and *negative*, a statistically significant effect in the wrong direction for the expected SA enhancement. Then, for the general class of dependent measure (e.g., performance-based, memory probe) a mini vote count was conducted. The overall result for the dependent measure type was positive if the number of positive results exceeded the number of negative results. The overall result was negative if the number of negative results exceeded the number of positive results. The overall result was no effect if there were no statistically significant results or if the positive and negative results canceled each other out. Thus, for each paper in the database there would be one result (i.e., positive, no effect, or negative) for each class of dependent measure reported in the paper (i.e., performance-based, memory probe, subjective rating, physiological, or workload). Any given paper might contribute one, two, three, or four tests of an interface manipulation on SA measures and possibly one test of the interface manipulation on a workload measure.

An important assumption behind the scoring approach was that the interface manipulations were generally effective in improving SA. This approach is undermined if the interface manipulation in an experiment failed to effect SA or if the direction of the effect was in the opposite direction than that predicted. In these circumstances, a SA metric that was sensitive to the real effect would be inaccurately scored as insensitive in the current evaluation. Consequently, this evaluation necessarily involves simultaneous tests of the effectiveness of the various SA manipulations as well as testing the sensitivity of the SA metrics.

## CHARACTERIZING THE DATABASE

Before reviewing the results of the meta-analysis conducted for this chapter, it would be worthwhile to highlight some of the general trends observed while assembling and scoring the database. These results are described in more detail in Vidulich (in press) and are reviewed here to provide some context for understanding the results of the meta-analysis.

### Number of Papers, Experiments, and Tests

In the final database there were 57 papers. Several papers reviewed multiple experiments so there was a total of 65 experiments in the database. Many experiments used more than one type of SA metric so there was a total of 104 tests of SA metrics. Also, 15 of the experiments included a mental workload metric.

Of the 104 tests of SA metrics, the most popular type of metric was task-performance. Sixty-one of the 65 experiments reported performance results, although not all of these were explicitly intended as SA measures. The next most popular approach for measuring SA in the database was memory probes. Twenty-two studies used a memory probe SA metric. Subjective ratings were used in 20 experiments. Physiological measures were used only once as a SA metric. Therefore, physiological measures were not included in the meta-analysis.

## Types of Interface Manipulations

The most common interface manipulation was a new format for information already available in the other interface. Thirty-eight experiments used this type of manipulation. For example, Wickens and Prevett (1995) contrasted an interface using an integrated 3D representation of the airspace to an interface that had all of the same information present but divided over several displays. In 22 experiments, the interface manipulation involved adding new information. For example, Zenyuh, Reising, Walchli, and Biers (1988) added stereoscopic depth information to a visual search display. In the remaining five experiments the interface manipulation involved automation. In one such study, Endsley and Kiris (1995) examined the effects of five different automation strategies on SA, performance, and mental workload in automobile navigation.

## Year of Publication

The publication trends suggest that SA has been growing in popularity. Out of the 65 experiments in the database, there was only one from the 1970s, 16 from the 1980s, and 48 in the first 7 years of the 1990s.

## Publication Type

The database was dominated by proceedings articles (48). There were also 10 journal articles and 7 technical reports. The relatively low proportion of experiments from journals (15.4%) might appear worrisome. However, a common concern in meta-analysis is the "file drawer problem" (see Rosenthal, 1984). The "file drawer problem" refers to the possible bias toward only publishing statistically significant results. This tendency might cause a review of the published literature to overestimate the strength of effect for a given variable. Thus, it is a good idea to assess research from sources that are less likely to be prone to such a publication bias.

## Task Type and Experimental Design

The majority of experiments in the database (39) used experimental tasks that were simulations of realistic, complex tasks or part-tasks that required preexisting expertise to perform. The overwhelming majority of experiments (57) used a within-subject experimental design.

## Subject Type and Sample Size

Fifty-one percent of the studies used subjects that possessed a preexisting expertise for performing the experimental task. Five studies used mixed samples of experts and novices. The range of sample sizes was relatively large, ranging from a low of 4 to a high of 80 subjects. The typical sample size was 12 to 16 subjects. The mean number of subjects was 16.1 and the median was 12. Twelve was also the modal number of subjects. Seventeen experiments used a sample size of 12.

## ANALYZING THE DATABASE

There were three main questions to be addressed in the analysis of the database: were alternative SA metrics sensitive to interface manipulations, was there evidence that the specified moderating variables affected the sensitivity of the SA measures, and was there any consistent relation between SA and mental workload. Each of these questions will be addressed in turn.

### Sensitivity of SA Measures

As previously discussed, the present meta-analysis was based on the "vote-counting" approach described by Bushman (1994). More specifically, the results in the database were used to calculate confidence intervals for the proportion of "successful" outcomes (see Bushman, 1994, p. 196, Formula 14–11). A "successful" outcome in this context refers to the interface manipulation that was intended to improve SA having a statistically significant effect (at the 0.05 level) in the predicted direction. Assuming that the experimenters have a good understanding of SA and how it is influenced by interface design, then, in general, it should be expected that the independent variables in these experiments should have affected the subjects' SA. If so, and if the SA measurement technique under consideration is sensitive to SA, then a high proportion of successful outcomes should occur.

Ninety-five percent confidence intervals were calculated for the dependent variables from each of the 61 performance-based tests, the 22 memory probe tests, and the 20 subjective rating tests. As a rule of thumb for this meta-analysis, a measure was described as sensitive if the 95% confidence interval around the proportion of that measure's positive outcomes was completely above the 50% level. This would imply that the data supported the contention that, in an experiment with a manipulation effect similar to those present in the database, the experimenter would have a better than 50% chance of achieving a significant result at the 0.05 level.

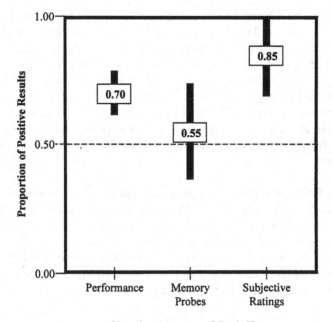

FIG. 11.1.   Ninety-five percent confidence intervals of the proportion of positive results (at a 0.05 level) for the three main types of SA metrics in the database: performance-based, memory probes, and subjective ratings. Confidence intervals that extend above 1.0 are truncated due to the logical impossibility of a proportion greater than 1.

At first consideration this may appear to be a very generous criteria for sensitivity, but given the low statistical power that is likely to be present in such experiments with typically only 12 to 16 subjects, this seemed like a reasonable criteria.

As can be seen in Fig. 11.1, performance-based measures and subjective ratings both passed the test, but memory probe measures failed. This result implies that experimenters were generally successful at creating an interface manipulation that had the expected effect on most SA measures. The next step was to examine the effects of moderating variables within specific SA measure types.

## Effects of Moderating Variables

*Performance-Based Measures.*   One question concerning the performance-based measures is whether the likelihood of an effect was related to the experimenter's intention to use performance as a SA measure. In fact, there was no suggestion of such a relation in the data. The proportion of successful outcomes when SA was being inferred from performance (0.700) was essentially identical to the proportion of successful outcomes when SA was not being inferred from performance (0.710). In both cases, the 95% confidence intervals were completely above the 50% level. This suggests that interface manipulations intended to improve SA also tended to improve performance regardless of whether the experimenter was using performance as a SA measure. The moderating variable of Metric Type was not useful due to the overwhelming tendency for experimenters to use measures of performance (53 out of 61 tests) rather than measures of effectiveness (8 tests).

*Memory Probe Measures.* Timing of memory probe presentation was not examined due to only 5 of the 22 memory probe studies using the retrospective approach. In contrast, the split between experiments that used a "wide" breadth of questions versus a "narrow" breadth of questions was almost equal; 10 versus 12, respectively. There was a striking difference in the sensitivity of the two approaches. As shown in Fig. 11.2, the proportion of successful attempts at detecting a change in SA with a narrow mem-

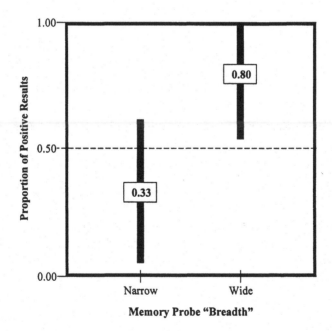

FIG. 11.2. Ninety-five percent confidence intervals of the proportion of positive results (at a 0.05 level) for the two levels of question "breadth" in memory probe measures: Narrow versus Wide. Confidence intervals that extend above 1.0 are truncated due to the logical impossibility of a proportion greater than 1.

ory probe procedure was very low, but the proportion of success with a wide memory probe procedure was much higher with the entire 95% confidence interval over the 50% level. Therefore, the memory probe results from the current database strongly suggest that using a wide variety of questions in a memory probe metric is advisable.

***Subjective Rating Measures.***   Despite the utility of categorizing subjective mental workload rating metrics by dimensionality, immediacy, and evaluation style (Tsang & Vidulich, 1994; Vidulich & Tsang, 1987), these categories were not useful in the present meta-analysis. Out of the 20 tests of subjective ratings, unidimensional ratings (5 examples), retrospective ratings (6), and relative ratings (2) were all too rare in the database to make comparisons to the alternatives reasonable.

By far the most common subjective technique in the database was the Situation Awareness Rating Technique (SART, Selcon & Taylor, 1990; Taylor, 1990); a multidimensional, immediate, and absolute rating technique. Twelve out of the 20 experiments that used subjective SA ratings used the SART technique. The data suggests that this was a good choice. Figure 11.3 shows the confidence intervals for the SART experiments con-

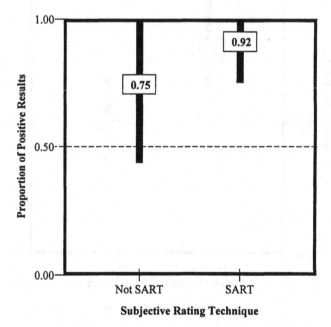

FIG. 11.3.   Ninety-five percent confidence intervals of the proportion of positive results (at a 0.05 level) for two categories of subjective rating techniques: Not SART versus SART. Confidence intervals that extend above 1.0 are truncated due to the logical impossibility of a proportion greater than 1.

trasted to the combined results of all other subjective rating techniques (i.e., non-SART). The low number of the non-SART studies and the variety of specific metrics lumped into that category makes it imprudent to draw any conclusions about the non-SART subjective measures. At the moment, the only specific subjective measure with enough data to reasonably evaluate was SART. SART was the most reliably sensitive SA measure in the current database.

## SA and Mental Workload

It is a common hope among interface designers that an improved interface will not only improve SA, but will do so while simultaneously reducing mental workload. Yet, the theoretical relation between SA and workload is questionable (Endsley, 1993). Fifteen studies measured both SA and workload, so it is reasonable to ask whether any relation was evident. Figure 11.4 shows the confidence intervals for the 15 studies improving SA and lowering workload. In both cases, the results are collapsed across the different SA or workload metrics used. The results suggest that the interface

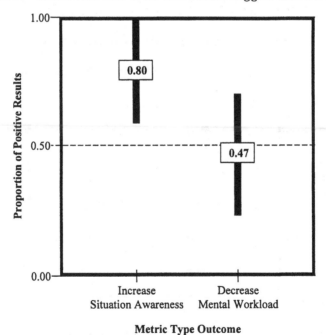

FIG. 11.4. Ninety-five percent confidence intervals of the proportion of positive results (at a 0.05 level) for improving SA (on the left) and reducing mental workload (on the right). Confidence intervals that extend above 1.0 are truncated due to the logical impossibility of a proportion greater than 1.

manipulations were generally successful in improving SA, but did not generally reduce mental workload.

## CONCLUSIONS AND FUTURE DIRECTIONS

### Conclusions From the Present Analysis

The results of the present analysis were limited. Although a database of 65 studies is a reasonable size, the generality of the present meta-analysis is constrained by limitations of the present database. One limitation was that the heavy reliance of researchers in this field upon within-subject designs limited the strength of the meta-analysis that could be done. A second limitation became evident as the database was subdivided to examine the effect of specific potential moderating variables. As the necessary subdivisions were done, the number of examples of studies in some categories was often too few to encourage meaningful interpretation of the results. Nevertheless, despite these limitations, some conclusions appear justified.

*The Efficacy of SA Manipulations on Performance.* As shown in Fig. 11.1, the manipulations of interface design undertaken to improve SA tended to improve task performance. This suggests that SA researchers are justified in believing that the SA construct represents an important phenomena in human–machine interaction. As Hardiman, Dudfield, Selcon, and Smith (1996) and Selcon, Hardiman, Croft, and Endsley (1996) have suggested, it might be worthwhile to consider SA a "meta-measure" in system evaluation. As Selcon et al. (1996) pointed out, it is seldom possible to evaluate all of the mission relevant uses of a display. Thus, a metric approach is needed that provides a generalized link to likely performance in future use. Hardiman et al. (1996) and Selcon et al. (1996) suggested that rather than focusing on task-specific performance, it might be preferable to examine meta-measures that encapsulate the cognitive reaction to performance with a given interface. One proposed meta-measure was mental workload. Presumably, if a task can be performed with one interface while experiencing less mental workload, this should provide for more robust transfer to using that interface in more challenging conditions. Another possible meta-measure is SA. Presumably the interface that provides a better understanding of the situation will provide the better chance of reacting appropriately to that situation. The fact that the current database suggests that manipulations designed to improve SA generally improve performance in a wide variety of tasks and by a wide variety of measures is consistent with the suggestion that SA could provide a useful metameasure of system interface design.

*The Sensitivity of SA Metrics.*   In addition to performance, subjective measures of SA tended to be sensitive to the interface manipulations, especially if the SART was used as the SA metric (see Figs. 11.1 and 11.3). Memory probe measures were generally not as sensitive to the interface manipulations (see Fig. 11.1), but there was evidence that the proper procedure (e.g., using a wide breadth of probe questions) might enhance the sensitivity of memory probes considerably (see Fig. 11.2). The resulting overall picture is that there are promising approaches for measuring SA, albeit no perfect solution. When they discussed guidelines for selecting mental workload metrics, Tsang and Wilson (1997) observed the following:

> Though the work is far from being complete, there is now a body of knowledge that can be put to practical use. From this, we have developed the following guidelines for selecting an appropriate strategy for workload assessment and evaluation. The philosophy underlying these guidelines is that conducting workload assessment, especially that of a fairly complex system, does require some knowledge and training in the area. (p. 435)

The body of knowledge regarding the use of SA metrics is even more incomplete than that of workload, but some practical implications are already emerging. It is important to realize, however, that just as a cookbook of easy recipes for mental workload assessment has not emerged from the workload research, it is unlikely that there will be a cookbook of simple SA measurement recipes. SA measurement will probably always require some expertise and careful application.

*SA and Mental Workload.*   As stated previously, the current database does not support a simple correlation between SA and mental workload. It is common for system designers to assert that their proposed interface will both improve SA and reduce mental workload, and there is no reason to think that a single interface redesign would necessarily be incapable of accomplishing both goals. However, the current database suggests that assertions regarding SA and mental workload must be considered to be separate assertions and must be assessed by appropriate measures. Improved SA should not, by itself, be used as evidence of lower mental workload (or higher mental workload either, for that matter). The simultaneous measurement of SA and mental workload in system evaluation has been suggested by numerous researchers (e.g., Endsley, 1993; Hardiman et al., 1996; Selcon et al., 1996; Tsang & Wilson, 1997).

### Future Directions

SA is a field of research that still appears to be growing in popularity. At the rate at which SA research is being conducted and reported, it would be unwise to assume that any present conclusions will be definitive. However, in addition to the conclusions reviewed, the present meta-analysis can be considered to have implications for future research in SA measurement and future applications of SA measurement.

*Future Directions in SA Measurement Research.* Although several approaches to SA measurement appeared to be sensitive in the existing database, it is clear that the existing database has limitations. Several potentially viable approaches to measuring SA have not been examined enough to reach even the most tentative conclusions. For example, would using retrospective memory probes or subjective ratings be as sensitive as memory probes collected during interruptions or ratings collected immediately following task completion? Perhaps not, but the current database does not allow for any convincing test of either assertion. Also, the effectiveness of interface manipulations intended to improve SA on overall measures of effectiveness in complex task environments has seldom been examined. Such gaps in the coverage of SA measurement issues can only be closed by further research.

Another point worth considering while performing future SA measurement research is the possible use of the results in future meta-analytic studies. Greater use of between-subject designs would allow more powerful meta-analytic techniques to be used. However, within-subject designs are likely to remain the standard approach for research in this area, if only because of the greater efficiency of such designs (e.g., fewer subjects means less "wasted" time spent in subject training). Yet, even if a within-subject design is used, the use of the results in a future meta-analysis could be enhanced by a more complete reporting of the statistical results. As pointed out by Dunlap et al. (1996), effect sizes for within-subject designs can be calculated if either the correlations across repeated measures or the relevant means and standard deviations are reported. Future use of meta-analysis in SA metric evaluation is desirable because it seems unlikely that there could be any single critical experiment that would establish or destroy the viability of a metric approach. It is more plausible that a metric will gain its strength by building a critical mass across many applications in diverse settings. This would allow a better judgment not only of the metric's overall value, but also about the parameters that make the metric useful in some settings but dubious in others. The assessment of whether such a critical mass has been reached should be a question well-suited to future meta-analyses.

*Future Directions in the Applications of SA Measurement.* The results of the present meta-analysis also have implications for researchers that are considering using currently available SA metric techniques in interface evaluations. Although it is to be hoped that better techniques and more complete guidelines for their application might be forthcoming from future research, it is advisable to use multiple SA metrics, if possible. In several of the database studies multiple metrics were collected. Often only a subset of the SA metrics were successful in detecting the interface manipulation effect. Inasmuch as the sensitivity of any given metric technique is never likely to be perfect, the use of multiple measures allows some assurance that the interface effect (if any) has the best possible chance to be detected.

The use of multiple measures is also advisable because it is wise to include mental workload assessment in parallel with any SA assessment. SA enhancement is a laudable goal for system designers, but only if it can be obtained without inducing unacceptable levels of mental workload. If a gain in SA can be accomplished with a reduction in mental workload, then that would be the best of both worlds.

## REFERENCES

References marked with an asterisk indicate studies included in the meta-analysis database.

*Adelstein, B. D., & Ellis, S. R. (1993). Effect of head-slaved visual image roll on spatial situation awareness. In *Proceedings of the Human Factors and Ergonomics Society 37th Annual Meeting* (Vol. 2, pp. 1350–1354). Santa Monica, CA: Human Factors & Ergonomics Society.

*Amar, M. J., Hansman, R. J., Vaneck, T. W., Chaudhry, A. I., & Hannon, D. J. (1995). A preliminary evaluation of electronic taxi charts with GPS derived position for airport surface situation awareness. In R. S. Jensen & L. R. Rakovan (Eds.), *Proceedings of the Eighth International Symposium on Aviation Psychology* (Vol. 1, pp. 499–504). Columbus: Aviation Psychology Laboratory, The Ohio State University.

*Andre, A. D., Wickens, C. D., Moorman, L., & Boschelli, M. M. (1991). Display formatting techniques for improving situation awareness in the aircraft cockpit. *The International Journal of Aviation Psychology, 1,* 205–218.

*Aretz, A. J. (1988). A model of electronic map interpretation. In *Proceedings of the Human Factors Society 32nd Annual Meeting* (Vol. 1, pp. 130–134). Santa Monica, CA: Human Factors Society.

*Aretz, A. J. (1991). The design of electronic map displays. *Human Factors, 33*(1), 85–101.

*Ballas, J. A., Heitmeyer, C. L., & Perez, M. A. (1991). Interface styles for adaptive automation. In R. S. Jensen (Ed.), *Proceedings of the Sixth International Symposium on Aviation Psychology* (Vol. 1, pp. 108–113). Columbus: The Ohio State University.

*Barfield, W., Rosenberg, C., & Furness, T. A. (1995). Situation awareness as a function of frame of reference, computer-graphics eyepoint elevation, and geometric field of view. *International Journal of Aviation Psychology, 5,* 233–256.

Bushman, B. J. (1994). Vote-counting procedures in meta-analysis. In H. Cooper, & L. V. Hedges (Eds.), *The handbook of research synthesis* (pp. 193–213). New York: Russell Sage Foundation.

*Busquets, A. M., Parrish, R. V., Williams, S. P., & Nold, D. E. (1994). Comparison of pilots' acceptance and spatial awareness when using EFIS vs. pictorial display formats for complex, curved landing approaches. In R. D. Gilson, D. J. Garland, & J. M. Koonce (Eds.), *Situational awareness in complex systems: Proceedings of a CAHFA Conference* (pp. 139–167). Daytona Beach, FL: Embry-Riddle Aeronautical University Press.

*Calhoun, G. L., Janson, W. P., & Valencia, G. (1988). Effectiveness of three-dimensional auditory directional cues. In *Proceedings of the Human Factors Society 32nd Annual Meeting* (Vol. 1, pp. 68–72). Santa Monica, CA: Human Factors Society.

*Carmody, M. A., & Gluckman, J. P. (1993). Task specific effects of automation and automation failure on performance, workload, and situational awareness. In R. S. Jensen & D. Neumeister (Eds.), *Proceedings of the Seventh International Symposium on Aviation Psychology* (Vol. 1, pp. 167–171). Columbus: The Ohio State University.

*Casper, P. A. (1997). A full mission success story: RPA simulation at CSRDF yields promising results. In *Proceedings of the Human Factors and Ergonomics Society 41st Annual Meeting* (Vol. 1, pp. 95–99). Santa Monica, CA: Human Factors and Ergonomics Society.

*Chandra, D., & Bussolari, S. R. (1991). An evaluation of flight path management automation in transport category aircraft. In R. S. Jensen (Ed.), *Proceedings of the Sixth International Symposium on Aviation Psychology* (Vol. 1, pp. 139–144). Columbus: The Ohio State University.

*Crabtree, M. S., Marcelo, R. A. Q., McCoy, A. L., & Vidulich, M. A. (1993). An examination of a subjective situational awareness measure during training on a tactical operations simulator. In R. S. Jensen & D. Neumeister (Eds.), *Proceedings of the Seventh International Symposium on Aviation Psychology* (Vol. 2, pp. 891–895). Columbus: The Ohio State University.

*Crick, J. L., Selcon, S. J., Piras, M., Shanks, C., Drewery, C., & Bunting, A. (1997). Validation of the explanatory concept for decision support in air-to-air combat. In *Proceedings of the Human Factors and Ergonomics Society 41st Annual Meeting* (Vol. 1, pp. 28–31). Santa Monica, CA: Human Factors and Ergonomics Society.

Dominguez, C. (1994). Can SA be defined? In M. Vidulich, C. Dominguez, E. Vogel, & G. McMillan (Eds.), *Situation awareness: Papers and annotated bibliography* (Technical Report No. AL/CF–TR–1994–0085, pp. 5–15). Wright-Patterson AFB, OH: Armstrong Laboratory.

*Dudfield, H. J., Davy, E., Hardiman, T., & Smith, F. (1995). The effectiveness of colour coding collimated displays: An experimental evaluation of performance benefits. In R. S. Jensen & L. R. Rakovan (Eds.), *Proceedings of the Eighth International Symposium on Aviation Psychology* (Vol. 1, pp. 58–63). Columbus: Aviation Psychology Laboratory, The Ohio State University.

Dunlap, W. P., Cortina, J. M., Vaslow, J. B., & Burke, M. J. (1996). Meta-analysis of experiments with matched groups or repeated measures designs. *Psychological Methods, 1*, 170–177.

*Eckel, J. S., Mollenhauer, P. C., & Patterson, M. J. (1989). Pilot evaluation of selected colors and scales using a digitized map display. In R. S. Jensen (Ed.), *Proceedings of the Fifth International Symposium on Aviation Psychology* (Vol. 1, pp. 313–318). Columbus: Aviation Psychology Laboratory, The Ohio State University.

*Elias, B. (1995). Dynamic auditory preview for visually guided target aiming. In *Proceedings of the Human Factors and Ergonomics Society 39th Annual Meeting* (Vol. 2, pp. 1415–1419). Santa Monica, CA: Human Factors and Ergonomics Society.

Endsley, M. R. (1993). Situation awareness and workload: Flip sides of the same coin. In R. S. Jensen & D. Neumeister (Eds.), *Proceedings of the Seventh International Symposium on Aviation Psychology* (Vol. 2, pp. 906–911). Columbus: Aviation Psychology Laboratory, The Ohio State University.

Endsley, M. R. (1995). Measurement of Situation Awareness in dynamic systems. *Human Factors, 37*(1), 65–84.

*Endsley, M. R., & Kiris, E. O. (1995). The out-of-the-loop performance problem and level of control in automation. *Human Factors, 37*(2), 381–394.

*Endsley, M. R., Mogford, R. H., Stein, E., & Hughes, W. J. (1997). Controller situation awareness in free flight. In *Proceedings of the Human Factors and Ergonomics Society 41st An-*

*nual Meeting* (Vol. 1, pp. 4–8). Santa Monica, CA: Human Factors and Ergonomics Society.

*Ericson, M. A., & Yee, W. D. (1991). Target acquisition performance using spatially correlated auditory information over headphones. In R. S. Jensen (Ed.), *Proceedings of the Sixth International Symposium on Aviation Psychology* (Vol. 1, pp. 589–594). Columbus: Aviation Psychology Laboratory, The Ohio State University.

*Guttman, J. (1986). *Evaluation of the F/A–18 Head–Up–Display for recover from unusual attitudes* (NADC–86157–60). Warminster, PA: Naval Air Development Center.

*Haas, M. W., Hettinger, L. J., Nelson, W. T., & Shaw, R. L. (1995). *Developing virtual interfaces for use in future fighter aircraft cockpits* (AL/CF–TR–1995–0154). Wright-Patterson AFB, OH: Armstrong Laboratory.

*Hardiman, T. D., Dudfield, H. J., Selcon, S. J., & Smith, F. J. (1996). In *AGARD Conference Proceedings 575—Situation Awareness: Limitations and Enhancement in the Aviation Environment* (pp. 15-1–15-7). Neuilly Sur Seine, France: AGARD.

*Hart, S. G., & Chappel, S. L. (1983). Influence of pilot workload and traffic information on pilot's situation awareness. In *Nineteenth Annual Conference on Manual Control* (pp. 4–26). Cambridge, MA: MIT Press.

*Hartz, J. O. (1995). The integration of peripheral vision supporting features into conventional display design. In R. S. Jensen & L. R. Rakovan (Eds.), *Proceedings of the Eighth International Symposium on Aviation Psychology* (Vol. 2, pp. 863–868). Columbus: Aviation Psychology Laboratory, The Ohio State University.

*Haskell, I. D., & Wickens, C. D. (1993). Two- and three-dimensional displays for aviation: A theoretical and empirical comparison. *International Journal of Aviation Psychology, 3*, 87–109.

*Hettinger, L. J., Brickman, B. J., Roe, M. M., Nelson, W. T., & Haas, M. W. (1996). Effects of virtually-augmented fighter cockpit displays on pilot performance, workload, and situation awareness. In *Proceedings of the Human Factors and Ergonomics Society 40th Annual Meeting* (Vol. 1, pp. 30–33). Santa Monica, CA: Human Factors and Ergonomics Society.

*Hughes, E. R., Hassoun, J. A., Ward, G. F., & Rueb, J. D. (1990). *An assessment of selected workload and situation awareness metrics in a part-mission simulation* (Technical Report No. ASD–TR–90–5009). Wright-Patterson AFB, OH: Aeronautical Systems Division.

*Johnston, J. C., Horlitz, K. L., & Edmiston, R. W. (1993). Improving situation awareness displays for air traffic controllers. In R. S. Jensen & D. Neumeister (Eds.), *Proceedings of the Seventh International Symposium on Aviation Psychology* (Vol. 1, pp. 328–334). Columbus: The Ohio State University.

*Kibbe, M. P. (1988). Information transfer from intelligent EW displays. In *Proceedings of the Human Factors Society 32nd Annual Meeting* (Vol. 1, pp. 107–110). Santa Monica, CA: Human Factors Society.

*Kinsley, S. A., Warner, N. W., & Gleisner, D. P. (1985). *A comparison of two pitch ladder formats and an ADI ball for recovery from unusual attitudes* (Technical Report No. NADC–86012–60). Warminster, PA: Naval Air Development Center.

*Kuchar, J. K., & Hansman, R. J. (1993). An exploratory study of plan-view terrain displays for air carrier operations. *The International Journal of Aviation Psychology, 3*, 39–54.

*Liang, C. C., Wickens, C. D., & Olmos, O. (1995). Perspective electronic map evaluation in visual flight. In R. S. Jensen & L. R. Rakovan (Eds.), *Proceedings of the Eighth International Symposium on Aviation Psychology* (Vol. 1, pp. 104–109). Columbus: Aviation Psychology Laboratory, The Ohio State University.

*Marshak, W. P., Kuperman, G., Ramsey, E. G., & Wilson, D. (1987). Situational awareness in map displays. In *Proceedings of the Human Factors Society 31st Annual Meeting* (Vol. 1, pp. 533–535). Santa Monica, CA: Human Factors Society.

*Melanson, D., Curry, R. E., Howell, J. D., & Connelly, M. E. (1973). The effects of communications and traffic situation displays on pilots awareness of traffic in the terminal area.

In *Proceedings of the Ninth Annual Conference on Manual Control* (pp. 25–39). Cambridge, MA: MIT Press.

*Morrison, J. G., Kelly, R. T., & Hutchins, S. G. (1996). Impact of naturalistic decision support on tactical situation awareness. In *Proceedings of the Human Factors and Ergonomics Society 40th Annual Meeting* (Vol. 1, pp. 199–203). Santa Monica, CA: Human Factors and Ergonomics Society.

*Neale, D. C. (1997). Factors influencing spatial awareness and orientation in desktop virtual environments. In *Proceedings of the Human Factors and Ergonomics Society 41st Annual Meeting* (Vol. 2, pp. 1278–1282). Santa Monica, CA: Human Factors and Ergonomics Society.

*Olson, J. L., Arbak, C. J., & Jauer, R. A. (1991). *Panoramic Cockpit Control and Display System—Volume 2: PCCADS 2000* (AFWAL–TR–88–1038). Wright-Patterson AFB, OH: Wright Laboratory.

*Regal, D., & Whittington, D. (1995). Guidance symbology for curved flight paths. In R. S. Jensen & L. R. Rakovan (Eds.), *Proceedings of the Eighth International Symposium on Aviation Psychology* (Vol. 1, pp. 74–79). Columbus: Aviation Psychology Laboratory, The Ohio State University.

*Reising, J., Barthelemy, K., & Hartsock, D. (1989). Pathway-in-the-sky evaluation. In R. S. Jensen (Ed.), *Proceedings of the Fifth International Symposium on Aviation Psychology* (Vol. 1, pp. 233–238). Columbus: Aviation Psychology Laboratory, The Ohio State University.

Rosenthal, R. (1984). *Meta-analytic procedures for social research*. Beverly Hills, CA: Sage.

*Selcon, S. J., Bunting, A., Coxell, A. W., Lal, R., & Dudfield, H. J. (1995). Explaining decision support: An experimental evaluation of an explanatory tool for data-fused displays. In R. S. Jensen & L. R. Rakovan (Eds.), *Proceedings of the Eighth International Symposium on Aviation Psychology* (Vol. 1, pp. 92–97). Columbus: Aviation Psychology Laboratory, The Ohio State University.

*Selcon, S. J., Hardiman, T. D., Croft, D. G., & Endsley, M. R. (1996). A test-battery approach to cognitive engineering: To meta-measure or not to meta-measure, that is the question! In *Proceedings of the Human Factors and Ergonomics Society 40th Annual Meeting* (Vol. 1, pp. 228–232). Santa Monica, CA: Human Factors and Ergonomics Society.

*Selcon, S. J., & Taylor, R. M. (1990). Evaluation of the situational awareness rating technique (SART) as a tool for aircrew systems design. In *Situational Awareness in Aerospace Operations* (AGARD–CP–478, pp. 5–1 to 5–8). Neuilly Sur Seine, France: NATO–AGARD.

*Selcon, S. J., Taylor, R. M., & Shadrake, R. A. (1992). Multi-modal cockpit warnings: Pictures, words, or both? In *Proceedings of the Human Factors Society 36th Annual Meeting* (Vol. 1, pp. 57–61). Santa Monica, CA: Human Factors Society.

*Shively, R. J., & Goodman, A. D. (1994). Effects of perceptual augmentation of visual displays: Dissociation of performance and situation awareness. In *Proceedings of the Human Factors and Ergonomics Society 38th Annual Meeting* (Vol. 2, pp. 1271–1274). Santa Monica, CA: Human Factors and Ergonomics Society.

Spick, M. (1988). *The ace factor: Air combat and the role of situational awareness*. Annapolis, MD: Naval Institute Press.

*Steiner, B. A., & Camacho, M. J. (1989). Situation awareness: Icons vs. alphanumerics. In *Proceedings of the Human Factors Society 33rd Annual Meeting* (Vol. 1, pp. 28–32). Santa Monica, CA: Human Factors Society.

*Steiner, B. A., & Dotson, D. A. (1990). The use of 3-D stereo display of tactical information. In *Proceedings of the Human Factors Society 34th Annual Meeting* (Vol. 1, pp. 36–40). Santa Monica, CA: Human Factors Society.

Taylor, R. M. (1990). Situational awareness rating technique (SART): The development of a tool for aircrew systems design. In *Situational Awareness in Aerospace Operations* (AGARD–CP–478, pp. 3–1 to 3–17). Neuilly Sur Seine, France: NATO–AGARD.

*Taylor, R. M., & Selcon, S. J. (1990). Cognitive quality and situational awareness with advanced aircraft attitude displays. In *Proceedings of the Human Factors Society 34th Annual Meeting* (Vol. 1, pp. 26–30). Santa Monica, CA: Human Factors Society.

Tsang, P. S., & Vidulich, M. A. (1994). The roles of immediacy and redundancy in relative subjective workload assessment. *Human Factors, 36*(3), 503–513.

Tsang, P. S., & Wilson, G. F. (1997). Mental workload. In G. Salvendy (Ed.), *Handbook of human factors and ergonomics* (2nd ed., pp. 417–449). New York: Wiley.

*Venturino, M., & Kunze, R. J. (1989). Spatial awareness with a helmet-mounted display. In *Proceedings of the Human Factors Society 33rd Annual Meeting* (Vol. 2, pp. 1388–1391). Santa Monica, CA: Human Factors Society.

Vidulich, M. A. (in press). *Interface Manipulations and Situation Awareness Metrics: A Bibliographic Database.* Air Force Research Laboratory Technical Report.

Vidulich, M., Dominguez, C., Vogel, E., & McMillan, G. (1994). *Situation awareness: Papers and annotated bibliography* (AL/CF–TR–1994–0085). Wright-Patterson AFB, OH: Armstrong Laboratory.

*Vidulich, M. A., McCoy, A. L., & Crabtree, M. S. (1996). Attentional control and situational awareness in a complex air combat simulation. In *AGARD Conference Proceedings 575—Situation Awareness: Limitations and Enhancement in the Aviation Environment* (pp. 18-1–18-5). Neuilly Sur Seine, France: AGARD.

*Vidulich, M. A., Stratton, M., Crabtree, M., & Wilson, G. (1994). Performance-based and physiological measures of situational awareness. *Aviation, Space, and Environmental Medicine, 65*(5, Suppl.), A7–A12.

Vidulich, M. A., & Tsang, P. S. (1987). Absolute magnitude estimation and relative judgment approaches to subjective workload assessment. In *Proceedings of the Human Factors Society 31st Annual Meeting* (Vol. 2, pp. 1057–1061). Santa Monica, CA: Human Factors Society.

Waddell, D. (1979, Winter). Situational awareness. *USAF Fighter Weapons Review*, pp. 3–6.

*Ward, G. F., & Hassoun, J. A. (1990). *The effects of head–up display (HUD) pitch ladder articulation, pitch number location and horizon line length on unusual attitude recoveries for the F-16* (Technical Report No. ASD–TR–90–5008). Wright-Patterson AFB, OH: Aeronautical Systems Division.

*Wells, M. J., & Venturino, M. (1989). The effects of increasing task complexity on the field-of-view requirements for a visually coupled system. In *Proceedings of the Human Factors Society 33rd Annual Meeting* (Vol. 1, pp. 91–95). Santa Monica, CA: Human Factors Society.

*Wells, M. J., Venturino, M., & Osgood, R. K. (1988). Using target replacement performance to measure spatial awareness in a helmet-mounted simulator. In *Proceedings of the Human Factors Society 32nd Annual Meeting* (Vol. 2, pp. 1429–1433). Santa Monica, CA: Human Factors Society.

*Wickens, C. D., Liang, C. C., Prevett, T., & Olmos, O. (1996). Electronic maps for terminal area navigation: Effects of frame of reference and dimensionality. *International Journal of Aviation Psychology, 6*, 241–271.

*Wickens, C. D., & Prevett, T. T. (1995). Exploring the dimensions of egocentricity in aircraft navigation displays. *Journal of Experimental Psychology: Applied, 1*, 110–135.

*Zenyuh, J. P., Reising, J. M., Walchli, S., & Biers, D. (1988). A comparison of a stereographic 3-D display versus a 2-D display using an advanced air-to-air format. In *Proceedings of the Human Factors Society 32nd Annual Meeting* (Vol. 1, pp. 53–57). Santa Monica, CA: Human Factors Society.

# Part III

# SPECIAL TOPICS
# IN SITUATION AWARENESS

# Chapter 12

# Individual Differences in Situation Awareness

Leo J. Gugerty
*Galaxy Scientific Corporation*

William C. Tirre
*US Air Force Research Laboratory*

One of the most important abilities needed to perform real-time tasks involves monitoring and comprehending the rapidly changing situations in these tasks; that is, maintaining *situation awareness* (SA). People differ in their ability to maintain SA. In a number of experiments, we have measured these individual differences in SA abilities and also measured participants' abilities at a variety of basic cognitive tasks. Discovering the basic cognitive abilities that correlate with SA abilities provided information about the cognitive processes used in maintaining SA. Thus, this research helped refine our understanding of the global construct of SA. Also, identifying tests of the cognitive correlates of SA can help in selecting individuals who are likely to perform well in real-time tasks.

A real-time task is defined as one where the external task environment changes continuously with some changes being beyond the operator's control, and where the operator must allocate attention among multiple subtasks (time-sharing or multitasking). SA is the activated knowledge used to perform a real-time task (i.e., the knowledge in working memory or easily available to working memory, cf. Ballas, Heitmeyer, & Perez, 1992). Situation assessment refers to the cognitive processes used to maintain SA. These processes involve comprehending a dynamic, multifaceted task situation.

One of the major models of real-time task performance, Klein's recognition-primed decision model, emphasizes the importance of SA (Kaempf, Klein, Thordsen, & Wolf, 1996; Klein, 1993). This model suggests that real-time operators do very little problem solving. That is, once they have recognized (or comprehended) a situation, they consider very few response alternatives. Their responses are usually determined by simple condition–action rules (Orasanu & Fischer, 1997). Thus, the quality of operators' responses depends largely on how well they have assessed the situation.

To support this claim, naturalistic studies demonstrated that the most frequent cause of errors in real-time tasks is errors in SA (Endsley, 1995a; Hartel, Smith, & Prince, 1991). For example, a study of 420 automobile accidents found that the most frequent causes were "recognition" errors such as improper lookout, inattention, and internal distraction; errors in response selection and execution were less frequent (Shinar, 1993).

We studied SA using the real-time task of driving. Figure 12.1 presents our analysis of some of the cognitive processes involved in situation assessment during driving and includes these processes in a simple framework for driving decision making. Drivers must maintain knowledge of route location for navigation, knowledge of nearby traffic for maneuvering (local scene comprehension), knowledge of spatial orientation (e.g., lane position) for path tracking, and knowledge of their vehicle's status. For one of these driving processes, local scene comprehension, Fig. 12.1 indicates a

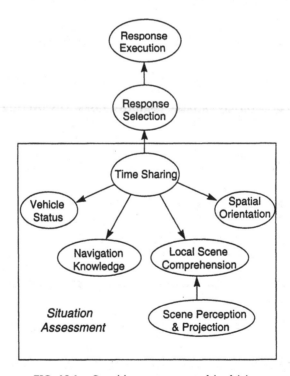

FIG. 12.1. Cognitive processes used in driving.

subprocess, scene perception and projection, that refers to a driver's ability to perceive the locations and speeds of nearby vehicles and project these into the future. In contrast to scene perception and projection, local scene comprehension focuses on a driver's ability to understand the meaning inherent in the nearby traffic (e.g., to identify potential hazards). Subprocesses for other driving processes, such as maintaining navigation knowledge, are not indicated in Fig. 12.1 because our analysis has not yet focused on these processes.

The time-sharing process in Fig. 12.1 refers to a driver's ability to allocate attention among the four situation-assessment processes. Finally, the situation-assessment processes are shown in the context of the later stages of driving decision making, response selection and execution.

Our investigation of individual differences in SA abilities focused on local scene comprehension, scene perception and projection, and time-sharing. Additionally, drivers' abilities at a variety of basic cognitive tasks were measured in order to see which of these abilities predicted SA abilities. The cognitive abilities we measured included working memory, temporal processing, visual processing, and perceptual-motor coordination. The reasoning behind the selection of these predictors is as follows:

• Working memory has a limited capacity that must be shared between current operations and temporary storage of intermediate results and freshly encoded data. It is the central bottleneck in information processing and its capacity varies considerably across individuals. Thus, SA is expected to be limited by working memory capacity.

• Visual processing is likely an important part of SA for driving in that safe drivers scan the environment by checking mirrors and the forward view to monitor the locations and actions of other vehicles.

• Temporal processing is probably a component of performance in dynamic visual environments such as driving. For example, temporal processing is probably involved in estimating velocities, maintaining a safe stopping distance between cars, and noticing and passing slower vehicles on the highway.

• Lastly, perceptual-motor coordination might be involved in SA, not because maintaining SA requires motor control, but because it requires time-sharing ability. In the SA tests used in this study, participants monitored multiple vehicles so that they could later perform a number of tasks: identifying hazardous vehicles, recalling vehicle locations, and making driving actions that avoid hazards. Monitoring multiple vehicles in order to perform multiple tasks required time-sharing. Similarly, the perceptual-motor coordination tasks we used required time-sharing because the participants had to view moving stimuli and coordinate movements of hand and feet.

## PREVIOUS RESEARCH ON INDIVIDUAL DIFFERENCES IN SA

Previous research examined the cognitive correlates of SA in aircraft pilots. For example, Carretta, Perry, and Ree (1996) reported a study in which a large battery of cognitive, perceptual, and perceptual-motor tasks were administered to 171 F-15 pilots. The dependent variable was the first unrotated principal component found on a set of peer and supervisory ratings of SA.[1] The dependent variable was predicted well by F-15 flying hours (experience), but the only individual difference variable that had any incremental validity was a general cognitive ability ($g$) composite.[2] When F-15 flying experience was partialled out of both the dependent variable and the predictor variables, significant correlations were found for working memory, divided-attention tests, and two perceptual-motor tests.

Objective, performance-based measurement of SA, perhaps in a flight simulator, would probably be much more informative. Along these lines, Endsley and Bolstad (1994) examined correlates of SA using the Situation Awareness Global Assessment Technique (SAGAT) and a battery of 18 cognitive, perceptual, and perceptual-motor tests. Because of the small sample ($N = 21$), strong conclusions about correlations with SAGAT are not warranted; however, it is interesting to note that a perceptual-motor tracking task correlated .72 with SAGAT. Endsley and Bolstad suggested that pilots with superior perceptual-motor abilities had spare attentional capacity that could be devoted to situation assessment; consequently they demonstrated better SA.

Other research has investigated SA in the context of artificial tasks that vary in the degree to which they resemble real-world tasks. In this literature, SA is often described as a high-level attention management ability distinct from more elementary cognitive processes (see e.g., Hopkin,

---

[1]Someone might question whether the Carretta et al. (1996) dependent variable actually reflected SA. One might doubt it is meaningful to ask a person to rate the quality of another person's unobservable mental events. An alternative interpretation of their dependent variable is that it reflected general airmanship (piloting skill) as perceived by other pilots who presumably had ample occasion to make informed judgments. One bit of data that might be used to support the SA interpretation of the rating scale is found in Bell and Waag (1995). They found a correlation of .6 ($N = 40$) between the SA rating scale used by Carretta et al. (1996) and ratings of SA performance in simulated fighter missions.

[2]Spearman (1904) presented the first formal theory of general cognitive ability or general intelligence. He argued that performance on ability tests was a function of a general intellectual factor underlying all cognitive performance to some extent and a specific factor unique to the task. Interestingly, Spearman (1923/1973) anticipated modern cognitive psychology when he speculated that general intelligence reflected individual differences in "mental energy" or attention.

1993). Along this line, O'Hare (1997) reported a study in which a small sample of adult males ($N$ = 24) was administered the WOMBAT Situational Awareness and Stress Tolerance Test (Roscoe, 1993; Roscoe & Corl, 1987) and tests from the Walter Reed Performance Assessment Battery that corresponded to the component tasks of the WOMBAT. The WOMBAT is a complex time-sharing task that requires the examinee to divide attention among multiple tasks on separate screens. To score well on the WOMBAT, examinees must be able to quickly assess which task has priority at any given moment and direct attention to it. O'Hare found that only one subtest, pattern-recognition,[3] was consistently correlated with WOMBAT performance through the 60-minute duration of the task. During the first 10 minutes of practice the correlation was .59 and during the last 10 minutes, .57 ($p$ < .01). O'Hare presented analyses that suggested WOMBAT performance became less dependent on computer game experience and elementary cognitive processing abilities with practice.

Our reanalysis of the data did not support this conclusion: in the final 10 minutes of task performance, both computer experience and pattern-recognition scores continued to make unique contributions to $R^2$. Given the small sample size and the small differences between predictor-criterion correlations for the first and final 10 minutes of practice, the prudent strategy would be to use the final 10 minutes of WOMBAT performance as the dependent variable. If one considers the .63 correlation between the first 10 and final 10 minutes of WOMBAT performance to be a reliability estimate, the disattenuated correlation between pattern recognition and WOMBAT performance is .72. If anything, this correlation indicates that SA (if we accept WOMBAT as a measure of SA) is related to visual recognition memory even after 60 minutes of practice at the SA task.

In complex real-time environments, correlation of task performance with cognitive ability might increase rather than decrease with practice. Rabbit, Banerji, and Szymanski (1989) reported a study in which 56 males were administered five 1-hour training sessions on the Space Fortress task over 5 days. They were also administered a standard intelligence (general cognitive ability) test. The Space Fortress task is a complex videogame that involves manipulating a spacecraft to attack a space fortress that is trying to defend itself. Perceptual-motor as well as purely cognitive abilities are required to perform the task. Performance improves slowly over hours of practice, but participants generally find it engaging. Space Fortress was designed to develop general workload-coping and attention-

---

[3]The pattern-recognition test presented a random arrangement of 16 asterisks for 1.5 seconds followed by a retention interval of 3.5 seconds. Then a second pattern is presented that has two randomly chosen asterisks changed in position. The examinee must decide if the study and test stimuli are the same or different.

management skills and there is some empirical validation that training transfers to real aircraft piloting (Gopher, Weil, & Bareket, 1992; Hart & Battiste, 1992). Rabbit et al. (1989) found that Space Fortress performance correlated .28 with general cognitive ability in the first hour of practice and .69 in the final or fifth hour of practice. The interpretation Rabbit et al. (1989) gave these data clearly links general cognitive ability to SA. High correlations between general cognitive ability test scores and game performance

> . . . may occur because people who can master most of the wide range of problems included in IQ tests can also more rapidly learn to master complex systems of rules, to attend selectively to the critical portions of complex scenarios, to make rapid and correct predictions of immanent events and to prioritize and update information in working memory. (p. 254)

Results with Space Fortress might be unfairly dismissed by some critics because the task is unrealistic, but ongoing research by W. C. Tirre that made use of a more realistic aircrew task has yielded similar results. He investigated the cognitive correlates of a synthetic task designed to simulate the B-1 defensive systems operator task (the DSO Analog task), but which did not require specialized knowledge. DSO Analog involved identifying threats (called enemies in the game) and selecting the appropriate defense. The subject's ship moved forward at a constant rate, but the subject was able to move it left or right. When an enemy came within range of the subject's ship, the enemy was able to shoot either missiles or lasers at the ship (depending on the type of enemy). Some enemies could also launch more enemies. If the subject applied the correct defense while an enemy was within range, the enemy could not attack the ship and would harmlessly pass by. The task required comprehending a complex set of rules, noticing when the automation misidentified enemies, and selecting appropriate actions in real time. As with Space Fortress, scores started out very low (even negative) and improved with hours of practice.

Tirre found that correlations of DSO performance with a working memory capacity composite score steadily climbed from .38 for the first hour to .53 for the fourth and final hour ($N = 130$). Likewise, correlations with a general cognitive ability score derived from paper-and-pencil tests climbed from .31 to .53. The reason for the increase in correlation between task performance and the ability measures is likely due to an general increase in DSO internal consistency. Internal consistency reliability increased from .76 to .92 with practice, which probably reflects a stabilization of strategies adopted by the examinees to perform the DSO task.

Joslyn and Hunt (1998) investigated the cognitive correlates of three realistic tasks that seemed to require SA because they required operators to monitor and classify a number of changing situations and allocate scarce resources to the situations based on their classifications. These tasks simulated aspects of the job of a public safety (911) dispatcher, a public safety call receiver, and an air traffic controller. In four studies, the performance of both college students and professional operators (911 dispatchers) on these realistic criterion tasks were predicted, with correlations ranging from .50 to .70, by performance on the abstract decision-making task (ADM). ADM, also developed by Joslyn and Hunt, was a dynamic, content-free task that required monitoring and classifying multiple events that overlapped in time. ADM probably measures what has been termed time-sharing ability because it required participants to prioritize and switch attention among multiple overlapping events. It also seems to measure working memory ability because participants must remember the rules for classifying events, the events that still need to be classified, and what features of those events have been identified. In a fifth study, Joslyn and Hunt (1998) found that ADM predicted both initial and well-practiced performance on the 911 dispatch task.

To summarize, previous research on the cognitive correlates of SA measured SA ability using questionnaire data (Carretta et al., 1996), performance on moderate and high-fidelity simulators (Endsley & Bolstad, 1994; Joslyn & Hunt, 1998), and synthetic laboratory tasks (O'Hare, 1997; Rabbit, Banerji, & Szymanski, 1989). Although some of these studies used too few participants to warrant strong conclusions, the general finding is that SA is related to working memory capacity, general intelligence ($g$), time-sharing ability, perceptual-motor ability, and visual-recognition ability.

In the next section, we describe the SA tests we used and then present the results of three studies of the cognitive correlates of SA. Our studies were similar to those in Endsley and Bolstad (1994), O'Hare (1997), Rabbit et al. (1989), and Joslyn and Hunt (1998) because SA was measured objectively. We measured SA in a low-fidelity driving simulator that was less complex than the high-fidelity simulator used in the Endsley and Bolstad (1994) study, but still represented a dynamic environment. The participants in our studies had a broader range of abilities than considered in either the Carretta et al. (1996) or Endsley and Bolstad (1994) studies.

## DESCRIPTIONS OF SITUATION AWARENESS TESTS

Our SA tests focused on the driving subtasks of local scene comprehension and scene perception and projection shown in Fig. 12.1. Each test required participants to monitor the movements of the nearby vehicles in a simulated driving task and to make judgments and actions based on these movements.

The driving task was performed on a PC-based driving simulator. The simulator showed 3-dimensional animated driving scenes in a window that filled a 17 inch computer screen. The participants saw the front view from the driver's perspective and the view reflected in the rearview, left-side-view, and right-sideview mirrors (see Fig. 12.2). All scenes showed traffic on a three-lane divided highway with all cars moving in the same direction. Participants watched animated scenes lasting from 18 to 35 seconds and were instructed to imagine that their simulated car was on autopilot. At the end of each scene, the participants' knowledge of the traffic vehicles was probed using a number of methods.

In the *recall probes*, the moving scene disappeared and participants indicated the locations of the traffic cars at the end of the scene on a bird's-eye view of the road, using the mouse. The bird's-eye view showed the road 20 car lengths ahead of the driver, 10 lengths behind, and the driver's car (in the correct lane). After participants finished recalling the car locations for a scene, they received feedback indicating the correct final car locations for that scene.

In the *time-to-passage probes*, the driver's car was in the center lane and was either overtaking or being overtaken by two cars in the right and left lanes. In either case, the moving scene stopped before the two traffic cars reached the driver. The participant then judged which of the two traffic cars would reach his or her car first.

In the *performance probes*, participants could make driving responses while viewing the moving scenes; that is, they could override the autopilot.

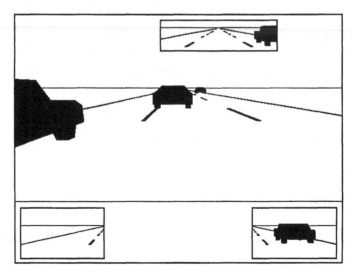

FIG. 12.2.   Three-dimensional scene from driving simulator. The actual scenes were in color.

On some trials, an incident would occur that required a driving response; for example, a car would move into the driver's lane ahead of the driver while moving slowly enough that it would hit the driver. Other hazards approached the driver from the rear. Participants could make four responses to avoid hazards: accelerate, decelerate, move to the lane on the left, or move to the lane on the right. They indicated these responses with the up, down, left, and right arrow keys, respectively. When participants pressed an arrow key during a hazard, the scene ended and feedback was displayed concerning the correctness of the response. Correct performance required the participant to avoid hazard cars without hitting any other traffic cars and to refrain from responding on catch (nonhazard) trials.

In the *scene-interpretation probes*, no driving (arrow-key) responses were required. At the end of the moving scene, the participants saw a bird's-eye view of the road that included the correct locations of the traffic cars and were asked a question that required them to identify potentially hazardous cars. The questions included: which car was driving erratically, which car was tailgating, which car is most likely to change lanes now, which car is driving fastest (or slowest) now, and which two cars are on a collision course now. Participants clicked on the car or cars that answered the question and then received feedback.

The time-to-passage probes were always done in a separate block of trials; however, the other three probes were usually done in the same block. A participant would watch a moving scene and make a driving response (performance probe) if necessary. Then, at the end of the scene, the participant might be asked to recall car locations or identify hazards. Thus, on the blocks with three types of probes, participants had to monitor the information required for all three probes on each trial. This required time-sharing ability.

We hypothesized that the recall probes and time-to-passage probes would assess participants' abilities to perceive speeds and distances and to project speed and distance information into the future so as to predict future vehicle locations. These are the abilities referred to as *scene perception and projection* in Fig. 12.1. The scene-interpretation probes and performance probes required participants to do more than perceive and project speeds and distances. In addition, participants had to comprehend the meaning of the speed and distance information in order to identify and avoid hazards. Thus, the latter two probes were hypothesized to reflect *local scene comprehension* abilities.

The specific ability measures used for each probe type were as follows. To evaluate the "goodness" of participants' recall data, a computer algorithm matched the cars recalled by participants with the actual locations of cars at the end of each trial and also identified omissions (i.e., nonrecalled cars) and intrusions (i.e., nonexistent cars that were recalled). Once the matching was done for a trial, the *composite recall error* was calculated based on the average distance between recalled cars and the actual cars they were matched to and on the number of omissions and intrusions (see Gugerty, 1997, for details concerning how this and other SA measures were calculated).

The time-to-passage probe data were scored in terms of the percentage of trials on which participants answered correctly. The scene-interpretation data were scored in the same manner.

Two SA measures were derived from the performance-probe data. The first was *hazard detection*. This was calculated using the $A'$ nonparametric, signal-detection measure of sensitivity (Grier, 1971). On each signal (hazard) trial, a response interval was defined as beginning when a car entered the driver's lane on a trajectory that would hit the driver and ending when it was too late for the driver to avoid the oncoming car. Following the procedure of Watson and Nichols (1976) for measuring sensitivity with continuous signal-detection tasks, we defined catch-trial response intervals that were equal in duration to those on hazard trials. A hit was defined as any arrow-key response, even an incorrect response, during the response interval of a hazard trial. A false alarm was any arrow-key response during the response interval of a catch trial. For all trials, responses before the response interval, which were infrequent, were ignored in this analysis.

We attempted to construct the hazard detection measure so that it assessed SA ability and not decision-making or response-execution ability even though any performance-based measure is affected by all stages of the perception-action cycle. When participants responded incorrectly to a hazard car, this showed that they were aware of the hazardous situation, but selected and executed an inappropriate avoidance response. Therefore, by defining even incorrect responses to hazards as hits in this measure, we hoped that it would reflect the participants' ability to detect hazards (an aspect of SA) more than their decision-action abilities.

The hazard-detection measure focuses on the participants' awareness of vehicles in front of and behind their car because the hazardous cars always entered the driver's lane from a side lane and then approached the driver. The second SA measure derived from the performance-probe data, *blocking-car detection*, focused on the participants' awareness of blocking cars to their immediate right and left. These cars were usually completely within the participants' blind spot. Participants could usually only know about blocking cars by remembering that a car had entered the blind spot and had not left it. On a trial where the hazard car approached from the front and there were blocking cars to the right and left, participants who tried to avoid the hazard car by moving right or left would miss one of the blocking cars but hit the other blocking car. In this case, participants were considered as detecting one of two blocking cars (i.e., a 50%

blocking-car detection score). Overall, blocking-car detection was estimated by the ratio of the total number of blocking cars avoided over the total number of blocking cars on hazard trials.

As in the hazard-detection measure, scoring high on the blocking-car detection measure does not depend on making a correct response in terms of global task performance. In the above example, participants would be credited with 50% blocking-car detection on a trial where they crashed. Thus, blocking-car detection should reflect participants' awareness of blocking cars more than their decision-action processes.

## EXPERIMENT 1

This study was conducted in conjunction with a larger factor analytic study of the Cognitive Abilities Measurement (CAM) battery (Kyllonen, 1994). CAM attempts to comprehensively measure the human abilities essential to the acquisition of knowledge and skill.

### Method

*Participants.* Participants were hired from temporary employment agencies. In the larger study, 230 participants of both sexes between the ages 18 to 30 were administered the CAM and the Armed Services Vocational Aptitude Battery (ASVAB) over 5 days. A subset ($N = 34$) was also administered the driving simulator task.

*Abilities Tests.* In the driving simulator session, participants completed 84 trials with both performance and recall probes and 84 trials with only recall probes.[4] CAM, a battery of 59 computer-administered tests, was created through the use of a taxonomy. The six rows of the taxonomy reflect the major abilities suggested by cognitive psychology (viz., working memory, processing speed, induction, declarative knowledge, declarative (associative) learning, and procedural (skill) learning). The columns reflect three information domains suggested by psychometric analyses (viz., verbal, quantitative, and spatial). Three or four tests were given for each cell of the taxonomy. The ASVAB is the entrance examination for the U.S. armed services. It is a paper and pencil test that includes ten subtests: arithmetic reasoning, math knowledge, numerical operations, coding speed, electronics information, auto and shop, mechanical comprehension, general science, paragraph comprehension, and word knowledge.

---

[4]The scene interpretation and time-to-passage probes had not been developed at the time of this study.

**Procedure.** Participants were tested in groups of about 20 who completed the tests over a week-long period.

### Results and Discussion

As discussed in Gugerty (1997, Experiment 1), participants' scores for the driving simulator tests were reliable and above chance. Test battery scores for the cognitive abilities tests were reduced to a manageable number in two steps. First, for the 230 participants in the full study, 18 cell scores for the 6 rows and 3 columns of the CAM taxonomy were formed (e.g., all spatial working memory tests were combined into one score). Second, the 18 CAM and 10 ASVAB scores were factor analyzed in separate runs. Factor analysis resulted in three CAM factors—$g$/working memory ($g$/WM), processing speed, and declarative knowledge—and two ASVAB factors—$g$ and clerical speed. Each factor set was orthogonally rotated using the quartimax procedure, which emphasizes the amount of variance explained by the first factor. The first CAM factor, $g$/WM, had its highest loadings on the working memory and procedural learning tasks. This finding is consistent with prior research (e.g., Kyllonen & Christal, 1990; Tirre & Pena, 1993) that indicated working memory and tasks with heavy working memory requirements might be the core of the psychometric phenomenon known as $g$. The first ASVAB factor had its highest loadings on tests requiring quantitative reasoning skills (viz., arithmetic reasoning and math knowledge, both of which benefit from education). As such, ASVAB $g$ reflects crystallized instead of fluid ability (Cattell, 1971).

The SA measures were highly correlated with the CAM $g$/WM factor and slightly less so with the ASVAB $g$ factor. Hazard detection was correlated with $g$/WM ($r = .60, p < .001$) and with ASVAB $g$ ($r = .39, p < .03$). Likewise, blocking-car detection correlated with CAM $g$/WM ($r = .74, p < .001$) and with ASVAB $g$ ($r = .76, p < .001$). Composite recall error correlated with $g$/WM ($r = -.73, p < .001$) and with ASVAB $g$ ($r = -.50, p < .003$). None of the remaining ability factors correlated significantly with the SA measures. These results are consistent with the hypothesis that SA depends critically on the working memory system (Endsley, 1995b).

### EXPERIMENT 2

The second experiment was conducted as part of a larger factor analytic study of cognitive, perceptual-motor, and temporal processing abilities. Each participant completed a session in the driving simulator. In addition, computer-based batteries of cognitive, perceptual-motor and temporal tests, and one paper-and-pencil battery (the Air Force Officer Qualifying

Test, AFOQT) were administered. The AFOQT includes 16 tests that measure general ability plus 7 group factors: spatial, verbal, quantitative, visual search, technical knowledge, and science and math (Goff & Tirre, 1995).

## Method

*Participants.* Participants were hired from temporary employment agencies. The 88 participants included 61 males and 27 females ranging in age from 18 to 30 years.

*Abilities Tests.* In the driving simulator session, participants completed 148 trials with only performance probes. In this experiment, we were interested in whether the frequency of hazards in the performance probes affected the participants' ability to detect hazards. There were two groups, one that experienced hazards on 75% of the trials and a second that experienced hazards on only 25% of the trials. Both groups first experienced practice in which hazards occurred on 50% of the trials. As it turns out, hazard frequency did not have a significant effect on any of the dependent variables (Gugerty & Tirre, 1996), so we collapsed across groups for the correlational analyses.

The computer-administered cognitive battery was a subset of the CAM 4.1 battery consisting of tests of working memory, processing speed, induction, declarative knowledge, declarative learning, and procedural learning (Kyllonen, 1994). We used only the working memory tests in our analysis. These consisted of a spatial test where participants memorized, combined, and then recalled stick figures; a quantitative test where participants studied lists of numbers, made computations with them, and then recalled them; and a verbal test where participants studied a series of short sentences and then created word lists based on the relations in the sentences.

The perceptual-motor battery consisted of 17 tests designed to measure 4 of the Fleishman factors: multilimb coordination, control precision, rate control, and response orientation (Fleishman & Quaintance, 1984). In the analysis for this experiment, we used the multilimb coordination tests. In the center-the-ball task, the participant adjusted the horizontal motion of a drifting circle with foot pedals and the vertical motion with the joystick so as to keep the circle over a target. In "pop the balloons," the participant used the foot pedals and joystick to move a sight over a moving target circle and then "popped" the balloon with a trigger press. In the Mashburn task, participants used the joystick and foot pedals to move and maintain three cursors so that each was at a separate target point. Each trial was completed when a 2-second period had elapsed in which all three cursors were at their target points. In other tasks, participants used both hands (two joysticks) to track a moving object and to move an object along a path.

In the temporal battery, participants estimated when a growing line or an ascending digital clock would reach a target, selected which of two growing lines or ascending clocks would reach a target first, and estimated when short time intervals had ended.

Visual search ability was estimated based on three subtests of the AFOQT: table-reading, block-counting, and scale-reading. Each of these subtests appears to require the participant to visually scan a printed stimulus (a table of numbers, a picture of stacked blocks, or a scale with markings) for a certain item or set of data. Each subtest requires attention to visual detail and might be regarded as an occulomotor test because eye movements must be carefully controlled.

*Procedure.*    Participants were tested in groups of about 20 who completed the test batteries over a week-long period. Each battery was given on a fixed day or days. The order of tests within a battery was randomized for each participant.

**Results and Discussion**

Because the participant to variable ratio was too small to warrant factor analysis of the total dataset, we created composite scores corresponding to factors found in previous analyses of the CAM 4.0, perceptual-motor, temporal, and AFOQT batteries in which sample size was no problem. Considering CAM first, we created a composite for working memory that corresponded to a factor found by Kyllonen (1993). For the perceptual-motor battery we simply created a multilimb coordination composite corresponding to Fleishman and Quaintance's (1984) factors because we had deliberately attempted to simulate original apparatus tests used by Fleishman in his factor analytic research. We made a separate temporal processing composite because Tirre (1997) found that a similar factor loaded by two of the temporal processing tests was distinct from four perceptual-motor factors including multilimb coordination. Based on Goff and Tirre (1995), the visual search composite was created from the table-reading, scale-reading, and block-counting subtests of the AFOQT.

The SA performance measures, hazard detection and blocking-car detection, were significantly intercorrelated ($r = .42, p < .001$). The correlations of the four ability predictors with the SA dependent variables (see Table 12.1) indicate that: hazard detection is correlated significantly with all predictors, with working memory having the highest correlation, and blocking-car detection is correlated significantly with three of the four predictors. However, substantial intercorrelations of the predictor vari-

TABLE 12.1
Correlations Between Predictor and SA Measures
for Experiment 2 ($N = 88$)

| | SA Measure | |
| --- | --- | --- |
| Predictor | Hazard Detection | Blocking-Car Detection |
| Working memory | .44*** | .47*** |
| Multilimb coordination | .31** | .37*** |
| Visual search | .22* | .21 |
| Temporal processing | .35*** | .31** |

*$p < .05$. **$p < .01$. ***$p < .001$.

ables and the small sample size suggest caution in interpreting these correlational patterns.

Multiple regression analyses with the dependent variables reflect the pattern displayed in Table 12.1. That is, working memory was the only significant predictor ($p < .05$) in equations for hazard detection ($R = .46$), and both working memory and multilimb coordination were significant predictors of blocking-car detection ($R = .51$). These two predictor variables were correlated ($r = .47$), and so we might expect a substantial common contribution from them in a regression equation. In a two-variable equation predicting hazard detection, the unique contributions of working memory and multilimb coordination to the explained variance were 11% and 1%, respectively; while the common contribution was 8%. The equation predicting blocking-car detection showed a similar pattern; the unique contributions of working memory and multilimb coordination to the explained variance were 11% and 3%, respectively; while the common contribution was 10%.

In summary, the findings suggest that SA ability, as measured in this study, depends primarily on working memory and multilimb coordination ability and to a lesser extent on temporal processing and visual search ability. As mentioned above, multilimb coordination may predict the nonmotor SA tasks because it taps into participants' time-sharing ability.

However, there are alternative interpretations of the Experiment 2 findings. The multilimb coordination measures all used dynamic visual stimuli, and thus the correlation between these tests and the SA tests could have been due to the dynamic visual processing required for both types of tests instead of any common time-sharing requirements. Similarly, because the temporal processing tests in Experiment 2 also used dynamic visual stimuli, they could have correlated with SA because of shared visual rather than temporal processing requirements.

In Experiment 3, we attempted to determine to what extent the correlation of both multilimb coordination and temporal processing with SA was due to the visual nature of these tests. For temporal processing, we did this by using auditory tests of temporal processing. If, using these auditory tests, we still find that temporal processing ability predicts SA, then we will have good evidence that this relation is based on temporal and not visual processing ability.

We could not use the same approach with multilimb coordination because we do not yet have available nonvisual tests of multilimb coordination. To determine whether multilimb coordination predicts SA due to its time-sharing or dynamic visual components, we added tests of dynamic visual processing to our battery of cognitive predictors. If we find that dynamic visual processing predicts SA independently of multilimb coordination, this will provide evidence that the multilimb coordination tests predict SA because of their time-sharing rather than visual requirements.

The new tests of dynamic visual processing may also help us understand the relatively low correlation of visual processing with SA in Experiment 2. Experiment 2 found a significant but low correlation between visual search and one SA measure (hazard detection). One possible reason for the low correlation between visual processing and the highly visual SA tasks could be that the visual tests we used required processing of static, unchanging stimuli (e.g., searching for a target in a visual array) whereas the SA tasks required processing of dynamic, changing visual stimuli. Researchers have found that visual-spatial processing of static and dynamic stimuli are distinguishable abilities (Hunt, Pelligrino, Frick, Farr, & Alderton, 1988). Thus, we hypothesized that the dynamic visual processing tests would predict SA independently of static visual processing and that the dynamic tests would be better predictors than the static ones.

## EXPERIMENT 3

In the third experiment, we used the SA tests from Experiments 1 and 2, the performance and recall probes, and added two tests to the SA battery. These were the time-to-passage and scene interpretation tests described earlier. In terms of the SA model in Fig. 12.1, this expanded SA battery included two measures of scene perception and projection and three of local scene comprehension. Also, because the scene interpretation task was done concurrently with the performance and recall tasks, the new battery increased the demands for time-sharing ability.

In terms of cognitive ability tests, we assessed the same four factors as in Experiment 2: working memory, visual, temporal, and perceptual-motor ability. For working memory and perceptual-motor ability, we used tests from the CAM, ASVAB, and Fleishman batteries used in Experiments 1 and 2. However, as mentioned previously, we used auditory instead of visual tests to assess temporal ability and, in addition to tests of static visual processing, added two tests of dynamic visual processing.

## Method

*Participants.* Participants were U.S. Air Force recruits tested during basic training. The 129 participants included 64 males and 65 females ranging in age from 17 to 35 years with a mean of 20.1 years.

*Abilities Tests.* In the driving simulator session, participants first completed 32 time-to-passage probes. Then they completed a block of 70 trials, including 30 with only performance probes, 20 with both performance and scene interpretation probes, and 20 with both performance and recall probes. The order of trials in this block were randomized separately for each participant. Because of this, participants had to be prepared for the performance, recall, and scene-interpretation tasks on each trial.

The multilimb coordination tests, center-the-ball and pop-the-balloons, were described in Experiment 2. The variable representing center-the-ball performance was mean tracking error distance. For pop-the-balloons, the variable was elapsed time to pop all balloons. The spatial working memory test was similar to the one used in Experiment 2. The participants' ASVAB scores were also obtained.

The first auditory temporal test focused on duration discrimination. On each trial, participants heard an 800 msec tone and either a 620, 692, 764, 836, 908, or 980 msec tone and judged which tone was longer. Tests were scored in terms of percent correct on the four easier discriminations because participants performed at chance for the two difficult discriminations. In the second test, rate extrapolation, participants heard a series of three or five beeps at one of three rates and then had to press a key when the tenth beep in the series would occur. Performance was scored in terms of the percentage of key presses that were within .5 beats of the tenth beat.

In the first static visual search task, spread-out-search, participants searched a three by three array of nine 2-digit numbers, each separated by 9.5 cm, to determine whether any of the numbers matched a target number. The second task was a computerized version of the AFOQT table-reading task. The third was the coding speed test taken from the participants' ASVAB scores. On each item of this test, participants had to determine which of five 4-digit number strings matched a word. To determine which number string matched the word, participants searched a key at the top of the page that consisted of 10 word-number pairs.

In the first dynamic visual test, direction detection, participants saw objects briefly moving across the screen and then, using a response wheel with 16 spokes, indicated the direction in which the object had been moving. In the second test, the road-sign test, participants saw very small letters, numbers, or figures that increased in size. As soon as the object could be identified, the participants pressed a key and then selected the object from a set of distracters.

*Procedure.* Participants were tested in groups of about 20 who completed the test batteries in a 3.5 hour period. The order of tests was: driving simulator time-to-passage, second driving simulator test, and then cognitive abilities tests. The order of tests within the cognitive abilities battery was randomized for each participant.

**Results and Discussion**

*SA Tests.* Table 12.2 shows how participants performed on the SA tests and the test reliabilities. In this experiment, we used a different measure for hazard detection. Instead of the sensitivity measure used in the previous experiments, hazard detection was estimated by the participants' rate of responding to hazards (hit rate). We ignored the false alarm rate and did not calculate sensitivity because the number of catch trials was too low to reliably estimate the false alarm rate. Also, three of the items in the scene interpretation test were changed slightly after the first data collection session ($N = 28$). The scene interpretation percent correct scores for these participants were estimated from the 17 items that remained unchanged.

The SA scores showed moderate intercorrelations. After switching signs so that positive scores always reflected better performance, the correlations ranged from .10 to .47 with a median of .34. Factor analysis of the SA scores did not reveal the 2-factor structure hypothesized in Fig. 12.1. Instead we found that a single factor adequately described the SA data ($\chi^2(5, N = 111) = 5.51$, $p = .36$ for the 1-factor model, 45% of the variance explained).[5] Because of this, we created a *composite SA score* that was the average of the standardized scores of the five SA tests with the sign of the standardized composite recall error switched.

*Cognitive Correlates of SA.* Our goal was to assess the relations between SA and each of the cognitive, perceptual, and perceptual-motor tasks and determine if the abilities measured by these tasks made unique contributions to the prediction of SA in the driving simulator. We began with a factor analysis of the tests intended to measure perceptual-motor processing, temporal processing, static visual processing, and dynamic visual processing. To enable the factor analysis program to make fine distinctions among the perceptual and motor tests, we included multiple in-

---

[5]If a second factor was extracted (eigenvalue = .94) and an oblique rotation was selected, the factors correlated about .6. For simplicity, we decided to accept the single factor model.

TABLE 12.2
Performance on SA Tests for Experiment 3

| Measure (units) | M | (SD) | Reliability |
|---|---|---|---|
| Time to passage (percent correct) | 76 | (11) | .63[a] |
| Composite recall error (car lengths) | 0.58 | (0.58) | .80[b] |
| Hazard detection (percent correct) | 86 | (12) | .80[b] |
| Blocking-car detection (percent correct) | 70 | (15) | na |
| Scene interpretation (percent correct) | 51 | (15) | .60[a] |

[a]Cronbach's alpha.
[b]Corrected even-odd reliability.

dicators (two to four) for some of the tests in the analysis. For example, road-sign recognition time scores for letter, figure, and number stimuli were kept separate, and duration discrimination scores for the four easy duration comparison conditions were kept separate.

We excluded certain ASVAB subtests and the spatial working memory test from this analysis and used them as indicators of $g$/WM in a separate factor analysis. We estimated $g$/WM as the first unrotated principal axis factor involved in arithmetic reasoning, math knowledge, paragraph comprehension, word knowledge, mechanical comprehension, and spatial working memory. Note that in this set of tests there were two tests for each of the quantitative, verbal, and spatial domains.

The factor analysis of the perceptual-motor, temporal processing, and visual processing tests was performed in the exploratory mode. To determine the number of factors, we looked for convergence among three criteria: the scree test, the Kaiser-Guttman eigenvalue test, and the maximum likelihood test. Both the scree and the eigenvalue test indicated five factors, but the maximum likelihood test indicated six ($\chi^2(86, N = 111) = 117.70, p = .013$ for five factor model). The five factor solution (see Table 12.3) was rotated using an oblique method. We interpreted the factors as multilimb coordination (errors and performance time), auditory duration discrimination (accuracy), auditory rate extrapolation (accuracy), dynamic visual processing (time), and visual search (accuracy and rate). As shown in Table 12.4, the factors were weakly intercorrelated (median $r = .15$), but the two temporal processing factors were correlated .32. The only factor showing substantial correlation with $g$/WM was multilimb coordination ($r = -.52$), a negative correlation because motor ability was indexed by errors and performance time. This correlation replicates previous research by Chaiken, Tirre, and Kyllonen (1996).

Our primary interest was in how $g$/WM and these five factors combined to predict SA. We computed factor scores using the Anderson-Rubin method which yields uncorrelated factor score estimates. Each of the factors was correlated significantly with the SA composite score (see Table

TABLE 12.3
Factor Analysis of the Perceptual-Motor,
Temporal Processing, and Visual Processing Tests ($N = 111$)

| | Multilimb Coordination | Auditory Duration Discrimination | Auditory Rate Extrapolation | Dynamic Visual Processing (Road Sign) | Visual Search |
|---|---|---|---|---|---|
| Center-the-ball (Block 3) mean error distance | .87 | | | | |
| Center-the-ball (Block 4) mean error distance | .86 | | | | |
| Pop-the-balloons (Block 2) time | .68 | | | | |
| Pop-the-balloons (Block 3) time | .75 | | | | |
| Auditory duration discrimination (800 vs. 620 msec) accuracy | | .73 | | | |
| Auditory duration discrimination (800 vs. 692 msec) accuracy | | .66 | | | |
| Auditory duration discrimination (800 vs. 908 msec) accuracy | | .67 | | | |
| Auditory duration discrimination (800 vs. 980 msec) accuracy | | .52 | | | |
| Auditory extrapolation (Rate 1) accuracy | | | .77 | | |
| Auditory extrapolation (Rate 2) accuracy | | | .84 | | |
| Auditory extrapolation (Rate 3) accuracy | | | .83 | | |
| Road sign recognition (letters) time | | | | .88 | |
| Road sign recognition (figures) time | | | | .95 | |
| Road sign recognition (numbers) time | | | | .92 | |
| Spread-out search time | | | | | -.38 |
| ASVAB coding speed rate | | | | | .37 |
| AFOQT table reading rate | | | | | .41 |
| Direction detection (even difficult blocks) accuracy | -.49 | | | | .50 |
| Direction detection (odd difficult blocks) accuracy | -.43 | | | | .51 |

Note. Loadings < .30 omitted.

## TABLE 12.4
Factor Analysis of the Perceptual-Motor, Temporal Processing, and Visual Processing Tests:
Factor Intercorrelations and Correlations With Other Tests ($N = 111$)

| | Multilimb Coordination | Auditory Duration Discrimination | Auditory Rate Extrapolation | Dynamic Visual Processing (Road Sign) | Visual Search |
|---|---|---|---|---|---|
| **Factor Intercorrelations** | | | | | |
| Multilimb coordination | 1.00 | | | | |
| Auditory duration discrimination | -.24 | 1.00 | | | |
| Auditory rate extrapolation | -.09 | .32 | 1.00 | | |
| Dynamic visual processing | .23 | -.19 | -.04 | 1.00 | |
| Visual search | -.25 | .12 | -.02 | -.07 | 1.00 |
| **Correlations with** | | | | | |
| General cognitive ability (g/WM) | -.52** | .20* | .09 | -.12 | .24* |
| Driving SA | -.54** | .30** | .18* | -.29** | .18* |

*$p < .05$. **$p < .01$. $r(g/WM, SA) = .445$.

269

12.4) with the highest correlations coming from multilimb coordination (errors and performance time; $-.54$) and $g$/WM (.45). We then regressed the composite SA score on $g$/WM and the five factors using the simultaneous inclusion method (see Table 12.5). The six predictors together accounted for 52.9% of the variance in the SA composite score ($R = .73$, $R_{adj} = .71$). All predictors except $g$/WM made significant unique contributions to the equation. The primary contest for criterion variance was between $g$/WM and multilimb coordination. When these two variables were used to predict the SA composite score in a 2-variable equation, multilimb-coordination explained 13% of the variance and $g$/WM explained 4%. The common contribution by $g$/WM and multilimb coordination was 16%.

The results of the regression analysis are consistent with the idea that SA is a product of several cognitive and perceptual processes—time-sharing (indicated through multilimb coordination), temporal processing, and visual processing. Experiment 3 also provided information about the reasons behind the correlations of temporal processing and multilimb coordination with SA that were found in Experiment 2. The fact that both auditory temporal processing tests made significant contributions to predicting SA suggests that temporal processing is an important part of SA and that the correlation between temporal processing and SA in Experiment 2 was not solely due to shared visual processing.

Concerning multilimb coordination, we hypothesized earlier that it is strongly related to SA either because both of these tasks require time-sharing capacity or because both tasks require dynamic visual processing. The evidence from Experiment 3 appears to favor the first hypothesis because multilimb coordination and both the dynamic and static visual processing factors all made unique contributions to the explained variance in SA. It should be noted that we do not possess direct evidence to support the interpretation of multilimb coordination as reflecting time-sharing capacity. However, Baddeley (1993) described working memory as involving the management of cognitive resources or time-sharing, and the substantial

TABLE 12.5
Regression Analysis: Prediction of Driving SA Composite ($N = 106$)

| Variable | Beta | $r$ | Part $r$ | Partial $r$ | $T$ |
|---|---|---|---|---|---|
| General cognitive ability ($g$/WM) | .02 | .45 | .02 | .02 | 0.23 |
| Multilimb coordination | −.53 | −.54 | −.44 | −.54 | −6.38** |
| Auditory duration discrimination | .30 | .30 | .29 | .39 | 4.17** |
| Auditory rate extrapolation | .18 | .18 | .18 | .25 | 2.59* |
| Dynamic visual processing | −.29 | −.29 | −.28 | −.38 | −4.12** |
| Visual search | .17 | .18 | .17 | .24 | 2.42* |

*Note.* $R = .73$. $R_{adj} = .71$. $F(6, 99) = 18.56$. $p < .0001$.
*$p < .05$. **$p < .001$.

correlation between $g$/WM and multilimb coordination is what we would expect if multilimb coordination involved time-sharing as well.

Finally, Experiment 3 provided mixed evidence on the question of whether dynamic visual processing predicts SA independently of static visual processing. One dynamic visual test, the road-sign test, was grouped by the factor analysis in a separate factor from the static visual tests and was found by the regression analysis to predict SA independently of the static tests.

The other dynamic visual test, direction detection, was grouped by the factor analysis with both dynamic tests (viz., pop-the-balloons and center-the-ball) and static visual search. This suggests that direction detection involves two visual components: a dynamic visual component distinct from the component involved in the road-sign test, but required by visual tracking tasks such as center-the-ball and pop-the-balloons and a static visual component similar to that required in visual search tasks. If we accept this interpretation of variance on the direction detection test, we can conclude that both dynamic and static visual processes are involved in SA.

We had also hypothesized that dynamic visual tests would be better predictors of SA than static visual tests. However, the correlation between SA and the dynamic visual processing (road-sign) factor ($r = .29$) was not significantly higher than that between SA and the visual search factor that included the static tests ($r = .18$, $t(108) = 0.88$, $p < .20$).

## GENERAL DISCUSSION

In these studies we measured SA using a low-fidelity driving simulator in which participants were required to recall car locations and to identify and avoid hazards. Cognitive abilities were measured using test batteries that focused on working memory capacity and processing speed in Experiment 1, and on working memory and visual, temporal, and perceptual-motor processing in Experiments 2 and 3.

Only working memory was found to be strongly related to SA in Experiment 1, but statistical power was low due to a small sample size and relationships with other predictors may have been obscured. In Experiment 2, we employed a larger sample and found that working memory was strongly predictive of both SA dependent variables employed. We also found that multilimb coordination added to working memory in predicting one SA dependent variable (blocking-car detection). Multilimb coordination, interpreted as an indicator for time-sharing ability, and working memory each made unique contributions to the explained variance in SA. However, their shared or common contribution was also substantial, indicating that there is a high degree of commonality between these variables as Baddeley (1993) suggested.

On the surface, the results of Experiment 3 might appear to contradict the tentative conclusion that working memory plays a substantial role in SA. In Experiment 3 we found that $g$/WM failed to make a significant unique contribution to explanation of SA variance. Unfortunately, we did not measure $g$/WM in quite the same manner as in Experiments 1 and 2, relying instead on the $g$ component found among selected ASVAB tests along with a spatial working memory measure. This decision probably weakened our measure of working memory to some degree. Multilimb-coordination overlapped substantially with $g$/WM and in regression equations may have "stolen" some of the variance $g$/WM would have explained on its own. The shared contribution of these two variables was larger than either variable's unique contribution.

Working memory, as it is the central bottleneck in controlled cognitive processing, would be expected to be involved in conscious processing leading to SA; but, as noted by other researchers, SA might also have an automatic component that might not be limited by working memory (Kennedy & Ordy, 1995; Orasanu & Fischer, 1997). This automatic component of SA merits further investigation.

Visual processing was found to predict SA ability in Experiments 2 and 3. The evidence suggests that both dynamic (e.g., road-sign recognition) and static (e.g., visual search) processing factors are involved in SA. Our understanding of the visual processing factors operating in the dynamic driving environment is admittedly incomplete, and we hope to refine our measurement of these abilities in future research. Temporal processing predicted SA ability in Experiments 2 and 3 regardless of whether temporal ability was measured by visual or auditory tests; and Experiment 3 demonstrated that temporal processing might be multicomponential itself. Two temporal factors were found in the factor analysis and each contributed significantly to the prediction of SA.

Our findings that SA ability is correlated with $g$, working memory ability, and perceptual-motor ability fit with the findings of Carretta et al. (1996) and Endsley and Bolstad (1994) as well as with our findings that the real-time DSO task is predicted by $g$. Our finding that SA ability is correlated with static visual processing fits with O'Hare's (1997) finding that SA is predicted by visual recognition ability. In addition, we also found that SA ability is correlated with dynamic visual processing ability and temporal processing ability.

As the next steps in our research program, we plan to pursue two streams simultaneously. One involves refining our measures of the cognitive correlates of SA and exploring relations among these. As suggested earlier, we believe that time-sharing, dynamic visual processing, and temporal processing abilities each need further research.

The second line of research involves refining our measurement of SA. The next version of the driving simulator will introduce interactive driving and multitasking to increase realism and generalizability to automobile driving. Research participants will be able to steer, brake, accelerate, and navigate their simulated cars through a simulated city environment. The multitasking requirement will be increased through introduction of navigation and communication tasks. Imagine the workload of a sales representative trying to find a location in an unfamiliar neighborhood who must respond to calls from the office on his cellular phone. We can simulate these conditions and measure the effect on SA.

An important part of refining our SA measures will involve validating that these measures predict performance of real-time tasks under realistic conditions. This validation is necessary to support the claims made earlier that maintaining SA is a crucial part of real-time task performance. In our case, this would require predicting performance in on-the-road driving. As mentioned earlier, Gopher et al. (1992) and Hart and Battiste (1992) found that practice on Space Fortress benefitted subsequent aircraft piloting performance. Similarly, performance on the WOMBAT test of SA differentiated between elite pilots and nonpilots (O'Hare; 1997).[6]

SA is a concept that first emerged in the aviation community and recently, human factors researchers have extended this concept to many domains of human performance. We initially chose driving as a simpler and more common analogue to flying, thinking that if we could first understand SA in driving, we would have a good foundation for understanding SA in flying. Our plans are to examine correlations between measures of SA in the driving simulator and measures of SA in other simulated environments (e.g., air combat and air traffic control). Confirmatory factor analysis will be used to test alternative structural models of the cognitive processes involved in SA. Also in this research, we will determine if similar predictors (cognitive correlates) emerge for SA in different environments.

## ACKNOWLEDGMENTS

This research project was performed at the U.S. Air Force Armstrong Laboratory, Cognitive Technologies Branch (HRMC) with support from Dr.

---

[6]To date we have not conducted a validation study of our SA measures. However, we have found that these SA measures correlate (maximum $r = .83$, median $r = .64$) with performance on the useful-field-of-view test (UFOV); (Chaparro, Groff, Tabor, Sifrit, & Gugerty, 1999); and the UFOV test is known to predict on-the-road driving performance (e.g., accidents) quite well (Ball, Owsley, Sloane, Roenker, & Bruni, 1993).

John Tangney at the Air Force Office of Scientific Research (grant PE 61102F, work unit 2313T147, AFOSR task 2313/BA (LAMP), LRIR #92AL009). Part of the research was performed under a fellowship from the U.S. National Research Council to Leo Gugerty. We would like to thank Jamie Burns, Marvin McFadin, In Soo Park, Roger Dubbs, Karen Rouf, and Terry Simmons for programming the driving simulator and other tests; Janice Hereford for data analysis; and Ralph Salyer and Candee Dietz for administering tests.

## REFERENCES

Baddeley, A. (1993). Working memory or working attention? In A. Baddeley & L. Weiskrantz (Eds.), *Attention: Selection, awareness, and control*. Oxford, England: Clarendon.

Ball, K., Owsley, C., Sloane, M. E., Roenker, D. L., & Bruni, J. R. (1993). Visual attention problems as a predictor of vehicle crashes in older drivers. *Investigative Ophthalmology & Visual Science, 34*(11), 3110–3123.

Ballas, J., Heitmeyer, C., & Perez, M. (1992). Evaluating two aspects of direct manipulation in advanced cockpits. In *Proceedings of the Computer-Human-Interaction Conference (CHI '92)*. New York: Association for Computing Machinery.

Bell, H. H., & Waag, W. L. (1995). Using observer ratings to assess situational awareness in tactical air environments. In D. J. Garland & M. R. Endsley (Eds.), *Experimental Analysis and measurement of situation awareness* (pp. 93–99). Daytona Beach, FL: Embry-Riddle Aeronautical University Press.

Carretta, T. R., Perry, D. C., Jr., & Ree, M. J. (1996). Prediction of situational awareness in F-15 pilots. *The International Journal of Aviation Psychology, 6*, 21–41.

Cattell, R. B. (1971). *Abilities: Structure, growth, and action*. Boston: Houghton Mifflin.

Chaiken, S., Tirre, W., & Kyllonen, P. (1996, August). *Cognitive and psychomotor determinants of complex real-time performance*. Paper presented at the Symposium: New Approaches for Psychomotor Assessment—Theory and Application, American Psychological Association, Toronto, Ontario, Canada.

Chaparro, A., Groff, L., Tabor, K., Sifrit, K. & Gugerty, L. (1999). Maintaining situational awareness: The role of visual attention. In *Proceedings of the Human Factors Society 43rd Annual Meeting, Vol. 2* (pp. 1343–1347). Santa Monica, CA: Human Factors Society.

Endsley, M. R. (1995a). A taxonomy of situation awareness errors. In R. Fuller, N. Johnston, & N. McDonald (Eds.), *Human factors in aviation operations* (pp. 287–292). Aldershot, England: Ashgate Publishing, Ltd.

Endsley, M. R. (1995b). Toward a theory of situation awareness in dynamic systems. *Human Factors, 37*(1), 32–64.

Endsley, M. R., & Bolstad, C. A. (1994). Individual differences in pilot situation awareness. *International Journal of Aviation Psychology, 4*, 241–264.

Fleishman, E. A., & Quaintance, M. K. (1984). *Taxonomies of human performance: The description of human tasks*. Orlando, FL: Academic Press.

Goff, G. N., & Tirre, W. C. (1995). *Confirmatory factor analysis of the Air Force Officer Qualifying Test, Forms O, P1 and P2*. Unpublished manuscript.

Gopher, D., Weil, M., & Bareket, T. (1992). The transfer of skill from a computer game trainer to actual flight. In *Proceedings of the Human Factors Society 36th Annual Meeting, Vol. 2* (pp. 1285–1290). Santa Monica, CA: Human Factors Society.

Grier, J. (1971). Nonparametric indexes for sensitivity and bias: Computing formulas. *Psychological Bulletin, 75*, 424–429.

Gugerty, L. (1997). Situation awareness during driving: Explicit and implicit knowledge in dynamic spatial memory. *Journal of Experimental Psychology: Applied, 3*, 1–26.

Gugerty, L., & Tirre, W. (1996). Situation awareness: A validation study and investigation of individual differences. In *Proceedings of the Human Factors Society 40th Annual Meeting, Vol. 1* (pp. 564–568). Santa Monica, CA: Human Factors and Ergonomics Society.

Hart, S., & Battiste, V. (1992). Field test of video game trainer. In *Proceedings of the Human Factors Society 36th Annual Meeting, Vol. 2* (pp. 1291–1295). Santa Monica, CA: Human Factors Society.

Hartel, C., Smith, K., & Prince, C. (1991). *Defining aircrew coordination: Searching for mishaps with meaning*. Paper presented at the sixth International Symposium on Aviation Psychology. Columbus, OH.

Hopkin, D. (1993, February). *Situation awareness in air traffic control*. Paper presented at the Center for Applied Human Factors in Aviation Conference Situation Awareness In Complex Systems, Orlando, FL.

Hunt, E., Pelligrino, J., Frick, R., Farr, S., & Alderton, D. (1988). The ability to reason about movement in the visual field. *Intelligence, 12*, 77–100.

Joslyn, S., & Hunt, E. (1998). Evaluating individual differences in response to time-pressure situations. *Journal of Experimental Psychology: Applied, 4*(1), 16–43.

Kaempf, G., Klein, G., Thordsen, M., & Wolf, S. (1996). Decision making in complex naval command-and-control environments. *Human Factors, 38*(2), 220–231.

Kennedy, R., & Ordy, J. M. (1995). Situation awareness: A cognitive neuroscience model based on specific neurobehavioral mechanisms. In D. J. Garland & M. R. Endsley (Eds.), *Experimental analysis and measurement of situation awareness* (pp. 49–56). Daytona Beach, FL: Embry-Riddle Aeronautical University Press.

Klein, G. (1993). A recognition-primed decision (RPD) model of rapid decision making. In G. Klein, J. Orasanu, R. Calderwood, & C. Zsambok (Eds.), *Decision making in action: Models and methods* (pp. 138–147). Norwood, NJ: Ablex.

Kyllonen, P. C. (1993). Aptitude testing based on information processing: A test of the four-sources model. *Journal of General Psychology, 120*, 375–405.

Kyllonen, P. C. (1994). CAM: A theoretical framework for cognitive abilities measurement. In D. Detterman (Ed.), *Current topics in human intelligence: Volume 4, Theories of intelligence* (pp. 307–359). Norwood, NJ: Ablex.

Kyllonen, P. C., & Christal, R. E. (1990). Reasoning ability is (little more than) working memory capacity? *Intelligence, 14*, 389–433.

O'Hare, D. (1997). Cognitive ability determinants of elite pilot performance. *Human Factors, 39*(4), 540–552.

Orasanu, J., & Fischer, U. (1997). Finding decisions in natural environments: Towards a theory of situated decision making. In C. Zsambok & G. Klein (Eds.), *Naturalistic decision making* (pp. 343–357) Mahwah, NJ: Lawrence Erlbaum Associates.

Rabbit, P., Banerji, N., & Szymanski, A. (1989). Space Fortress as an IQ test? Predictions of learning and of practice performance in a complex interactive video-game. *Acta Psychologica, 71*, 243–257.

Roscoe, S. N. (1993). Predicting and enhancing flight deck performance. In R. Telfer & P. Moore (Eds.), *Aviation training: Pilot, instructor and organization* (pp. 195–208). Aldershot, England: Avebury.

Roscoe, S. N., & Corl, L. (1987). Wondrous original method for basic airmanship testing. In R. S. Jensen (Ed.), *Proceedings of the Fourth International Symposium on Aviation Psychology* (pp. 493–499). Columbus: The Ohio State University.

Shinar, D. (1993). Traffic safety and individual differences in drivers' attention and information processing capacity. *Alcohol, Drugs and Driving, 9*, 219–237.

Spearman, C. E. (1904). "General intelligence" objectively determined and measured. *American Journal of Psychology, 15,* 201–293.

Spearman, C. E. (1973). *The nature of intelligence and the principles of cognition.* North Stratford, NH: Ayer Company Publishers. (Original work published 1923)

Tirre, W. C. (1997). Steps toward an improved pilot selection battery. In R. Dillon (Ed.), *Handbook on Testing* (pp. 220–255) Westport, CT: Greenwood.

Tirre, W. C., & Pena, M. C. (1993). Components of quantitative reasoning: General and group factors. *Intelligence, 17,* 501–522.

Watson, C., & Nichols, T. (1976). Detectability of auditory signals presented without defined observation intervals. *Journal of the Acoustical Society of America, 59,* 655–667.

# Chapter 13

# Situation Awareness and Aging

Cheryl A. Bolstad
*North Carolina State University*

Thomas M. Hess
*North Carolina State University*

During the 20th century, we have been witnessing the "graying of America." Specifically, there has been a significant demographic shift in the age structure of the United States, with the average age of the population steadily rising. For example, the average life expectancy of 47.3 years in 1900 has risen to our current level of about 75 years. In addition, older adults (over age 65) now comprise over 13% of the population, as compared to 4% in 1900, with this proportion projected to increase to approximately 20% by the year 2030 (Moody, 1994). By the year 2020, population growth will be concentrated among those individuals over the age of 50 as the "baby boomers" become senior citizens. Along with this shift in the population structure is an increasing concern about the capabilities of and problems faced by older adults.

Importantly, as our population is aging, our world is also becoming increasingly technical and complex. This means more demands are being placed on individuals while performing everyday tasks. Whereas this may be manageable for younger adults, older individuals may experience difficulties as complex environments and new technologies tax the limits of their cognitive systems. For example, consider the act of driving a car. As automobiles become more sophisticated and highways become more congested, greater demands are being placed on vehicle operators. An older individual who is not used to such sophistication and complexity may have difficulty understanding the dynamics in a specific situation as well as making effective, timely decisions. It is this understanding that has become known as situation awareness (SA). SA applies to complex dynamical systems, and such systems are a part of the everyday lives of most older adults.

As has been noted elsewhere in this volume, SA is a widely known and used construct in the field of aviation, particularly in military arenas. It has been traced back to World War I, where it was recognized as a crucial component for crews in military aircraft (Press, 1986). Currently, the study of this important construct is going beyond the field of aviation, with situation awareness now being examined in such diverse areas as air-traffic control, nuclear power plant operation, and anesthesiology (Collier & Follesø, 1995; Endsley, 1995; Gaba, Howard, & Small, 1995; Sollenberger & Stein, 1995).

Although we are unaware of any existing work in this area, the SA construct may also prove useful in studying the specific problems faced by older adults. For example, aging-related changes in cognitive abilities that affect attention and memory may result in older individuals experiencing difficulties in performing certain essential tasks in work-related settings. Examination of older adults' performance within the SA framework might prove to be beneficial in understanding the specific problems associated with aging by providing clues regarding potential difficulties and how they might be remediated.

We would also argue that usefulness of the SA framework in examining aging is not just limited to work-related activities; benefits might also be found by investigating problems older adults encounter in the context of everyday activities. For example, SA is as important to the older adult trying to cross a busy street as it is to a pilot trying to shoot down the enemy. In both situations, the individual's life is placed in jeopardy, but good SA should lower the probability of any negative outcomes. Thus, the formation of SA may be critical to continued well-being and adaptive functioning of older adults.

Unique problems may present themselves as researchers attempt to incorporate knowledge about the aging of cognitive abilities into their models. Presently, most of the theoretical work on SA and its measurement relies on studies in which the subject population consists almost entirely of young adults (e.g., college students, military pilots). Due to cognitive changes that may begin occurring during the middle years of a person's lifetime, however, one could reasonably question the generalizability of these theoretical accounts and assessment techniques to the majority of individuals who comprise the remainder of the adult lifespan. In addition, use of these models in designing situations to promote SA may not have the same benefits for younger and older adults, because such design is typically based on data obtained from younger individuals and thus does not take into account normative age-related changes in cognitive abilities. Future SA research and model building needs to address these previously nonstudied populations of both middle-aged and older adults. This is especially true now that SA research is being conducted in nonmilitary contexts, where the target populations are likely to be more heterogeneous with respect to age than those in the military.

The primary goal of this chapter is to examine SA acquisition within the context of what we currently know about aging-related cognitive changes. We begin our examination of SA and aging by presenting a brief definition of SA acquisition, followed by a short discussion of some of the major theoretical views and empirical findings regarding cognitive aging that we view as relevant to the current topic. We then attempt an integration of these two sections by discussing the potential impact of aging at each stage of SA acquisition. Lastly, we will address some general issues regarding the measurement of SA across the adult lifespan.

## COMPONENTS OF SA ACQUISITION

The term *SA* has been used to describe both the process of how SA is formed and the product of this development. Due to space constraints, this chapter will focus on the acquisition of SA, not the "resultant, elusive thing we call SA itself" (Tenney, Adams, Pew, Huggins, & Rogers, 1992). We assume, however, that any age-related changes that affect the acquisition of SA may in turn also affect the product known as SA and vice versa.

While reviewing the literature, one finds many different definitions characterizing the process of SA acquisition. For present purposes we follow Endsley (1988) who stated that, "Situation awareness is the perception of the elements within a volume of time and space, the comprehension of their meaning and the projection of their status in the near future" (p. 97) to guide our discussion. This definition focuses on the cognitive nature of SA, with an emphasis on three stages of information processing involved in SA acquisition. Specifically, the individual must (1) perceive and extract the necessary information from their environment, (2) integrate this information with existing knowledge in order to form a coherent and useful picture of the situation, and (3) project what will happen based on their understanding of this information. These stages are evident in several theories that have been proposed to describe the formation of SA (e.g., Adams, Tenney, & Pew, 1995; Endsley, 1995; Gaba et al., 1995; Salas, Prince, Baker, & Shreshta, 1995). Whereas each theory takes a slightly different perspective, they all have in common the notion that acquisition is based on the same information-processing structures associated with other aspects of cognitive skills. As we consider the impact of aging, it is useful to think of how age-related changes in these information-processing structures might affect each of the three stages mentioned above.

## GENERAL AGING-RELATED PHENOMENON

Cognitive skills are not static across the lifespan, but instead change in response to normative and nonnormative forces associated with biological, psychological, social, and historical systems. In examining the impact of aging on SA, it is useful to understand the nature of these changes. Naturally, a comprehensive discussion of age-related cognitive changes is beyond the scope of this chapter, and the interested reader is referred to several recent sources for a more complete overview of cognitive change in adulthood (Blanchard-Fields & Hess, 1996; Craik & Salthouse, 1992; Kausler, 1991). Instead, we focus on four general themes in the aging literature that are representative of major empirical and theoretical traditions and that we have judged to be relevant to understanding aging and SA acquisition. It is important to note that, contrary to the dominant stereotype of aging in our culture, these ideas are not just about cognitive decline. Rather, they are representative of the lifespan developmental view that aging is characterized by both gains and losses (Baltes, 1987; Hess & Blanchard-Fields, 1996). It is also important to note that the aging-related changes described in this chapter are normative, and that there is a great deal of interindividual variability in the timing, order, and even occurrence of these changes.

### Slowing

One of the most robust findings in the aging literature is that older adults consistently perform more slowly than younger adults on many cognitive and psychomotor tasks. Within most domains, this slowing can be characterized as proportional; that is, across task conditions, the older adults' performance can often be represented as a linear function (with a slope > 1) of the young adults' performance in the same conditions. Salthouse (1996) argued that age-related changes in speed may account for much of the age-related variability in cognitive performance across a variety of tasks. There is disagreement, however, as to whether this slowing represents the primary factor behind much of the observed aging-related decline in performance or whether it is a manifestation of a more general, as yet unidentified aging process (Lindenberger & Baltes, 1994; Salthouse, Hancock, Meinz, & Hambrick, 1996). There is also some disagreement as to whether the slowing is general or domain-specific (e.g., Fisk & Fisher, 1994). Irrespective of these controversies, however, is the fact that aging is associated with slowing and that the impact of slowing increases with task complexity.

## Decreasing Processing Resources

Much of the work on aging and cognition has been done from within the information processing framework, with a focus on the role of capacity limitations in determining age effects on performance. One of the more prominent views of cognitive aging to arise from this work is that aging is associated with reductions in processing resources (Craik & Byrd, 1982; Salthouse, 1991). The exact nature of these resources is open to debate, with some scientists suggesting that a direct relationship exists between the amount of processing resources that are available and the capacity of a person's working memory, whereas others believe processing resources are directly linked to the amount of mental energy one possesses. Regardless of the nature of this entity, a processing resource deficit is typically conceptualized in terms of the ability to efficiently process information within working memory and is most evident in situations where one must, for example, initiate specific mental operations, such as those involved in encoding (Craik, 1986), consciously examine the contents of memory (Jennings & Jacoby, 1993), or perform complex operations, including division of attention (McDowd & Craik, 1988). Given that most of the active processes for SA are thought to occur in working memory, such age-related changes should be of major concern.

Whereas the research points to a reduction in processing resources for older adults, with a concomitant impact on cognitive performance, this general finding should be tempered. When tasks do not tap into working-memory resources, functioning is fairly stable with age. For example, recent research suggests that aging-related deficits are much more severe in explicit memory than in implicit memory (see Howard, 1996, for review). This appears to be due to the fact that the deliberate retention processes associated with explicit memory, such as conscious retrieval of previously stored information, place more demands on processing resources than do the relatively unconscious retrieval processes associated with implicit memory.

## Disinhibition

A more recent theoretical view suggests that aging is associated with an increased inability to inhibit nonselected information (Hasher & Zacks, 1988). For example, studies of negative priming have shown that older adults experience greater interference in a visual search task when a distracter on the previous trial subsequently becomes the target (e.g., Hasher, Stolzfus, Zacks, & Rympa, 1991). This suggests that responses associated with previously nonselected target information are more likely to experience suppression in younger adults than in older adults. It has been hypothesized that this age-related difficulty in suppressing or inhibiting unattended information may have the effect of decreasing the efficiency of working-memory functions as precious capacity is taken up by irrelevant information in older adults (Hasher & Zacks, 1988).

More important, such effects are not just limited to the visual domain. Research has also shown that aging has a negative impact on the ability to actively filter or select out information from multiple sources in the auditory domain. This is especially true in tasks that place a heavy demand on the limited attentional capacities of a person. For example, a familiar auditory selection task includes the "cocktail party phenomenon," in which a person can selectively attend to many of the conversations going on around the room while ignoring others. Relative to younger adults, older adults demonstrate a greater susceptibility to interference in such situations (Barr & Giambra, 1990).

Although there is a reasonable amount of support for the notion of decreased inhibitory functioning in later adulthood, there is evidence that this effect is moderated by certain conditions. For example, the previously identified negative priming effects appear to be domain specific, being observed in situations involving identity information, but not spatial location (e.g., Connelly & Hasher, 1993). Studies have also shown that aging effects on selective attention are substantially reduced when the individual does not have to search the environment for information. Specifically, older adults exhibit disproportionate increases in time to search for a visual target as the number of items in the visual array increases; when search is not involved, however, age is not associated with array-size effects (cf. Plude, Schwartz, & Murphy, 1996). Finally, it should also be noted that there is some controversy in the field as to whether the age-related deficits in inhibitory functions represent a distinct phenomenon or are simply manifestations of other more general aging related-related processes, such as generalized slowing (e.g., Salthouse & Meinz, 1995).

### Moderating Role Of Experience

Up to this point, the dominant perspective presented regarding aging—at least with respect to information processing—is one of decline. There is, however, substantial evidence that experience can modify the extent to which age effects are observed in certain tasks. This is illustrated nicely in the theory of crystallized and fluid intelligence (Horn & Cattell, 1967), in which intellectual ability is composed of two major components that are distinguished primarily in terms of the extent to which each is related to experience. (See also Baltes' [1987] distinction between the mechanics and pragmatics of cognition.) Fluid intelligence represents performance in novel situations, where prior experience is of little value and performance is dependent primarily on the integrity of the nervous system.

Fluid ability is thought to increase with development through childhood, reaching a peak in young adulthood before beginning a systematic decline beginning in the third decade of life. In contrast, crystallized intelligence represents performance influenced by acquired knowledge, such as the verbal and numerical skills that are formed during schooling. Crystallized skills also increase throughout childhood, but then exhibit stability or modest growth throughout much of the adult lifespan. This suggests that when older adults can draw on their experience (i.e. crystallized intelligence), age differences in performance should be substantially reduced.

This conclusion is supported in studies of expertise. It appears that the extensive practice and knowledge base associated with expertise may help to alleviate some of the age-related changes described above (for more details on how expertise impacts SA, see chap. 16, this volume). Many studies have shown that expert older adults can outperform novice younger adults and in some situations can perform as well as younger experts (e.g., Charness, 1981a, 1987; Morrow, Leirer, Fitzsimmons, & Altieri, 1994; Salthouse, 1984). A primary reason for this is that experts are able to perform many of the tasks in their domain of expertise with little drain on precious processing resources. This is due to (1) their ability to draw on readily accessible knowledge structures to organize incoming information and formulate responses, (2) their use of operations with a high degree of automaticity, and (3) the fact that extensive practice often results in elimination of computational steps that may slow processing. Through these processes, the negative impact of aging on working memory functioning may be overcome to some degree.

An example of the benefits associated with expertise can be seen clearly in a recent study by Clancy and Hoyer (1994). They found that search times for older (ages 40–64) and younger (21–28) medical technologists were similar for a visual search task that was representative of their job (e.g., comparing different images of bacteria morphology). When same-aged control subjects (i.e., nontechnologists) were tested in the same task, search times increased dramatically with age. In addition, when the medical technologists were compared on a standard laboratory-type search task (e.g., letter search), the younger technologists easily out-performed the older ones. Thus, expertise was able to overcome some of the typical search deficits found in older subjects that are usually attributed to slowing or reduced processing resources.

It also appears that experience may offset the negative impact of aging by the development of compensatory mechanisms established through extensive practice in response to aging-related limitations. These mechanisms enable older adults to perform at levels similar to younger adults, compensating for the age-related changes they are experiencing in their basic cognitive processes. For example, studies of transcription typing have shown that older expert typists maintain their speed by increasing their preview span; that is, they look farther ahead in the document they are typing than do younger typists, presumably to allow more preparation time for translating the visual image into a keystroke (Bosman, 1993; Salthouse, 1984). Similarly, Charness (1981a, 1987) has suggested that older experts can produce performance levels similar to younger experts in the areas of chess and bridge by developing better and more efficient ways to search through their memory.

Thus, experience appears to allow older adults to maintain high levels of performance through automatization of specific operations, an increased knowledge base, and/or compensatory mechanisms. If SA is measured in areas where older adults have considerable expertise, the difference between their measured SA and that of younger adults may be minimal. Note, however, that expertise does not always result in elimination of age effects on performance. For example, Charness (1981b) found that older chess experts still exhibited worse memory for chess patterns than young experts when the time to view the pattern was limited. Thus, in certain situations, deficits in basic processes and an inability to enact appropriate compensatory mechanisms may limit the ability of older adults to take advantage of their higher levels of knowledge.

It should also be noted that, although typically thought to have beneficial effects for performance in older adults, extensive experience may also have negative effects that may mimic those that have been hypothesized to be based in deteriorating physiological structures. Charness and Bieman-Copeland (1992) identify three such effects. First, the more knowledge a person has about a specific task or domain, the more time it will take them to access specific information. In other words, it takes longer to search through memory as the amount of information increases. Second, increased knowledge may also increase the likelihood that attention will be captured by environmental events through automatic activation of existing conceptual structures. This may partially account for aging-related problems in disinhibition. Last, the speed of adding new information to existing structures is negatively correlated with the amount of information within those structures. Unfortunately, few studies have been done to compare the experience versus decrement-based views in accounting for aging effects in cognitive performance.

## AGING EFFECTS ON SA ACQUISITION

Little research has been conducted on the effects of aging on SA. Thus, in this section, we highlight the potential role of aging at each of the three general stages of SA acquisition mentioned earlier, using the four described themes: slowing, decreased processing resources, disinhibition, and the role of experience. Once again, note that there are tremendous individual differences in the aging process, and thus our discussion focuses on aging in general.

## STAGE 1: PERCEPTION AND REGISTRATION OF INFORMATION

As mentioned above, the first step toward the formation of SA is the perception of information within a given situation (Salas et al., 1995). Endsley (1995) referred to this state as Level 1 SA: "The perception of the elements in the environment within a volume of time and space" (p. 36). During this stage, individuals extract important components from the environment as well as information about the status and dynamics of this information. Attentional processes obviously play a major role in this stage as the individual searches the environment, selectively attending to important/essential information while disregarding nonessential data. This process may be influenced by specific processing objectives, expectations, and/or prior experience (Dominguez, 1994). The information derived during this stage is then encoded and stored in memory, and serves as the data for the second stage of SA acquisition.

Naturally, one critical aspect of perception involves the extent to which our senses are effectively registering information. Although we focus on cognitive aspects of the aging process, it should be recognized that aging is also associated with specific changes in sensory functions, including those associated with the visual, auditory, and kinesthetic systems. For example, in the visual system, changes in the lens and retina result in reductions in near visual acuity, blue-green color discrimination, dark adaptation, and contrast sensitivity (Schieber & Baldwin, 1996). Such changes would obviously affect the quality, quantity and type of information reaching our sensory memory and thus ultimately impact the formation of SA. The interested reader is referred to Schieber (1992) and Schieber and Baldwin (1996) for further details.

### Perceptual/Attentional Processes

*Slowing.* With respect to cognitive skills, it appears that aging could have a major effect on Stage 1 SA acquisition through a number of mechanisms. Age-related declines in attentional operations could conceivably affect both the efficiency of the search processes as well as the amount and type of information that eventually enters into working memory during this stage. As mentioned earlier, one characteristic of the aging phenomenon is the slowing of most cognitive processes. Thus, older adults would, in general, be somewhat slower in terms of the speed with which they both orient to events in the environment and scan the environment for relevant information. Cognitive slowing may be particularly disadvantageous in complex situations where cues in the environment are changing at a relatively rapid, externally paced rate, straining older adults' capabilities to keep pace with the information flow. The impact of this change in processing speed may be negligible, however, in situations where (1) circumstances are relatively stable, (2) the places to be scanned in the environment are limited in number, and (3) the time to respond to changing circumstances is not severely restricted.

*Decreased Processing Resources.*   Aging-related reductions in processing resources might also limit the amount and complexity of information that can be dealt with at any given time. This effect can be seen in a variety of cognitive tasks where age differences in performance tend to increase along with the difficulty of the task, such as performance in divided attention tasks with meaningless stimuli (Tun & Wingfield, 1997). The aging-related limitations associated with processing resources is very apparent when information is presented in multiple modalities (Wickens, Braune, & Stokes, 1987). Unfortunately, in most complex, dynamical situations, information is presented in multiple modalities at the same time. Even young adults' attentional resources are limited when attention must be directed to more than one modality, such as visual and auditory (Endsley, 1990); however, older adults appear to be at an even greater disadvantage in these situations.

Not only do older adults suffer from attentional limitations when information is presented in multiple modalities, but they are also disproportionately affected when multiple tasks are performed at the same time. Whereas researchers have found that attentional resources can be shared across many tasks for younger adults (Damos & Wickens, 1980; Schneider & Fisk, 1982), this skill appears to be more limited in older adults (Korteling, 1993; Lorsbach & Simpson, 1988). Age differences are especially apparent in cases where the two tasks are high in complexity or uncorrelated to one another (incoherent similarity), thereby putting additional strain on processing resources (Korteling, 1993; Lorsbach & Simpson, 1988). In fact, McDowd and Craik (1988) have argued that the negative effect of aging may have to do with the complexity of dual-task performance rather than with having to perform two tasks simultaneously.

What is interesting is that there have been some studies that show no differences in the time-sharing abilities of older and younger individuals (e.g., Wickens, Braune, & Stokes, 1987). This may be due to the fact that individuals are not actually performing two discrete tasks, as they are able to share some of the same resources between them. These studies do consistently demonstrate, however, that older adults perform more slowly than their younger counterparts. Older adults may also be able to perform dual tasks at a high level of proficiency if the skills used in one of the tasks is automatized, thereby placing little strain on cognitive resources (Hasher & Zacks, 1978). Thus, the complexity and type of task being used to measure time-sharing abilities can affect the extent to which age differences are observed in performance in dual-task situations.

Whereas much of the aging-related variation in attentional mechanisms appears to reflect general changes in speed and/or processing resources rather than a qualitative change in the way in which the cognitive system operates, aging does appear to have a disproportionate impact in certain situations. For example, whereas older adults are generally slower than the young in standard visual search tasks, selective attention appears to be particularly affected by aging when the individual must search the environment for relevant information or when target information competes with other information in the environment (for a review, see Plude et al., 1996).

We assume, as do others, that the first stage of SA places great demands on attentional resources. Given the just-discussed changes in attentional capacities, it can be hypothesized that older adults will not be able to achieve or maintain as high a level of SA as younger individuals based on the affects of aging on the first level of SA.

It should be noted, however, that the impact of some of these attentional deficits associated with basic cognitive processes is moderated by certain conditions. For example, Craik (1986) has hypothesized that the limitations associated with reduced processing resources can be made smaller with appropriate environmental supports to guide processing. This can be seen in studies where distinctive cues and preexisting information regarding spatial location are made available, resulting in attenuation or even elimination of visual selection differences between older and younger subjects (Plude & Doussard-Roosevelt, 1989). Presumably, such information enhances the selection of relevant information and inhibits the processing of nonrelevant information by focusing attention and eliminating search.

***Experience.*** It should also be noted that many of the attentional problems attributed to older adults in the literature typically occur within novel contexts (e.g., standard laboratory tasks). In situations where the subject has expertise, such effects may be greatly reduced or eliminated, as demonstrated in several studies of expertise that use domain-specific tasks (e.g., Clancy & Hoyer, 1994; Morrow et al., 1994). Experts have the ability to activate appropriate schemas from long-term memory to aid in performance in domain-specific tasks. These schemas allow them to focus their attention on the appropriate information as well as help direct their attention to where information may be presented through use of probabilistic information. In this latter case, probabilistic knowledge regarding the occurrence and location of events may help to offset the problems with search that have been observed in older adults (Plude et al., 1996). For example, in the Clancy and Hoyer (1994) study cited earlier, older medical technologists had similar search times as their younger counterparts when the task was representative of their job, presumably due too their ability to rely on their many schemas of bacteria morphology in order to aid their search.

Consistent with the foregoing, it has been suggested (e.g., Endsley, 1995) that schemas play a major role during the first stage of SA acquisition by organizing attentional processes. Schemas are thought to guide what information is attended to and what is ignored, such that important or even unusual information is remembered, while trivial, irrelevant information is not processed.

Importantly, one aspect of cognition that appears to be minimally affected by aging is schema activation and use (Hess, 1990). Assuming no variation in the nature of the schema itself, young and older adults differ little in terms of the impact of schemas on processing. For example, in several studies of schema processing involving script-based narratives, visual scenes, and person descriptions, Hess and colleagues (Hess, Donley, & Vandermaas, 1989; Hess & Slaughter, 1990; Hess & Tate, 1991) found that patterns of study times were similar across age groups in response to different stimulus elements, with schema consistent items being examined more quickly than schema inconsistent ones.

There is some evidence that schemas may be particularly important in supporting the processing of older adults. For example, age differences in recall tend to be greater for schema-inconsistent than for schema-consistent information (e.g., Hess & Tate, 1991; Hess et al., 1989; but also see Arbuckle, Vanderleck, Harsany, & Lapidus, 1990), and older adults appear to have more difficulty in schema activation when the cues in the environment deviate somewhat (e.g., different spatial arrangement) from schema-based expectations (e.g., Hess & Slaughter, 1990). One way to interpret these effects is in terms of reduced processing resources, whereby aging effects are attenuated when resource requirements are minimized through use of existing schemas. When the individual must organize and integrate without the benefit of previous experience, resulting in demands being placed on working memory, age differences in performance are increased.

**Registration in Memory**

Once information has been selected from the environment, aging-related problems may also exist in the registration of this information in memory. Presumably, working memory processes play an important role here as well, and changes in speed, processing resources, and/or inhibitory functions may all affect the efficiency of processing within working memory.

One impact of aging-related changes in these processes appears to be deficits in the encoding of contextual information associated with events in the environment. As an example, aging has a negative effect on source memory, with older adults having more problems than young adults in identifying the context in which an event occurred (for review, see Hess & Pullen, 1996). This is particularly true when one must discriminate between sources within the same context, such as in reality monitoring (e.g., did I do $X$ or just think about doing it?).

Contextual information plays an important role in retention by helping one to locate a particular event in time and space, and such information appears to be particularly important as the individual attempts to attach meaning to the information that exists in memory during the next stage of SA acquisition. Older adults appear to engage in less extensive processing and/or have reduced ability to inhibit task extraneous thoughts, which then serve to limit the informativeness of the available contextual information.

As with attentional processes, the negative impact of aging on episodic memory functions may be moderated by environmental supports whereby external guidance is provided during the encoding and retrieval of information. For instance, while driving down a busy highway, if an older driver encounters a sign saying "slow down construction ahead" they are more likely to take note of the speed of surrounding cars and thus notice those not obeying the speed limit. This sign may also act to retrieve information from memory regarding safe driving in a road construction area, such as watching out for workman and possible lane shifts. Likewise, existing knowledge and expertise may also reduce age differences by providing readily accessible encoding structures that reduce demands on cognitive resources and assist in organizing incoming information in memory.

**Summary**

In summary, aging would appear to have the potential of disrupting the initial stage of SA acquisition by decreasing the efficiency with which individuals can extract information from the environment and accurately store it in memory. This may result in older adults creating less complete and/or qualitatively different representations of their environment than younger adults. This is especially true when there is a lot of information present, when information is presented in multiple modalities, and when more than one task is performed at the same time. More important, however, the negative impact of aging on basic cognitive functions may be moderated by the specific circumstances. Thus, aging effects will be less prominent in simple environments, situations with a great deal of environmental support, unpaced situations, and tasks in which the individual can draw on his or her expertise.

## STAGE 2: COMPREHENSION

Once individuals have perceived the information in the environment, the next step is to integrate the information within working memory (Salas et al., 1995). Relevant information is brought into consciousness, thus allowing the person to consider all elements and organize the information into a coherent picture (Dominguez, 1994). In this stage, the operator gains an understanding of the significance of the elements within his or her environment. Endsley (1988, 1990, 1995) refers to this state as Level 2 SA: ". . . the comprehension of their (elements) meaning." Many authors refer to this product of SA as a mental model, which is a representation of the current situation as the individual perceives it. Elements within the environment may activate specific schemas that reside in long-term memory (Dominguez, 1994) that help to organize and update this mental model by giving meaning to individual elements and their interrelationships (Salas et al., 1995).

As in the previous stage, age differences in processing resources could have an impact on SA acquisition during this second stage. Because the bulk of SA processing is thought to occur in working memory (Endsley, 1995), we can assume that aging-related reductions in the efficiency of working memory processes may negatively affect a person's ability to create an accurate mental model. For example, reduced processing resources may limit the amount of information that can be considered at any given time in working memory. Hasher and Zacks (1988) have also suggested that age-related reductions in inhibitory processes may limit that amount of information being considered by effectively limiting the capacity of working memory due to the entry of uninhibited, task-extraneous elements in the system. In addition to limiting capacity, the introduction of such elements may also result in inappropriate information being incorporated into the mental model.

Reduced processing resources might also have a negative effect on Stage 2 SA acquisition due to the impact on information that was encoded into memory during Stage 1. Three circumstances come to mind here. First, older adults may have problems retrieving this information from memory, depending on the level of effort required. Studies have shown that there is an age-related impairment in the processing component concerned with the retrieval of information from memory (Salthouse, 1992). For example, using a secondary task to measure capacity use, Craik and McDowd (1987) demonstrated that older adults experience a disproportionate increase in capacity usage when recall is required as opposed to recognition in a memory task. Presumably, the effort associated with the greater use of self-initiated retrieval mechanisms in recall is particularly problematic as processing resources decline in later adulthood. Memory research in general has shown that age has a more negative effect on retention in situations involving conscious recollection processes (e.g., direct tests) than in those involving automatic retrieval (e.g., indirect tests; Howard, 1996).

Part of the retrieval problem experienced by older adults may have to do with the incomplete encoding of contextual information in the first place, which brings us to the second circumstance relating to Stage 1 effects on Stage 2. Specifically, once information is activated in memory, older adults may have difficulty identifying the source of the information and, thereby, discerning both where the information came from and what actually occurred versus what was inferred from existing schemas. An example of this type of problem is clearly demonstrated in studies of false fame (e.g., Jennings & Jacoby, 1993), where older adults are more likely than younger adults to misattribute fame to nonfamous but previously viewed names because they have difficulty identifying the source of their feelings of familiarity. Interestingly, it might be hypothesized that experts would be more susceptible to false familiarity problems associated with aging than would be novices during SA acquisition. Experts possess many mental models of various situations and may have difficulty discerning between what actually occurred and what was activated in their memory by these models.

Such problems may prove particularly troublesome in situations where it is important not only to be aware of information essential to SA, but also to know the origins of such information. Through the operation of automatic encoding and retrieval processes, older adults may have much of the same information available to them as younger adults. Uncertainty about the validity of the information used in SA formation may result in the construction of more tenuous mental models that may have a negative impact on decision making. The automatic processing associated with many scenarios may result in problems for individuals of all ages, but these problems may be exacerbated with age as the ability to consciously control and monitor working memory processes suffers.

Finally, it is apparent that Stage 2 SA acquisition will be affected by se-lection processes operating in Stage 1. Due to age-related cognitive changes, individuals may have either selected inappropriate information or been unable to select the appropriate information during the initial stage, thus affecting the nature of the information available to the individ-ual and the subsequent accuracy of their SA. This is also true for the infor-mation being processed during the final stage of SA acquisition.

### Summary

In summary, aging may be associated with problems during Stage 2 SA ac-quisition, as reductions in processing resources and inhibition problems affect the ability of the individual to create an accurate mental model of the situation in working memory. Problems here include the ability to util-ize effortful processing operations to deal with large amounts of and/or complex information, as well as difficulties in retrieving and utilizing in-formation (e.g., source) registered during Stage 1.

## STAGE 3: PROJECTION

Several definitions allude to the temporal nature of SA, such that the proc-esses used to form SA need to be in continuous operation in order for the in-dividual to maintain a coherent picture of the situation. Awareness of this temporal dimension is critical in maintaining SA in dynamic situations (Dominguez, 1994). One must not only make decisions based on mental models reflecting past and current states; this information must also be used to direct further perceptions and to anticipate how the situation will change in the future. Endsley (1995) refers to this step as Level 3 SA. She be-lieves through knowing the status and dynamic of the elements and com-prehending the situation, one can have the ability to project future actions of the elements, which are used in the decision making process.

### Decreased Processing Resources

Temporal awareness involves the continuous extraction and updating of information from working memory. Once more, aging-related changes in processing resources that affect the efficiency of working memory opera-tions may have the greatest effect on temporal awareness. As has been al-ready noted, these problems in working-memory functioning are most ev-ident in tasks that require a great deal of deliberate, conscious processing

(Smith & Earles, 1996). Age-related limitations in working-memory processes may lead to difficulties in gaining a coherent picture of the environment as well as in making the inferences necessary for future projections because it can be reasonably assumed that at least some aspects of these functions require deliberate processing. In this latter situation, it might be hypothesized that age differences in inferences will be minimal in situations where individuals can tap into well-developed knowledge structures established through extensive experience in a domain (e.g., elements A and B have always been associated with outcome X). When such knowledge is not available and inference generation must proceed in a more bottom-up fashion, then age differences might be more prevalent due to the increased demands on working memory processes. Also note that the projection accuracy is only as good as the information used in that process (e.g. the level 2 SA). Thus, any age-related processes that affect the amount and quality of information contained in working memory may negatively impact Stage 3 processing.

Lastly, because SA is an active, ongoing process that requires processing and integration across several stages, reductions in processing resources may play a significant role in determining SA in older individuals. Specifically, SA acquisition requires that individuals continually perform operations associated with three different stages as they perceive changes in the environment, update their mental models, and prepare for action. In essence, SA acquisition involves performance of multiple tasks, with the degree of overlap being dependent on the type of situation with which one is dealing. In some cases, the overlap may be minimal, putting little strain on cognitive resources for both younger and older adults. In more complex environments where information inflow and adjustments to this information represents a continuous process, however, older adults may be at a particular disadvantage due to their problems in dual-task performance.

## Experience

As before, some of the negative impact of aging might be alleviated by experience, since many operations may be automatized. Thus, older adults with extensive experience in a particular domain may be able to compensate for some of their working memory declines as they develop automaticity of a skill (Bosman & Charness, 1996).

Experience may also benefit older adults if they have well developed scripts and schemas that they can use to circumvent processing limitations. In the same way that schemas can reduce the negative consequences of reduced processing resources in later adulthood during earlier stages of

SA acquisitions by directing attention and providing an organizational framework for incoming information, Stage 3 processing may be promoted to the extent that these same schemas allow prediction of future states and provide direction for actions to be taken.

## AGING AND SKILL ACQUISITION

Assuming that aging-related changes in cognitive skills will result in poorer SA acquisition in later adulthood, an important question relates to the extent to which training might overcome these potential problems. We have noted all along that expertise does seem to be especially beneficial to older adults. When we speak of expertise, however, we are typically speaking of skills and knowledge that have been acquired over a substantial period of time, usually beginning in young adulthood.

The important question concerns whether older adults acquire new skills through training and how does this affect their SA acquisition. Many studies have examined age differences in skill acquisition. In general, these studies have found that older adults can acquire new skills, but that age differences in performance continue to exist, and in some cases become exacerbated. For example, in one series of studies, a testing-the-limits approach was used in training the use of a mnemonic strategy (i.e., the method of loci) in young and older adults (Kliegl, Smith, & Baltes, 1989; Lindenberger, Kliegl, & Baltes, 1992). After extensive practice, it was found that older adults exhibited dramatic improvements in use of their new skill, and performed at a level that was significantly greater than untrained younger adults. It was also found, however, that age differences in performance actually increased with training; that is, younger adults benefitted more from training than older adults, suggesting that cognitive reserve capacity (i.e., the difference between actual and optimal levels of performance) decreases with age.

In general, older adults do show significant improvement in skilled performance with extensive practice, and this improvement can be seen for a number of different skills (see Bosman & Charness, 1996, for review). The implications of training for the study of aging and SA acquisition are encouraging in suggesting that initial aging-related skill limitations may be substantially reduced with practice. Although the benefit may be less in some realms than in others, it would appear that there is the potential for older adults to improve levels of performance to help achieve acceptable levels of SA.

It is also important to note that whereas these training studies employ many more task trials than used in typical studies of cognitive skills, the degree of practice still does not approach that which may be obtained in everyday life (e.g., working at a job 40 hours per week). Thus, it is possible that such studies underestimate the potential impact of extensive experience in a novel domain for older adults.

## MEASURING SA PROCESSES IN OLDER ADULTS

In this final section, we address issues and concerns that might affect the measurement of SA acquisition for older adults. One thing to keep in mind when measuring SA acquisition is that any measure is actually a measurement of a person's ability to acquire SA during the 3 initial stages mentioned earlier and any of these stages can be studied alone or in combination. For example, if one is interested in assessing Stage 1 SA acquisition in a specific situation, a measure should be selected that is appropriate to both the stage of acquisition being studied as well as the context in which SA is being examined.

For instance, there is some promising work on using eye saccade latencies to determine the impact of expectancy on individuals (Stern, Wang, & Schroeder, 1995). Such a technique may prove to be useful for measuring the final stage of SA acquisition (i.e., projection) in situations involving visual displays, but may be less valid for assessing Stage 2 acquisition.

The appropriateness of the measurement technique must also be considered when examining different-aged individuals. Many measurement techniques have been proposed for SA, ranging from simple physiological measures involving heart rate and EEG to more psychological techniques utilizing concurrent memory probes and subjective rating measures (for reviews, see Fracker, 1991; Garland & Endsley, 1995). It is possible, however, that some of these measures may not take into account aging-related changes in physiological, perceptual, and cognitive systems that may impact assessment and negatively affect validity.

Some of the physiological measures being studied to assess SA use heart rate, EEG, eye blinks, eye saccade latencies and pupil changes (Stern, Wang, & Schroeder, 1995; Wilson, 1995). As with any testing in which older adults are used, a prescreening for physiological health should be obtained to rule out any preexisting health problems, such as cataracts, dry or watery eyes and heart problems. These health problems can impact the physiological measures of SA. For instance, dry or watery eyes may make older adults blink more or less often than younger adults (Rossman, 1989), thus affecting the eye blink measurement technique. Likewise, normative age-related changes may also impact on assessment. For example, the use of heart rate as a measurement technique with healthy older adults may be problematic due to reductions in maximal heart rate with age (Lakatta, 1990). Whereas reaction time is a very useful measure for most psychological research, it also slows with age. In general, a simple reaction time measure for a 60 year old is 20% slower than for a 20 year old (Small, 1987). Thus, measurement techniques that employ reaction time should be evaluated closely to see if any age differences in performance are due to a general slowing of reaction time rather than a process-specific slowing.

In all of these cases, the main problems in assessment will arise if individuals are compared on absolute measures (e.g., response time, heart rate) without taking baseline rates into account. If the performance of older adults is compared to age-based norms or adjusted for range of response (e.g., resting vs. maximal heart rate), then many measurement problems can be overcome.

In the proceeding sections, we have also documented some of the important age-related cognitive changes that could potentially impact SA acquisition and hence impact SA measurement. As mentioned earlier, experience may have a disproportionate impact on SA acquisition in older adults. Specifically, whereas younger adults may experience some benefit in performance due to expertise, this appears to be even greater for older adults. For example, the difference in performance between domain-general versus domain-specific laboratory tasks is greater for older adults than for younger adults (e.g., Clancy & Hoyer, 1994).

So what does this mean for the measurement of SA? If standardized tests assessing individual differences in basic skills thought to be associated with the formation of SA are used in selection or performance assessment, older experts may be penalized by their poor performance on domain-general tasks. That is, standardized tests might be better predictors of SA in younger than in older adults, at least in their area of expertise due to the potential for greater dissasociation between general and domain specific skills with age. This suggests that domain-specific assessment techniques would be preferred in many situations, especially those associated with evaluating ongoing performance (e.g., yearly job evaluations).

However, it also needs to be recognized that certain situations exist where absolute standards rather than relative ones (e.g., based on age-group norms) should be the criteria for evaluation. If absolute standards are of interest, then domain general tasks would be more appropriate. For example, if an individual has to switch attention between tasks (e.g., such as processing traffic signals to watching for oncoming cars) or hold a specific amount of information in mind while performing a task (e.g., the speed and distance of the surrounding cars while navigating in traffic), it will be more important to know if the criteria can be met than how the individual performs relative to others in his or her age group. On the other hand, if one wants to know how SA acquisition varies within 60 year olds, then domain specific tasks can be used. Thus, the situation needs to be considered in determining the appropriateness of absolute and relative standards in assessing SA acquisition.

Aging-related cognitive limitations may also come into play to the extent that the measure taps into performance aspects that are unrelated to SA acquisition. For example, the use of explicit memory tests that put demands on processing resources, such as retrospective event recall and concurrent memory probes to measure SA, may have a disproportionately negative impact on SA assessment in older individuals to the extent that these demands are independent of the skill being assessed. If this occurs, older adults may exhibit a lower SA not due to that fact that their ability to acquire SA has diminished, but because the measure being used consumes more of their limited processing resources.

When possible, researchers should attempt to limit the ability-independent test demands so that more accurate assessments of older (and younger) adults' abilities can be obtained. One such means might be through employment of indirect assessment techniques that do not require conscious examination of the contents of memory. Studies have shown that indirect memory tests are associated with much smaller aging effects than are direct memory tests (e.g., Hultsch, Masson, & Small, 1991). Older subjects may have difficulty explicitly recalling that an event has occurred, but their actions may demonstrate that the event was registered in memory. Test demands might also be reduced—along with potential age-related biases in measurement—through the use of recognition rather than recall tests due to the latter type of test requiring fewer processing resources (e.g., Craik & McDowd, 1987).

Obviously, in any situation, one would want to make sure that the test was appropriate to the situation before adopting these or any other recommendations. Thus, for example, one would not want to employ an indirect test to examine SA acquisition in a context in which conscious monitoring of the contents of memory is important for developing accurate mental models. Our point is simply that assessment of SA in a sample of individuals that varies widely in age needs to take into account that age differences in cognitive as well as other skills (e.g., perceptual) may result in unintended measurement biases, and that these biases may result in questionable validity of the resulting measures as well as inappropriate conclusions regarding the abilities of older adults.

## SUMMARY AND CONCLUSION

In this chapter, we have examined the impact of aging on SA acquisition. We argued that aging-related changes in cognitive ability may have a dramatic impact on SA formation in many work-related and everyday situations, and that use of the SA framework may prove fruitful in understanding and dealing with the potential problems faced by older adults in our society. Unfortunately, current models of and research on SA have neglected the aging component. In the absence of empirical studies, we have attempted to illustrate the potential impact of aging by examining the three stages of SA acquisition within the context of current theory and research on cognition and aging. Based on this examination, it appears that normative age-related changes in processing resources, speed, and inhibitory functions may have a negative impact on SA. This conclusion is tempered, however, by the facts that (1) older experts may show minimal decline in their areas of expertise, (2) extensive training in older adults may overcome some of the negative effects associated with aging, and (3) the cognitive changes discussed may vary widely in their form and progression across individuals. We have also suggested that measurement of SA in older adults may be somewhat problematic due to changes in physiological, perceptual, and cognitive functions. In assessing SA acquisition in older adults, and when comparing age-related variations in SA factors, these changes need to be taken into account in order to reduce age biases in measurement.

Although the literature supports the notion that SA will decline with age, there has been very little research to support or disprove this claim. One preliminary analysis of SA scores for experienced pilots found no correlation with age (Bolstad & Endsley, 1991), although it should be emphasized that the small number and the select nature of the subjects in this study may have precluded such effects. In fact, the lack of age effects in this sample may be used as evidence in support of the notion that expertise will minimize the impact of aging on SA.

The potential benefits of a lifetime of experience can be seen in a more recent examination of SA and driving (Bolstad, 1996). In this study, the performance of young (18 to 39 years), middle-aged (40 to 59) and older drivers (60 to 84 years) was compared using a subjective measure of SA. The individuals (n = 61) were administered a 40-item questionnaire designed to look at Level 1 and Level 2 SA as defined by Endsley (1990). Young drivers had less than 5 years of driving experience and older drivers had more than 20 years experience.

The expectations were that older drivers would experience more driving difficulties associated with potential SA problems than younger drivers. In fact, younger adults reported experiencing the most problems with forgetfulness, failure to stop, and navigating in unfamiliar areas whereas middle-aged drivers had the fewest difficulties.

These results could be attributed to expertise with the driving task. Younger adults may not have enough practice with the task of driving to allow them to develop driving components that are performed automatically. In contrast, older drivers may have developed compensatory mechanisms to overcome many of the changes associated with aging. These older drivers did report using more compensatory mechanisms, such as avoiding the freeway during rush hour and not driving at night than the other age groups. As in other studies of expertise, older adults reported failing to perform at levels comparable to younger experts (middle-aged drivers). Whereas they performed better than younger drivers, they reported more driving difficulties and lack of SA than the middle-aged adults.

These two studies by no means represent a systematic examination of SA and aging. They do, however, provide some support for our hypotheses and are consistent with research on aging and cognition. In addition, the results suggest that the impact of aging on SA acquisition is not simple, and needs to be understood within the context of the individual and the situation.

## ACKNOWLEDGMENTS

Preparation of this chapter was facilitated by a grant (AG05552) from the National Institute on Aging.

## REFERENCES

Adams, M. J., Tenney, Y. J., & Pew, R. W. (1995). Situation awareness and the cognitive management of complex systems. *Human Factors, 37*, 85–104.

Arbuckle, T. Y., Vanderleck, V. F., Harsany, M., & Lapidus, S. (1990). Adult age differences in memory in relation to availability and accessibility of knowledge-based schema. *Journal of Experimental Psychology: Learning, Memory and Cognition, 16*, 305–315.

Baltes, P. B. (1987). Theoretical propositions of life-span developmental psychology: One the dynamics between growth and decline. *Developmental Psychology, 23*, 611–626.

Barr, R. A., & Giambra, L. M. (1990). Age-related decrement in auditory selective attention. *Psychology and Aging, 5*, 597–599.

Blanchard-Fields, F., & Hess, T. M. (Eds.). (1996). *Perspectives on cognitive change in adulthood and aging*. New York, NY: McGraw-Hill.

Bolstad, C. A. (1996). [Situation Awareness and Driving: Potential Issues]. Unpublished report.

Bolstad, C. A., & Endsley, M. R. (1991). [Situation awareness attribute data]. Unpublished raw data.

Bosman, E. A. (1993). Age-related differences in the motoric aspects of transcription typing skill. *Psychology and Aging, 1*, 87–102.

Bosman, E. A., & Charness, N. (1996). Age-related differences in skilled performance and skill acquisition. In F. Blanchard-Fields & T. M. Hess (Eds.), *Perspectives on cognitive change in adulthood and aging* (pp. 428–453). New York, NY: McGraw-Hill.

Charness, N. (1981a). Aging and skilled problem solving. *Journal of Experimental Psychology: General, 110*, 21–38.

Charness, N. (1981b). Search in chess: Age and skill differences. *Journal of Experimental Psychology: Human Perception and Performance, 7*, 467–476.

Charness, N. (1987). Component processes in bridge bidding and novel problem-solving tasks. *Canadian Journal of Psychology, 41,* 223–243.

Charness, N., & Bieman-Copeland, S. (1992). The learning perspective: Adulthood. In R. J. Sternberg & C. A. Berg (Eds.), *Intellectual development* (pp. 301–327). New York, NY: Cambridge University Press.

Clancy, S. M., & Hoyer, W. J. (1994). Age and skill in visual search. *Developmental Psychology, 30,* 545–551.

Collier, S. G., & Follesø, K. (1995). SACRI: A measure of situation awareness for nuclear power plant control rooms. *Proceedings of an International Conference: Experimental Analysis and Measurement of Situation Awareness* (pp. 115–122). Daytona Beach, FL.

Connelly, S. L., & Hasher, L. (1993). Aging and the inhibition of spatial location. *Journal of Experimental Psychology: Human Perception and Performance, 19,* 1238–1250.

Craik, F. I. M. (1986). A functional account of age differences in memory. In F. Klix & H. Hagendorf (Eds.), *Human memory and cognitive capabilities* (pp. 409–442). New York, NY: Elsevier.

Craik, F. I. M., & Byrd, M. (1982). Aging and cognitive deficits: The role of attentional resources. In F. I. M. Craik & S. Trehub (Eds.), *Aging and cognitive processes* (pp. 191–211). New York, NY: Plenum.

Craik, F. I. M., & McDowd, J. M. (1987). Age differences in recall and recognition. *Journal of Experimental Psychology: Learning, Memory and Cognition, 13,* 474–479.

Craik, F. I. M., & Salthouse, T. A. (Eds.). (1992). *The handbook and aging and cognition.* Hillsdale, NJ: Lawrence Erlbaum Associates.

Damos, D., & Wickens, C. D. (1980). The acquisition and transfer of time-sharing skills, *Acta Psychologica, 6,* 569–577.

Dominguez, C. (1994). Can SA be defined? In M. Vidulich, C. Dominguez, E. Vogel, & G. McMillan, *Situation awareness: Papers and annotated bibliography* (AL/CF Publication No. TR–1994–0085), pp. 5–15). Armstrong Laboratory: Wright Patterson Air Force base.

Endsley, M. R. (1988). Design and evaluation for situation awareness enhancement. *Proceedings of the Human Factors Society 32nd Annual Meeting,* (pp. 97–101). Santa Monica, CA: Human Factors Society.

Endsley, M. R. (1990). *Situation awareness in dynamic human decision making: Theory and measurement* (NOR–DOC 90–49). Hawthorne, CA: Northrop Corporation.

Endsley, M. R. (1995). Towards a theory of situation awareness. *Human Factors, 37*(1), 32–64.

Fisk, A. D., & Fisher, D. L. (1994). Brinley plots and theories of aging: The explicit, muddled, and implicit debates. *Journal of Gerontology: Psychological Sciences, 49,* P81–P89.

Fracker, M. L. (1991). *Measures of situation awareness: Review and future directions* (AL–TR–1991–0128). Wright-Patterson AFB, OH: Armstong Laboratory.

Gaba, D. M., Howard, S. K., & Small, S. D. (1995). Situation awareness in anesthesiology. *Human Factors, 37,* 20–31.

Garland, D. J., & Endsley, M. R. (Eds.). (1995). *Proceedings of an International Conference: Excremental Analysis and Measurement of Situation Awareness.* Daytona Beach, FL: Embry-Riddle Aeronautical University Press.

Hasher, L., Stolzfus, E. R., Zacks, R. T., & Rympa, B. (1991). Age and inhibition. *Journal of Experimental Psychology: Learning, Memory and Cognition, 17,* 163–169.

Hasher, L., & Zacks, R. T. (1978). Automatic and effortful processes in memory. *Journal of Experimental Psychology: General, 108,* 356–388.

Hasher, L., & Zacks, R. T. (1988). Working memory, comprehension, and aging: A review and a new view. In G. H. Bower (Ed.), *The psychology of learning and motivation, Vol. 22* (pp. 193–225). Orlando, FL: Academic Press.

Hess, T. M. (1990). Aging and schematic influences on memory. In T. M. Hess (Ed.), *Aging and cognition: Knowledge organization and utilization* (pp. 93–160). Amsterdam: North-Holland.

Hess F., & Blanchard-Fields, T. M. (1996). Introduction to the study of cognitive change in adulthood. In F. Blanchard-Fields & T. M. Hess (Eds.), *Perspectives on cognitive change in adulthood and aging* (pp. 3–24). New York, NY: McGraw-Hill.

Hess, T. M., Donley, J., & Vandermaas, M. O. (1989). Age-related changes in the processing and retention of script information. *Experimental Aging Research, 15,* 89–96.

Hess, T. M., & Pullen, S, M. (1996). Memory in context. In F. Blanchard-Fields & T. M. Hess (Eds.), *Perspectives on cognitive change in adulthood and aging* (pp. 387–427). New York, NY: McGraw-Hill.

Hess, T. M., & Slaughter, S. J. (1990). Schematic knowledge influences on memory for scene information in younger and older adults. *Developmental Psychology, 26,* 855–865.

Hess, T. M., & Tate, C. S. (1991). Adult age differences in explanations and memory for behavioral information. *Psychology and Aging, 6,* 86–92.

Horn, J., & Cattell. R. (1967). Age differences in fluid and crystallized intelligence. *Acta Psychologica, 26,* 107–129.

Howard, D. V. (1996). The aging of implicit and explicit memory. In F. Blanchard-Fields & T. M. Hess (Eds.), *Perspectives on cognitive change in adulthood and aging* (pp. 221–254). New York, NY: McGraw-Hill.

Hultsch, D. F., Masson, M. E. J., & Small, B. J. (1991). Adult age difference in direct and indirect tests of memory. *Journal of Gerontology: Psychological Sciences, 46,* P22–P30.

Jennings, J. M., & Jacoby, L. L. (1993). Automatic versus intentional uses of memory: Aging, attention and control. *Psychology and Aging, 8*(2), 283–293.

Kausler, D. H. (1991). *Experimental psychology, cognition, and human aging.* New York, NY: Springer-Verlag.

Kliegl, R., Smith, J., & Baltes, P. B. (1989). Testing-the-limits and the study of adult age differences in cognitive plasticity of mnemonic skill. *Developmental Psychology, 25,* 247–256.

Korteling, J. E. (1993). Effects of age and task similarity on dual-task performance. *Human Factors, 35*(1), 99–114.

Lakatta, E. G. (1990). Heart and circulation. In E. L. Schneider & J. W. Rowe (Eds.), *Handbook of the biology of aging,* 3rd ed. (pp. 181–217). San Diego: Academic Press.

Lindenberger, U., & Baltes, P. B. (1994). Sensory functioning and intelligence in old age: A strong connection. *Psychology and Aging, 9,* 339–355.

Lindenberger, U., Kliegl, R., & Baltes, P. B. (1992). Professional expertise does not eliminate age differences in imaginary-based memory performance during adulthood. *Psychology and Aging, 7,* 585–593.

Lorsbach, T. C., & Simpson, G. B. (1988). Dual-task performance as a function of adult age and task complexity. *Psychology and Aging, 3,* 210–212.

McDowd, J. M., & Craik, F. I. M. (1988). Effects of aging and task difficulty on divided attention performance. *Journal of Experimental Psychology: Human Perception and Performance, 14,* 267–280.

Moody, H. (1994). *Aging: Concepts and controversies.* Thousand Oaks, CA: Pine Forge Press.

Morrow, D. G., Leirer, V. O., Fitzsimmons, C., & Altieri, P. A. (1994). When expertise reduces age differences in performance. *Psychology and Aging, 9,* 134–148.

Plude. D. J., & Doussard-Roosevelt, J. A. (1989). Aging, selective attention, and feature integration. *Psychology and Aging, 1,* 4–10.

Plude, D. J., Schwartz, L. K., & Murphy, L. J. (1996). Active selection and inhibition in the aging of attention. In F. Blanchard-Fields & T. M. Hess (Eds.), *Perspectives on cognitive change in adulthood and aging* (pp. 165–191). New York: McGraw-Hill.

Press, M. (1986). *Situation awareness: Let's get serious about the clue-bird.* Unpublished manuscript.

Rossman, I. (1989). Looking forward: The complete medical guide to successful aging. New York, NY: E. P. Dutton.

Salas, E., Prince, C., Baker, D. P., & Shreshta, L. (1995). Situation awareness in team performance: Implications for measurement and training. *Human Factors, 37*(1), 123–136.

Salthouse, T. A. (1984). Effects of age and skill in typing. *Journal of Experimental Psychology: General, 113*, 345–371.

Salthouse, T. A. (1991). *Theoretical perspectives on cognitive aging.* Hillsdale, NJ: Lawrence Erlbaum Associates.

Salthouse, T. A. (1992). The information processing perspective on cognitive aging. In R. J. Sternberg & C. A. Berg (Eds.), *Intellectual development.* New York, NY: Cambridge University Press.

Salthouse, T. A. (1996). The processing speed theory of adult age differences in cognition. *Psychological Review, 103*, 403–428.

Salthouse, T. A., Hancock, H. E., Meinz, E. J., & Hambrick, D. Z. (1996). Interrelations of age, visual acuity, and cognitive functioning. *Journal of Gerontology: Psychological Sciences, 51B*, P317–P330.

Salthouse, T. A., & Meinz, E. J. (1995). Aging, inhibition, working memory, and speed. *Journal of Gerontology: Psychological Sciences, 50B*, P297–P306.

Schieber, F. (1992). Aging and the senses. In J. E. Birren, R. B. Sloane & G. D. Cohen (Eds.), *Handbook of mental health and aging,* 2nd ed., (pp. 251–305). San Diego, CA: Academic Press.

Schieber, F., & Baldwin, C. L. (1996). Vision, audition, and aging research. In F. Blanchard-Fields & T. M. Hess (Eds.), *Perspectives on Cognitive Change in Adulthood and Aging* (pp. 122–162). New York, NY: McGraw-Hill.

Schneider, W., & Fisk, A. D. (1982). Concurrent automatic and controlled visual search: Can processing occur without costs? *Journal of Experimental Psychology: Learning, Memory and Cognition, 8*, 261–278.

Small, A. M. (1987). Design for older people. In G. Salvendy (Ed.), *Handbook of human factors* (pp. 495–504). New York, NY: Wiley.

Smith, E. A. L., & Earles, J. L. (1996). Memory changes in normal aging. In F. Blanchard-Fields & T. M. Hess (Eds.), *Perspectives on cognitive change in adulthood and aging* (pp. 192–220). New York, NY: McGraw-Hill.

Sollenberger, R. L., & Stein, E. S. (1995). A simulation study of air traffic controllers' situation awareness. *Proceedings of an International Conference: Experimental Analysis and Measurement of Situation Awareness* (pp. 211–217). Daytona Beach, FL: Embry-Riddle Aeronautical University Press.

Stern, J. A., Wang, L., & Schroeder, D. (1995). Physiological measurement techniques: What the heart and eye can tell us about aspects of situational awareness. *Proceedings of an International Conference: Experimental Analysis and Measurement of Situation Awareness* (pp. 155–162). Daytona Beach, FL: Embry-Riddle Aeronautical University Press.

Tenney, Y. J., Adams, M. J., Pew, R. W., Huggins, A. W. F., & Rogers, W, H. (1992, July). *A principled approach to the measurement of situation awareness in commercial aviation* (NASA Contractor Report 4551). Washington, DC: National Aeronautics and Space Administration.

Tun, P. A., & Wingfield, A. (1997). Language and communication. In A. D. Fisk & W. A. Rogers (Eds.), *Handbook of human factors and the older adults* (pp. 125–149). San Diego, CA: Academic Press.

Wickens, C. D., Braune, R., & Stokes, A. (1987). Age differences in the speed and capacity of information processing: 1. A dual-task approach. *Psychology and Aging, 2*, 70–78.

Wilson, G. F. (1995). Psychophysiological assessment of SA. *Proceedings of an International Conference: Experimental Analysis and Measurement of Situation Awareness* (pp. 141–145). Daytona Beach, FL: Embry-Riddle Aeronautical University Press.

# Chapter 14

# Situation Awareness, Automaticity, and Training

Wayne L. Shebilske
*Wright State University*

Barry P. Goettl
*U.S. Air Force Armstrong Laboratory*

Daniel J. Garland
*SA Technologies, Inc.*

Separate lines of research on training and *situation awareness* (SA) are converging to challenge dominant assumptions regarding automatic and controlled cognitive processing dynamics as complex skills develop. Specifically, prominent learning models suggest that controlled cognitive processes dominate early in training and become less important later in training as skill develops and automatic cognitive processes take over (e.g., Ackerman, 1987, 1988, 1990; Anderson, 1982, 1983; Fitts & Posner, 1967; Rasmussen, 1983; Schneider & Shiffrin, 1977; Shiffrin & Schneider, 1977). Shebilske, Goettl, and Regian (1999) noted that these single cycle models have promoted an assumption of one progressive alternation from high to low importance of controlled processing during skill development. Shebilske, Goettl, and Regian (1999) also proposed that an interactive iterative model more accurately describes processing dynamics. This model applies learning phases (controlled, intermediate, and automatic) to interactive task parts rather than to whole tasks and applies phases iteratively within components. As skill develops, this iterative interactivity increases the importance of controlled processes in two ways. First, controlled processes take on more intellectually challenging strategic problems as automatic processes take over lower order functions. Second, controlled cognitive processes take on the additional responsibilities of monitoring the outcome of automatic cognitive processes and correcting their failures.

Our early steps beyond single cycle models lead us to a rich literature that has documented SA's fundamental role in skilled performance (e.g.,

Endsley, 1995, 1999; Endsley & Bolstad, 1994; Endsley & Kaber, 1999; Garland & Endsley, 1996; Gilson, Garland, & Koonce, 1994). SA's definition, "the perception of the elements in the environment within a volume of time and space, the comprehension of their meaning and the projection of their status in the near future" (Endsley, 1988, p. 789), suggests that SA is a phenomenon mediated by controlled cognitive processes as opposed to automatic cognitive processes. Explicit control processes include conceptual meaning with specific parameters (e.g., time and location) as well as attention management strategies. This chapter reviews the theoretical and empirical foundations for understanding these dimensions of explicit control processes.

Further steps into the SA literature lead to the possibility of an intriguing parallel between overemphasizing cognitive automation during training and overemphasizing computer-automation in complex systems. Corrington and Shebilske (1995) suggested that too much emphasis on automation during training might reduce the effective operation of controlled cognitive processes in posttraining situations (cf. Shebilske, Goettl, & Regian, 1999). Researchers have documented a similar trade-off between computer automation and SA. Specifically, research suggests that computer automation of decision-making functions may reduce an operator's awareness of the system and of environmental dynamics (e.g., Endsley & Kaber, 1999; Endsley & Kiris, 1995; Endsley, Onal, & Kaber, 1997; Kaber & Endsley, 1997a, 1997b, 1997c; Kaber, Endsley, & Cassady, 1997). We review these results, interpret them with respect to the controlled and automatic cognitive processes behind SA, and apply this interpretation to training.

The steps beyond single cycle models are important from a human factors engineering perspective. Human factors engineers improve the fit between humans and machines through systematic design of the system itself, performance support systems, and training. Their approach is human-centered (Norman, 1986, 1992) in the sense that they prioritize changing machines to fit people as opposed to the other way around and they take into account human abilities when designing performance support systems and training. Present steps beyond single cycle models are aimed at improving human-centered shared control between operators and computer automation as well as human-centered training to enhance SA.

## SA'S FUNDAMENTAL ROLE IN SKILLED PERFORMANCE

Research documents the vital role SA plays in skilled performance, particularly in aviation (e.g., Endsley, 1995, 1999; Endsley & Bolstad, 1994). In piloting an aircraft or controlling air traffic, SA refers to the pilot's and controller's ability to recognize and comprehend the state of the aircraft systems, the state of the airspace and its accompanying air traffic, the components of the flight path, and the understanding of their respective predicted states.

Research on decision making (e.g., Klein, 1993; Klein, Orasanu, Calderwood, & Zsambok, 1993; Orasanu, 1993) and situation models (Van Dijk & Kintch, 1983) pointed to the importance of SA in effective pilot and controller decision making and effective and accurate assessment of the operational environment. The quality of the decisions made is contingent on an understanding of the situation parameters and their relations, and the acquisition and application of accurate and timely information. The ability of the flightcrew to maintain accurate and up-to-date SA is critical to the resultant decision-making processes and the safety of the aviation system (Strauch, 1996).

Aviation accidents all too often provide stark evidence of faulty decision-making processes resulting from inadequate situation assessment and awareness (see Garland, Wise, & Hopkin, 1999). For example, on February 1, 1991, at 6:07 p.m. (PST), a USAir Boeing 737–300, while landing on a runway at Los Angeles International Airport, collided with a SkyWest Fairchild Metroliner that was positioned on the same runway awaiting clearance for takeoff. The Boeing 737 and a large portion of the Metroliner that was crushed beneath the Boeing 737's left wing continued down the runway, veered left, and impacted a vacant building, resting 1,200 feet from the point of collision (National Transportation Safety Board [NTSB], 1991). As a result of the collision and postcrash fire both airplanes were destroyed. All 10 passengers and 2 crewmembers aboard the Metroliner and 20 passengers and 2 crewmembers aboard the Boeing 737 were fatally injured (Garland, Stein, & Muller, 1999).

John Lauber, working for the NTSB at the time, noted that:

> On the surface, the accident was simple: the local controller forgot that she had cleared SkyWest into position on the runway, on which she subsequently cleared USAir to land. Controller error "caused" this accident. But one must ask, Was this a simple controller error accident, or do other factors need to be considered? Most importantly, what must be done to minimize the probability of another accident of this kind? . . . "Forgetting" is not necessarily the same as "not attending to duty," and all the signs are that she was attending to her duties that day. Secondly, clearing an aircraft into position on a runway and then "forgetting" about that aircraft is not a unique event. A search of NASA's Aviation Safety Reporting System data base revealed that several instances had been reported, and a very similar incident occurred not long ago at San Diego. A fundamental characteristic of human performance is that "forgetting" is all too easy. Short-term memory . . . is highly vulnerable to intervening events disrupting it. Add a distraction here and a little time

pressure there, and presto, people forget—even very important things. (Lauber, 1993, pp. 24–25).

This accident and Lauber's assessment point to the fundamental need for developing and maintaining an accurate and up-to-date situation model of the rapidly changing environment. This situation model forms the central organizing feature from which all decision making and actions result. A fundamental and critical role for the operator is to develop and maintain an accurate mental and situation model. The following are two tragic examples depicting the criticality (and its volatility) of SA for aircrews. The following two examples are adapted from Krause (1995, pp. 135–139).

The first example involves an electrical problem on an aircraft. On October 11, 1983, Springfield, Illinois was experiencing instrument meteorological conditions with a cloud base at 2,000 feet MSL and cloud tops at 10,000 MSL. Visibility below 2,000 feet was one mile in the rain with scattered thunderstorms reported in the area. The Air Illinois Hawker Siddley 748-2A departed Springfield, Illinois around 8:20 p.m. (CDT). At 8:22 p.m., the crew radioed Springfield departure control and reported they had just experienced a slight electrical problem and were continuing to their destination of Carbondale, Illinois, about 40 minutes away.

Flight 710 was cleared to maintain 3,000 feet and the controller asked if he could be of any assistance. The crew responded, "we're doing okay, thanks."

At 8:23 p.m., the first officer told the captain that "the left [generator] is totally dead, the [generator] is putting out voltage, but I can't get a load on it." Seconds later, the first officer reported, ". . . zero voltage and amps on the left side, the right [generator] is putting out . . . [volts] . . . but I can't get it to come on the line." The first officer reported to the captain that the battery voltage was decreasing "pretty fast."

At 8:26 p.m., Flight 710 called into Kansas City Center with an "unusual request" for clearance to descend to 2,000 feet, ". . . even if we have to go VFR." However, because 2,000 feet was below Kansas City Center's lowest usable altitude and because the controller couldn't guarantee radar contact, the captain decided to remain at 3,000 feet.

At 8:28 p.m., the captain said, "Beacon's off . . . ," ". . . nav lights are off." The first officer reminded the captain that Carbondale had a 2,000 foot ceiling and the visibility was two miles with light rain and fog.

The captain instructed the flight attendant to inform the passengers that the airplane has experienced ". . . a bit of an electrical problem . . ." and that they were proceeding to Carbondale, 27 minutes away.

The first officer reported to the captain that ". . . when we . . . started losing the left one [generator], I reached up and hit the right [isolate button] trying to isolate the right side . . . cause I assumed the problem was the right side but they [the generators] both still went off."

At 8:51 p.m., the first officer told the captain, "I don't know if we have enough juice to get out of this." The captain then responded, "Watch my altitude, I'm going to go down to twenty-four hundred [feet]." At 8:53 p.m., the first officer reported, "We're losing everything . . . down to thirteen volts. . . ." A minute later the aircraft was at 2,400 feet and the captain asked the first officer, "Do you have any instruments? Do you have a horizon [attitude director indicator]?"

Sometime after 8:54 p.m., Flight 710 crashed 40 nautical miles (nm) north of its destination airport, killing 3 crewmembers and 7 passengers.

The NTSB investigation indicated that a probable cause of the accident was the captain's decision to continue the flight after the loss of direct current electrical power from both airplane generators. The flightcrew's inability to assess the airplane's battery endurance after the loss of generator power was the result of the formation of a faulty situation model that led to a loss of SA. The flightcrew's inadequate SA caused them to incorrectly weigh the magnitude of the risks involved in continuing to the destination airport.

The second example occurred on June 8 1992, when the crew of the GP Express Beech 99 departed from Atlanta, Georgia for the 50 minute flight to Anniston, Alabama. The captain had been hired 9 days prior to the accident—the day of the accident was his first day on the job. The first officer had been flying the Beech 99 for about 5 weeks and it was his first day flying the airline's southern route structure.

After departure, Flight 861 was cleared to maintain 6,000 feet. At 8:41 a.m., Atlanta Center cleared Flight 861 to ". . . descend at pilot's discretion, maintain five thousand [feet]." The captain asked the first officer, "Does he want us to resume own navigation?" The captain then said, "I heard him say that. As far as I'm concerned, I'm still on vectors two eight zero [heading]." The first officer responded, "Yeah, two eight zero's fine. Because we're on course anyway, so let's just hold it."

The captain then remarked that he thought they were ". . . slowly drifting off [course]." The captain asked, "What's the course?" The first officer replied, ". . . zero eight five inbound." The captain then concluded that ". . . we're way off course."

In fact, 085 degrees was the outbound course from the VOR that they had been tracking. The inbound course, to which the aircraft was headed, was the reciprocal of 085, or 265 degrees.

Atlanta Center then informed the crew that radar contact was terminated and they were to go over to Birmingham approach control.

At 8:43 a.m., Birmingham approach instructed the flight to descend and maintain 4,000 feet and to proceed direct to the [Talledega] VOR.

Birmingham approach also told them to expect a visual approach into Anniston. If the visual approach was not possible, the crew was to set up for an Instrument Landing System approach to runway five from over the BOGGA approach fix.

Several minutes later the controller notified them to "Proceed direct BOGGA, maintain four thousand 'til BOGGA, cleared . . . ILS runway five approach." Rather than ask the controller the distance to BOGGA, the first officer mentally computed the distance as being 5 miles.

By the time the first officer had tuned in the localizer, the airplane had gone ". . . right through it." The crew thought they were ". . . right over BOGGA . . . four and a half [miles] out. . . ." The first officer told the captain to, ". . . go ahead and drop your gear . . . speed checks."

The captain then asked, "What's the minimum altitude I can descend to 'til I'm established?" The response was ". . . twenty-two hundred [feet]."

By 8:50 a.m., the weather was moving in their direction and was reported to be only 2 miles from the localizer outer marker. The first officer notified Birmingham approach that the "procedure turn inbound [was] complete." But less than a minute later, the captain told the first officer ". . . we gotta go missed [approach] on this." The first officer replied ". . . there you go . . . there, you're gonna shoot right through it again . . . keep 'er goin' . . . you're okay." Within seconds, the captain called out, ". . . there's the glideslope." This was followed by the first officer saying "We can continue our descent on down. We're way high."

Immediately after they started their descent, they lost the glideslope. The captain then decided to go around. He asked the first officer "What's our missed approach point now?" His reply was "Twelve hundred [feet] . . . coming up. . . ." One second later was the sound of impact.

The aircraft crashed a little more than 7 miles from the airport at the 1,800 level. The captain and two passengers were killed and the first officer and the other two passengers received serious injuries.

The NTSB determined that the probable cause of the accident was the failure of the flightcrew to use approved instrument flight procedures that resulted in a loss of SA and terrain clearance.

These two accidents illustrate the significance and criticality of accurate SA to flight operations.

## ANALYZING THE CONTROLLED AND AUTOMATIC COGNITIVE PROCESSES BEHIND SA

The phenomenon of SA must be explained by psychological constructs that have an identity separate from the phenomenon itself (cf. Flach, 1995). Researchers have therefore drawn on established psychological constructs to explain SA phenomena. For example, two studies (Endsley,

1995, 1999) utilized Van Dijk and Kintch's (1983) construct, situation model. Building on existing theoretical foundations is consistent with the scientific rule of parsimony according to which the best explanations add the fewest new constructs.

## Situation Model and Related Constructs

Van Dijk and Kintch (1983) explained that situation models are necessary to account for the phenomenon of comprehending events whether one experiences the events directly, reads about them, or hears about them. Those interested in SA for directly experienced events can therefore draw on their rich arguments even though Van Dijk and Kintch concentrated on reading. Consider, for example, their arguments concerning reference, coherence, situational parameters, and perspective.

*Reference.* A situation model is needed to close the gap between input data's conceptual meaning and its reference to elements such as properties, relations, and facts in a possible world. Words such as "my friend" and "the dean" have conceptual meanings and they make reference to things in the world. The importance of the distinction is especially clear when words with different conceptual meanings refer to the same individual such as my friend, Steve, the dean. The situation model is needed to represent this same individual. Similarly, cockpit displays have conceptual meaning and make reference to things in the real world. The situation model is required to represent the things referenced, such as a plane's current speed and altitude.

*Coherence.* The situation model is also needed to represent the relations between the referenced things that are different than the relations in the input data. This difference means that two different sets of relations determine coherence in the situation model and in the input data. When the input is English words, the text representation is bound by the grammar learned in English classes and by conditional or temporal and causal relations among propositions. In contrast, relations among the referenced things in the situation model are richly elaborated structures of a person's internal knowledge contained in episodic and semantic memory. This distinction is clearest when references are less familiar to one person than another. For instance, two people can form a text representation and a situation model of "my friend, Steve, the dean." The first author's text representation would be the same as the reader's, but the situation models would be very different. The knowledge of friends, Steve, deans, and the first author determines the situation model. The author's knowledge of Steve and self is richer than the reader's so the resulting situation model is more richly elaborated with inferences supported by knowledge. The author's situation model includes, for example, that Steve is head and shoulders taller, about a year older, and a social psychologist who enriches his job by deftly applying psychological principles.

Similarly, suppose two pilots heard about an approaching storm. They would form equivalent text representations if they both understood the language, but they would form different situation models if they had different experiences with storms. Different experiences lead to different situation models with different inferences about the consequences of going through the storm and about alternative routes around the storm.

This example can be generalized beyond language by thinking of the text representation as a local representation determined primarily by the input data and the situation model as a global representation influenced extensively by internal knowledge. This distinction between local and global representations remains critical when the input data is experience mediated by instruments, such as seeing the storm on a radar screen, or direct experience, such as seeing the storm out the window. In all cases, the input data determines a local representation and internal knowledge enables a more richly elaborated situation model. Two pilots who know the rules for interpreting radar form similar local representations of a storm, but they form different situation models if they had different experiences with storms. Ericcson and Kintsch (1995) developed a process model of long-term working memory to explain how coherence is achieved between a local representation and internal knowledge.

*Situation Parameters.* Situation models are also needed to represent specific parameters (e.g., time and location). Authors sometimes force readers to infer these parameters from context information. The reader might infer from contextual details that a novel is about London in Dickens' day even though the text never mentions the location or time. It is difficult or impossible to account for such parameters in a text representation, but it is easy to account for such parameters in a situation model that is "indeed a 'model of the situation' " (p. 339, Van Dijk & Kintch, 1983).

*Perspective.* A text representation is poorer than a situation model at taking into account multiple perspectives that occur when authors describe events from the points of view of different people. A situation model provides a stable reference for taking into account these multiple perspectives. Generalizing beyond language, the incorporation of multiple perspectives is especially interesting when teammates must collaborate when they view events from different points of view.

Consider a paratrooper airdrop mission in which the air commander responsible for the aircraft must collaborate with the ground commander responsible for the paratroopers. Suppose that the best altitude and approach for an airdrop is different from these two perspectives and that the mission requires the least overall risk to both the aircraft and the paratroopers. The team's success will depend on the extent to which the air commander and the ground commander can integrate their situation model (Rentsch & Hall, 1994). The challenge is similar to that of the proverbial three blind men each feeling different parts of an elephant. The only way they can approach a realistic model is to share and integrate their private models. The required integration for the air and ground commanders is enabled by the ability of situation models to accommodate multiple perspectives and by the ability of each team member to form a model of the other team member's model.

### Distinguishing Situation Models and Mental Models

The representation of situation parameters in situation models provides another way to think of the situation model. It can be thought of as a parameterized mental model. Thinking of the situation model in this way brings to bear established literature on mental models that are general representations of how things work (Johnson-Laird, 1983; Norman, 1988). A mental model is a general framework for understanding many scenarios, each with its own specific parameters; a situation model represents one's SA and one's understanding of a specific situation. Pilots have mental models for flying through and around storms. They can run many different scenarios through their mind or they can construct situation models when input data specifies specific parameters, such as a specific plane, a certain kind of storm, or specific destinations. The literature on mental models provides valuable information about situation models. For instance, theorists remind scientists that they cannot directly observe another's mental model. Researchers infer another person's mental model by measuring observable behaviors. These inferences must take into account individual differences such as the experiential differences discussed earlier. The result of the scientist's inferences is not the original mental model, but a mental model of a mental model. SA researchers inherit this important lesson when they employ the established psychological construct of the situation model.

The challenges of inferring and distinguishing situation models and mental models are illustrated in the scenario depicted in Fig. 14.1. This figure shows a four-way intersection with traffic lights. Driver 1 is an inexperienced driver approaching the intersection from the south and intends to turn right using the turning lane. As the driver approaches the intersection the light turns amber. Driver 1 notices this and, as the driver rounds the turn, checks back to the left to make sure that the east bound traffic is

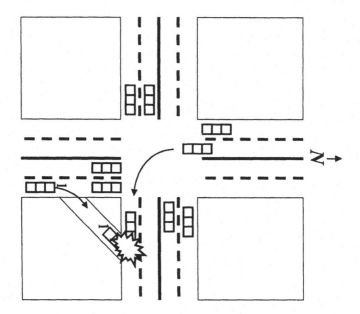

FIG. 14.1.  The text discusses difficulties in inferring a driver's mental model and situation model from observed behavior in the illustrated traffic accident.

still stopped at the light. Because of inexperience, Driver 1 does not appreciate the need to gather information about approaching cars and fails to notice the car turning left from the southbound lane into the eastbound lane. The cars collide. This accident can be attributed to poor SA. Driver 1 checked the eastbound traffic, but did not check for traffic that was turning onto the eastbound lane and crashed into the turning car. Compare this driver's behavior with several other hypothetical drivers in the same situation.

Driver 2, a cautious driver, always comes to a full stop when turning a corner. Although, Driver 2 did not see the turning car, the driver avoided a wreck simply by stopping at the corner. It cannot be inferred that this driver had better SA than Driver 1 despite a better outcome. Driver 2 avoided the wreck by stopping, not because the driver was more aware of the situation.

Consider Driver 3 who saw Driver 1 wreck. A few days later Driver 3 experienced a similar situation at the same intersection. This driver remembered how Driver 1 wrecked and adapted a model of the intersection to include turning cars. This driver was lazy, however, and although the driver had an accurate mental model of the intersection, the driver did not check for turning traffic. Consequently, this driver replicated the accident of Driver 1 and the accident can be attributed to poor SA. However, Drivers 1 and 3 demonstrate an important difference. Both drivers had poor SA, but Driver 1 developed a faulty situation model because of a poor mental model whereas Driver 3 developed a shabby situation model despite an accurate metal model. This difference was supplied as part of the example, it was not inferred from the outcome.

Finally, consider Driver 4, who was able to observe Drivers 1 and 3 and encountered the same situation a few days later. Driver 4 incorporated the information into a mental model of the intersection and used it to scan the situation. Because this driver had an appropriate mental model of the situation and used the model as a framework for attending to useful sources of information, an appropriate situation model was formed and the wreck was avoided.

These scenarios illustrate the challenges associated with inferring mental models and situation models form observed behavior. With inappropriate rules for drawing inferences from outcomes, one might conclude that Drivers 1 and 3 had poor SA and that Drivers 2 and 4 had good SA. This erroneous inference does not distinguish Drivers 1 and 3, who are quite different. The inference is also erroneous because it does not distinguish Drivers 2 and 4. Rules for inferring Situation Models and Mental Models must take into account attention management strategies. Driver 4 had a good mental model of the intersection as well as a good situation model and used them in an effective attention management strategy to avoid an accident; Driver 2 avoided the accident simply because of cautious behavior, not necessarily good SA.

Perhaps the most interesting difference is between Drivers 1 and 4. Both drivers use their mental model of the intersection to scan the environment to develop a situation model. The difference between the drivers is that Driver 4 has a more complete mental model of the intersection, forms a more accurate situation model, and avoids the wreck. Driver 1 wrecks because the novel event violates expectations. The inappropriate expectations came from a flawed mental model, that, in turn, led to a flawed situation model. The mental model itself did not include information about the unexpected event. To the driver it was a novel situation. If the driver had noticed the turning car by chance, it might have been accommodated in the formation of a more appropriate mental model. The improved mental model would have fostered more appropriate situation models in the future.

This example illustrates the general principle that a better mental model of the human–machine system and the environment fosters better situation models. However, mental models are rarely, if ever, perfect because they do not anticipate all possible events or interactions between multiple events in all situations (e.g., Suchman, 1987). Humans are included in computer automated systems to deal with the novel, unexpected events that cannot be predicted by programmers of the automated systems. Humans must also be attentive to unexpected events in order to accommodate to flaws in their own mental models. Accordingly, accurate situation models depend on both mental models and attention management strategies.

## Attention Management Strategies

The interactive iterative model highlights the changing interplay between automatic and controlled processes as humans reach mastery skill levels on complex tasks. It suggests that attention management strategies are central to understanding how experts respond to expected, unexpected, and novel events. Throughout skill acquisition a human's expectations provide a framework for attention management strategies. Early in training, there is relatively little information about the task; all events are new and unexpected. Consequently learners show marked inefficiencies in attention management. As skill develops, however, the expectations of the human become embodied in a mental model. This cognitive representation includes representations of the task, the system, the human's own capabilities, and the environment in which these elements interact. Thus, to understand how the role of attention mechanisms changes as operators of complex systems develop skill, one must understand how the operator's mental model changes with experience and how these changes influence automatic and controlled processes.

Four changes are relevant. First, the operator grows a richer database of possible events and possible corrective actions. Execution of the corrective actions can reach automatic cognitive control and can be performed rapidly and accurately with nothing more than a conscious intention. This type of automatic, preprogrammed response is what Rasmussen (1986) called *skill-based behavior*. Likewise, detection of possible events can reach the level of automatic control and these events become strongly associated with the corrective actions. In this way, events can reach a level of automatic detection and correction and be executed before they reach the level of conscious awareness. That this information can be accessed automatically implies that at least a portion of the model is encoded implicitly and therefore inaccessible to verbal processing.

A second change in an operator's mental model and its relation with automatic and controlled cognitive processes is that the mental model grows a rich database of overall event rates as well as event transition rates. That is, the operator can estimate the probability that certain events will occur and can predict what events are likely to follow the current event (i.e., event transition). This information influences both automatic and controlled processes. Automatic processes are manifested as anticipatory behaviors. For example, a strong temporal association between event A and event B can lead an experienced operator to begin searching for event B the moment event A is detected. This is illustrated in Moray's model of visual sampling behavior derived from eye movement patterns characteristic of fighter controllers (Moray, 1984). This patterning is also illustrated by several researchers who have showed evidence of predictable patterns of eye movements in pilots during normal operational situations (Bellenkes, Wickens, & Kramer, 1997; Fitts, Jones, & Milton, 1950). Finally, Moray and Rotenberg (1989) showed how eye movement patterns change before and after unexpected events.

Controlled processes may be invoked in the form of "metaattention" behaviors or in the form of executive processes when expectations are violated. In the first case, if event A is associated with numerous dangerous, unpredictable events, the detection of the event may lead to increased vigilance or a general increase in the level of caution. These metaattention behaviors may change the way the human samples the environment.

In the second case, unexpected events can invoke controlled processes when automatic processes are unable to handle the unexpected events. Because the mental model provides probabilistic expectations of future events, mismatches sometimes occur. Sometimes the unexpected events are familiar enough to be associated with rules indicating the proper corrective actions. This level of performance is what Rasmussen (1986) characterizes as *rule-based behavior*. When unexpected events involve events that are unfamiliar or for which there is no associated response, automatic processes are interrupted and task operation must be taken over by controlled processes. This is when attention management skills are most crucial. The interactive iterative model of skill acquisition assumes that controlled processes must monitor the automatic processes and detect when unexpected events exceed the capacity of automatic processes to control the situation. Once these events are detected, the controlled processes must quickly take over the task and find an appropriate corrective action that either resolves the problem or brings the task back within the control parameters of the automatic processes. During this period the task becomes extremely cognitively demanding because the controlled processes are slow, serial, and consume attention resources. This type of corrective behavior is similar to what Rasmussen calls *knowledge-based behavior*.

A third relevant change in an operator's mental model and its relation to automatic and controlled cognitive processes is that the set of truly novel events become fewer, but the remaining ones become more complex. Novel situations require slow, demanding, controlled processes so the reduction in novel events increases automatic cognitive processing. An experienced operator pushes the limit of optimal performance because previously novel events are no longer novel and no longer capable of disrupting automatic processes. However, as near optimal performers put themselves in increasingly complex situations, the remaining set of novel events tends to be composed of an increasingly higher concentration of complex interactions between multiple events. For example, a top gun in a dogfight automatically performs many operations that require thoughtful controlled processes during a first solo flight. However, the dogfight includes complex interactions between events that are not present during a first solo flight. These emergent interactions are particularly hard to deal with because they are novel and complex. Little, if any, information is available for prescribing the appropriate types of actions to deal with these complex novel situations.

The increased concentration of complex novel events is countered somewhat by a fourth change in the mental model and its relation to automatic and controlled cognitive processes. The expert's mental model promotes better diagnostic activities by providing a richer understanding of the functional properties of the system, the task goals and requirements, the operator's own capabilities, and relevant environmental factors. These enrichments are largely verbally coded and are accessible to conscious processes. As such, the enriched mental model can be used to diagnose the state of the system based on the pattern of "symptoms" observed. In addition, it can be used to plan corrective actions and make predictions about the effectiveness of those actions. These behaviors also match Rasmussen's (1986) characterization of knowledge-based behavior.

These four changes in the mental model and its relation to automatic and controlled cognitive processes as skill develops help us understand why computer automation can reduce an expert's SA.

Some of the changes in an operator's mental model increase the importance of automatic cognitive processes whereas others increase the importance of controlled cognitive processes. Specifically, the first change makes automatic cognitive processes fundamental parts of effective skilled performance. The fourth change makes controlled cognitive processes critical for successful skilled performance. The second and third changes increase the importance of both automatic and controlled cognitive processes. Successful attention management strategies must take into account the increasingly important role of both automatic and controlled cognitive processes as must effective human-centered designs.

## HUMAN-CENTERED SHARED CONTROL BETWEEN OPERATORS AND COMPUTER-AUTOMATION

As mentioned earlier, computer automation of decision-making functions in complex systems may reduce an operator's SA with respect to the system and the environment (e.g., Endsley & Kaber, 1999; Endsley & Kiris,

1995; Endsley, Onal, & Kaber, 1997; Kaber & Endsley, 1997a, 1997b, 1997c; Kaber, Endsley, & Cassady, 1997). Endsley and Kaber (1999) formulated taxonomic levels of automation (LOA) including low levels, such as assistance in implementation and high levels, such as supervisory control. Endsley, Onal, and Kaber (1997) used this taxonomy to analyze the consequences of malfunctions of system automation in teleoperations involving simulated movements of hazardous materials. They found that high LOA preceding a malfunction of computer automation led to significantly worse SA and corrective actions. They found better SA and corrective actions when the malfunctions occurred after lower levels of automation that involved the human in decision making.

Why does high LOA diminish SA and corrective actions after malfunctions? A partial answer can be found by considering how a high LOA changes an operator's mental model and its relation with automatic and controlled cognitive processes. The previous analysis suggests that a high LOA will yield inadequate mental models for both automatic and controlled cognitive processes during corrective actions. The mental model's representation of possible events, event rates, and the rates of transitions between events support automatic cognitive processing in conjunction with the computer automated system because the representation is appropriate for this situation. However, the mental model does not support automatic cognitive processes for corrective actions when the computer automation fails because the represented database will not fit the situation.

Similarly, the mental model of functional properties of the system, the task goals and requirements, the operator's own capabilities, and relevant environmental factors fit the system when the computer automation is functioning, but not when it is malfunctioning. Consequently, the mental model provides poor support for controlled cognitive processing during malfunctions of computer automation. These consequences for automatic and controlled cognitive processing have a major impact on the potential advantages and disadvantages of computer automation. Therefore, understanding them is necessary for effective human-centered shared control between humans and computer automation.

## HUMAN-CENTERED TRAINING TO ENHANCE SA

Computer automation and training are both interventions aimed at improving performance. It has been argued that computer automation hits its mark more effectively if it takes into account how the intervention changes an operator's mental model and its relation with automatic and controlled cognitive processes. We conclude this chapter by arguing that the same considerations improve human-centered training interventions to enhance SA.

Because an operator's mental model of a complex environment can never be fully complete, human-centered training should prepare skilled operators to encounter complex novel events. Such training should focus on attention management strategies because these strategies are at the heart of responding to unexpected events. The previous analysis suggests that these strategies change in response to the increasingly important role of both automatic and controlled cognitive processes as skill develops. Accordingly, human-centered training must promote effective development of both automatic and controlled processes. Many systematic training designs emphasize the promotion of automatic cognitive processes as skill develops (Gordon, 1994). This emphasis should be extended to include controlled cognitive processes, including detection of system failures, attention switching, and knowing when and how to apply metaattention skills.

Researchers have shown that aspects of attention related to sampling and time-sharing can be trained (Damos & Wickens, 1980; Gopher & Brickner, 1980; Senders, 1964; Sheridan, 1972). This research was the foundation for several protocols that showed promise as techniques for training-controlled cognitive processes for complex tasks. Gopher, Weil, and Seigel (1989) developed one protocol, multiple emphasis on components (MEC), in which participants learn a complex task composed of several components by practicing individual components within the context of the whole task. Their laboratory task was Space Fortress that represents some of aviation's information processing demands, such as short- and long-term memory loading, high workload demands, dynamic attention allocation, decision making, prioritization, resource management, continuous motor control, and discrete motor responses. Components of the task include using a joystick to control the velocity and position of a simulated space ship on a computer monitor, identifying and responding to simulated mines that appear on the monitor, collecting bonuses when opportunities occur, and battling a simulated fortress. Trainees practice the whole task but emphasize individual components. Trainees are instructed to perform the other components but without sacrificing performance on the emphasized one. Periodically throughout training, the component selected for emphasis changes so that all components receive emphasis at some point in training.

This approach has been shown to be an effective training approach (Gopher, Weil, & Bareket, 1994; Gopher et al., 1989). Its main benefit appears to be in training effective attention management skills. Gopher et al. (1989) showed that trainees using the MEC protocol were not as disrupted by the addition of secondary tasks as were trainees not using the MEC procedure. Gopher et al. (1994) showed that training using emphasis instructions transferred to the actual flight performance of Israeli Air Force flight school cadets. Gopher et al. (1994) argued that the emphasis-change protocol trained skills needed for coping with the attention management demands of flight training.

Shebilske, Regian, Arthur, and Jordan (1992) developed an Active Interlocked Modeling (AIM) dyad protocol that also enables trainees to focus attention on part of a complex task in the context of the whole task. It requires trainees to perform each half of a task alternately while being interlocked with a teammate who performs the other half. This protocol was compared with an Individual protocol in which trainees practiced individually. Despite having only half as much hands-on experience during practice sessions on Space Fortress, the AIM-dyad trainees performed as well as Individual trainees on tests performed alone by each trainee under identical conditions. Individual and AIM-dyad protocols have yielded similar performance, yet they have not done so by means of similar cognitive processes according to regression analyses testing the relation between cognitive abilities and Space Fortress performance (Day, Arthur, & Shebilske, 1997). One of several telling results was that for the AIM-dyad protocol, relative to the Individual protocol, the magnitude of the correlation for reasoning ability is higher than that for psychomotor ability (see also Shebilske, Jordan, Goettl, & Day, in press). Reasoning ability was measured by a computerized version of the Raven's progressive matrixes test, and psychomotor ability was measured by a computerized joystick-video-display aiming task (Mané & Donchin, 1989). This result suggested that increased reasoning compensates for reduced hands-on practice for trainees in the AIM-dyad protocol. An adaptation of this protocol has been successful at Aer Lingus Irish National Airline (Johnston, Regian, & Shebilske, 1995). In both contexts, trainees seem to enhance their controlled processing skills.

A more direct technique for enhancing controlled processing is explicit task-related elaborative processing during training. After Johnston had demonstrated beneficial interlesson training effects in protocols designed for Aer Lingus Airline, Johnston, Regian, and Shebilske (1995) expressed a need to analyze such effects in laboratory conditions. In response, Shebilske, Goettl, Corrington, and Day (1998) found that an explicit processing condition between training lessons was better than a filler control condition for training Space Fortress. Increased opportunities for explicit processing facilitate explicit controlled cognitive processing and hinder implicit automatic cognitive processing (Seger, 1994).

Therefore, single cycle models predict that beneficial effects of explicit processing should occur early in training. In contrast, the iterative interactive model predicts that the benefit should occur later in training when controlled processes increase in importance. Shebilske et al. (1998) found that the benefit occurred during the last six training lessons, but not during the first three; this supported the interactive iterative model. The benefit was robust and occurred during acquisition as well as during tests of retention, transfer from joystick to keyboard control, and interference from a secondary tapping task. These results suggested that pilot trainers should develop explicit task-related processing activities during inter-lesson intervals.

A foundation for explaining the benefits of training protocols that emphasize controlled cognitive processes is provided by our analysis of how an operator's mental model changes with experience and how these changes influence automatic and controlled cognitive processes. The analysis suggests that these benefits are mediated by improvements in a trainee's SA that in turn are mediated by improvements in the trainee's mental model and situation model. We will test this hypothesis in future research that will move us beyond our present theoretical integration to an empirical integration of SA and the interactive iterative model of complex skill acquisition.

## REFERENCES

Ackerman, P. L. (1987). Individual differences in skill learning: An integration of psychometric and information processing perspectives. *Psychological Bulletin, 102*, 3–27.

Ackerman, P. L. (1988). Determinants of individual differences during skill acquisition: Cognitive abilities and information processing. *Journal of Experimental Psychology: General, 117* (3), 288–318.

Ackerman, P. L. (1990). A correlational analysis of skill specificity: Learning, abilities, and individual differences. *Journal of Experimental Psychology: Learning, Memory, and Cognition, 16*(5), 883–901.

Anderson, J. R. (1982). Acquisition of cognitive skill. *Psychological Review, 89*, 369–406.

Anderson, J. R. (1983). *The architecture of cognition*. Cambridge, MA: Harvard University Press.

Bellenkes, A. H., Wickens, C. D., & Kramer, A. F. (1997). Visual scanning and pilot expertise: The role of attentional flexibility and mental model development. *Aviation, Space, and Environmental Medicine, 68*, 569–579.

Corrington, K. A., & Shebilske, W. L. (1995). Complex skill acquisition: Generalizing laboratory-based principles to football. *Applied Research in Coaching and Athletics Annual, 95*, 54–69.

Damos, D., & Wickens, C. D. (1980). The acquisition and transfer of time-sharing skills. *Acta Psychologica, 6*, 569–577.

Day, E. A., Arthur, W., Jr., & Shebilske, W. L. (1997). Ability determinants of complex skill acquisition: Effects of training protocol. *Acta Psychologica, 97*, 145–165.

Endsley, M. R. (1988). Situation awareness global assessment technique (SAGAT). In *Proceedings of the National Aerospace and Electronics Conference* (pp. 789–795). New York: IEEE.

Endsley, M. R. (1995). Towards a theory of situation awareness in dynamic systems. *Human Factors, 37*(1), 32–64.

Endsley, M. R. (1999). Situation awareness in aviation systems. In D. J. Garland, J. A. Wise, & V. D. Hopkin (Eds.), *Handbook of aviation human factors* (pp. 257–276). Mahwah, NJ: Lawrence Erlbaum Associates.

Endsley, M. R., & Bolstad, C. A. (1994). Individual differences in pilot situation awareness. *International Journal of Aviation Psychology, 4*, 241–264.

Endsley, M. R., & Kaber, D. B. (1999). Level of automation effects on performance, situation awareness and workload in a dynamic control task. *Ergonomics, 42*(3), 462–492.

Endsley, M. R., & Kiris, E. O. (1995). The out-of-the-loop performance problem and level of control in automation. *Human Factors, 37*(2), 381–394.

Endsley, M. R., Onal, E., & Kaber, D. B. (1997). The impact of intermediate levels of automation on situation awareness and performance in dynamic control systems. In D. I. Gertman, D. L. Schurman, & H. S. Blackman (Eds.), *Global perspectives of human factors in power generation. Proceedings of the 1997 IEEE Sixth Conference on Human Factors and Power Plants.* (pp. 7–7/7–12). New York: IEEE.

Ericcson, A., & Kintsch, W. (1995). Long-term working memory. *Psychological Review, 102*, 211–245.

Fitts, P., Jones, R. E., & Milton, E. (1950). Eye movements of aircraft pilots during instrument landing approaches. *Aeronautical Engineering Review, 9*, 24–29.

Fitts, P., & Posner, M. I. (1967). *Human performance.* Monterey, CA: Brooks/Cole.

Flach, J. M. (1995). Situation awareness: Proceed with caution. *Human Factors, 37*(1), 149–157.

Garland, D. J., & Endsley, M. R. (1996). *Experimental analysis and measurement of situation awareness.* Daytona Beach, FL: Embry-Riddle Aeronautical University Press.

Garland, D. J., Stein, E. S., & Muller, J. K. (1999). Air traffic controller memory: Capabilities, limitations, and volatility. In D. J. Garland, J. A. Wise, & V. D. Hopkin (Eds.), *Handbook of aviation human factors* (pp. 455–496). Mahwah, NJ: Lawrence Erlbaum Associates.

Garland, D. J., Wise, J. A., & Hopkin, V. D. (1999). *Handbook of Aviation Human Factors.* Mahwah, NJ: Lawrence Erlbaum Associates.

Gilson, R. D., Garland, D. J., & Koonce, J. M. (1994). *Situational awareness in complex systems.* Daytona Beach, FL: Embry-Riddle Aeronautical University Press.

Gopher, D., & Brickner, M. (1980). On the training of time-sharing skills: An attention viewpoint. In G. Corrick, M. Hazeltine, & R. Durst (Eds.), *In Proceedings of the 24th Annual Meeting of the Human Factors Society* (Vol. 1, pp. 259–263). Santa Monica, CA: The Human Factors Society.

Gopher, D., Weil, M., & Bareket, T. (1994). Transfer of skill from a computer game trainer to flight. *Human Factors, 36*(3), 387–405.

Gopher, D., Weil, M., & Siegel, D. (1989). Practice under changing priorities: An interactionist perspective. *Acta Psychologica, 71*, 147–178.

Gordon, S. E. (1994). *Systematic training program design: maximizing effectiveness and minimizing liability.* Englewood Cliffs, NJ: Prentice Hall.

Johnson-Laird, P. N. (1983). *Mental models.* Cambridge, England: Cambridge University Press.

Johnston, A. N., Regian, J. W., & Shebilske, W. L. (1995). Observational learning and training of complex skills in laboratory and applied settings. In N. Johnston, R. Fuller, & N. McDonald (Eds.), *Proceedings of the 21st Conference of the European Association for Aviation Psychology: Vol. 2. Aviation psychology: Training and selection* (pp. 316–323). Aldershot, England: Avebury Aviation.

Kaber, D. B., & Endsley, M. R. (1997a). The combined effect of level of automation and adaptive automation on human performance with complex, dynamic control systems. In *Proceedings of the Human Factors and Ergonomics Society 41st Annual Meeting* (Vol. 1, pp. 205–209). Santa Monica, CA: The Human Factors and Ergonomics Society.

Kaber, D. B., & Endsley, M. R. (1997b). Level of automation and adaptive automation effects on performance in a dynamic control task. In P. Seppala, T. Luopajarvi, C. H. Nygard, & M. Mattila (Eds.), *Proceedings of the 13th Triennial Congress of the International Ergonomics Association* (pp. 202–204). Helsinki: Finnish Institute of Occupational Health.

Kaber, D. B., & Endsley, M. R. (1997c). Out-of-the-loop performance problems and the use of intermediate levels of automation for improved control system functioning and safety. In *Proceedings of the 31st Annual Loss Prevention Seminar* (p. 38d). Houston, TX: American Institute of Chemical Engineers.

Kaber, D. B., Endsley, M. R., & Cassady, C. R. (1997). Level of automation for minimizing human out-of-the-loop performance effects on quality. In P. Seppala, T. Luopajarvi, C. H. Nygard, & M. Mattila (Eds.), *Proceedings of the 13th Triennial Congress of the International Ergonomics Association* (pp. 205–207). Helsinki: Finnish Institute of Occupational Health.

Klein, G. (1993). *Naturalistic decision making: Implications for design*. Wright-Patterson AFB, OH: Crew System Ergonomics Information Analysis Center.

Klein, G., Orasanu, J., Calderwood, R., & Zsambok, C. (1993). *Decision making in action: Models and methods*. Norwood, NJ: Ablex.

Krause, S. S. (1995). *Avoiding mid-air collisions*. Blue Ridge Summit, PA: TAB Books.

Lauber, J. K. (1993, July 16). Human performance issues in air traffic control. *Air Line Pilot*, 23–25.

Mané, A. M., & Donchin, E. (1989). The space fortress game. *Acta Psychologica, 71*, 17–22.

Moray, N. (1984). Attention to dynamic visual displays in man-machine systems. In R. Parasuraman & D. R. Davies (Eds.), *Varieties of attention* (pp. 485–513). Orlando, FL: Academic Press.

Moray, N., & Rotenberg, I. (1989). Fault management in process control: Eye movements and action. *Ergonomics, 32*, 1319–1342.

National Transportation Safety Board. (1991). *Aircraft accident report. Runway collision of USAir Flight 1493, Boeing 737 and SkyWest Flight 5569 Fairchild Metroliner Los Angeles International Airport Los Angeles, California, February 1, 1991* (Rep. No. NTSB–AAR–91–8). Washington, DC: Author.

Norman, D. A. (1986). Cognitive engineering. In D. A. Norman & S. W. Draper (Eds.), *User centered system design: New perspectives on human-computer interaction* (pp. 31–61). Hillsdale, NJ: Lawrence Erlbaum Associates.

Norman, D. A. (1988). *The psychology of everyday things*. Garden City, NY: Doubleday.

Norman, D. A (1992). *Turning signals are the facial expressions of automobiles*. Reading, MA: Addison-Wesley.

Orasanu, J. M. (1993). Decision making in the cockpit. In E. L. Wiener, R. L. Helmreich, & B. G. Kanki (Eds.), *Cockpit resource management*. New York: Academic Press.

Rasmussen, J. (1983). Skills, rules, knowledge: Signals, signs, and symbols and other distinctions in human performance models. *IEEE Transactions: Systems, Man, and Cybernetics, 13*(3), 257–267.

Rasmussen, J. (1986). *Information processing and human-machine interaction: An approach to cognitive engineering*. Amsterdam: North-Holland.

Rentsch, J. R., & Hall, R. J. (1994). Members of great teams think alike: A model of team effectiveness and schema similarity among team members. In M. M. Beyerlein & D. A. Johnson (Eds.), *Advances in interdisciplinary studies of work teams. Vol. 1. Series on self management of work teams* (pp. 223–262). Greenwich, CT: JAI.

Schneider, W., & Shiffrin, R. M. (1977). Controlled and automatic human information processing: I. Detection, search, and attention. *Psychological Review, 84*, 1–66.

Seger, C. A. (1994). Implicit learning. *Psychological Bulletin, 115*(2), 163–196.

Senders, J. (1964). The human operator as a monitor and controller of multidegree freedom systems. *IEEE Transactions on Human Factors in Electronics, HFE-5*, 2–6.

Sheridan, T. (1972). On how often the supervisor should sample. *IEEE Transactions on Systems, Science, and Cybernetics, SSC-6*, 140–145.

Shebilske, W. L., Goettl, B. P., Corrington, K., & Day, E. (1998). *Inter-lesson spacing and task related processing during complex skill acquisition*. Manuscript submitted for publication.

Shebilske, W. L., Goettl, B. P., & Regian, J. W. (1999). Individual and group protocols for training complex skills in laboratory and applied settings. In D. Gopher & A. Koriat (Eds.), *Attention and Performance XVII: Cognitive regulation of performance: Interaction of theory and application.* Mahwah, NJ: Lawrence Erlbaum Associates.

Shebilske, W. L., Jordan, J. A., Goettl, B. P., & Day, E. (in press). Cognitive and social influences in a training-team protocol for complex skills. *Journal of Experimental Psychology: Applied.*

Shebilske, W. L., Regian, J. W., Arthur, W., & Jordan, J. A. (1992). A dyadic protocol for training complex skills. *Human Factors, 34*(3), 369–374.

Shiffrin, R. M., & Schneider, W. (1977). Controlled and automatic human information processing: II. Perceptual learning, automatic attending, and a general theory. *Psychological Review, 84,* 127–190.

Strauch, B. (1996). Post hoc assessment of situation assessment in aircraft accident/incident investigation. In D. J. Garland & M. R. Endsley (Eds.), *Experimental analysis and measurement of situation awareness* (pp. 163–169). Daytona Beach, FL: Embry-Riddle Aeronautical University Press.

Suchman, L. A. (1987). *Plans and situated actions: the problem of human machine communications.* Cambridge, England: Cambridge University Press.

Van Dijk, T. A, & Kintch, W. (1983). *Strategies of discourse comprehension.* New York: Academic Press.

# Chapter 15

# Team Situation Awareness, Errors, and Crew Resource Management: Research Integration for Training Guidance[1]

Carolyn Prince
Eduardo Salas
*University of Central Florida*

Three current aviation topics, error, Crew Resource Management (CRM), and *situation awareness* (SA), each with its own body of research and theory, are bound in such a way that changes in the knowledge of one inevitably affects treatment of the other two. Until recently, most of the research and interest in these subjects was at the level of the individual. Now, each is being investigated at the team level. Already, information from this focus beyond the individual has caused a change in CRM training and yielded sufficient information to develop guidance for crew SA training.

SA is part of every aviator's vocabulary no matter the level of flight experience (Prince & Salas, 1999). Although originally associated with the military cockpit, SA is now recognized as necessary to all manned flight; without exception, cockpit environments present crew members with a continuous need to recognize and understand the requirements of a dynamic situation. Depending on the definition used, the concept encompasses components of sensory information, perception, memory, comprehension, and information integration. Because there are many elements important (or potentially important) to a crew's performance, maintaining the SA needed for a flight's dynamic conditions can be very demanding on

[1]The views expressed in this chapter are those of the authors and do not represent the official positions of the organization with which they are affiliated. Portions of this work were conducted under the NAWCTSD/UCF/FAA Partnership for Aviation Team Training.

**325**

the pilot's cognitive resources. When elements are not perceived, are misperceived, or misunderstood, the result is a "flawed" SA that can lead to error (Adams, Tenney, & Pew, 1995). Thus, defining SA to include perceiving, understanding, and then explaining some errors as having occurred through misperceiving and misunderstanding illuminates the natural relation between SA and errors.

For more than 20 years, human error has been recognized as the leading causal factor in accidents (Lauber, 1988). Its reduction is a goal for regulatory agencies, aviation organizations, and individual pilots. CRM training was introduced as one method of trying to achieve this goal. The objective of error reduction is evident even in the earliest CRM training materials where causal factors in accidents, such as the failure of crew members to alert the captain to a problem, were identified and discussed (Helmreich, 1996). Because errors were often traced to problems with one or more crew members' awareness, SA became a common CRM program topic (Helmreich & Foushee, 1993; Kanki & Palmer, 1993). The training of this cognitively complex subject in CRM programs was not designed to improve individual SA (e.g., improve instrument scans, increase understanding of certain signals, or learn to project the current situation into the flight's future); instead, it was focused on the need for crew members to contribute to the awareness of others.

The study of these subjects was originally at the level of the individual. Research and theory developed since 1990 have established a wider interpretation and understanding of error, CRM, and SA. Knowledge about human error in dynamic work environments has increased as researchers look at systems and subsystems for error causes (Moray, 1994; Rasmussen, 1996; Reason, 1994). Already, CRM programs have changed considerably from their initial form; this maturation is likely to continue due to research on team process training both inside and outside aviation (e.g., Prince & Salas, 1999; Salas, Cannon-Bowers, & Blickensderfer, 1997). In the area of SA, the statement, "notions of situation awareness must, at some point, be expanded beyond the individual to the aggregate set of individuals and systems responsible for the performance of a mission" (p. 103, Adams et al., 1995) is being realized. The expansion is occurring; team SA is a research issue studied on its own (Salas, Prince, Baker, & Shrestha, 1995).

In this chapter, we briefly review the traditional approaches to understanding errors, CRM, and SA, point to their interconnections, and describe how these efforts have helped shape training programs. Next, we present a sample of the current approaches to these topics, along with some of the research underlying the present state of knowledge. Finally, we integrate information from all three into guidance for training crew SA.

## HISTORIC CONNECTIONS

SA, error, and CRM training research each have a history that spans more than two decades. Much of what is currently being trained in CRM and crew SA is based on an early phase of this history. In that phase, error and SA, as constructs involving cognitive processes, were studied at the level of the individual in order to understand and explain the processes involved. Although CRM is primarily a group concept, much of the group research applied to the original CRM training programs centered on the input factors of the individual team members (Foushee & Helmreich, 1988).

### SA And Errors

An error, according to Senders (1994), is "failure to perform an intended action that was correct given the circumstances" (p. 166). Senders pointed out that even though there is a connection between errors and accidents, it is not necessarily a one to one relation. Errors may (or may not) result in an adverse consequence, but an accident is not the only possible result from the occurrence of an error.

Further expanding the concept of error, Norman (1981) distinguished three types: slips, violations, and mistakes. A slip occurs when an individual correctly assesses the situation and required action, but makes a wrong movement, such as hitting the wrong key when putting information into the Flight Management System. A violation is a purposeful error where the person takes a nonstandard action and has a reason for doing so. Both slips and violations can be made by someone who is well aware of the situation and the necessary or accepted response to it. A mistake is a different kind of error where the person does not form the right intention and therefore chooses an incorrect action. This happens because of problems with the environment (e.g., not supplying sufficient information, masking information) or with the individual (e.g., inadequate training, memory lapse, failure to focus attention). "Mistakes" is the error category most often associated with a problem with SA.

In studying human error in flight operations, Nagel (1988) looked at the breakdown of SA and characterized it as occurring through faulty acquisition of information or problems in information processing. Nagel depicted the pilot's maintenance of SA as placing demands on the visual, aural, and vestibular senses, all at the same time. That is, information that tells about the plane's location in relation to the earth and to other aircraft, its performance state, power setting, and configuration, all must enter through the senses. These same senses carry messages and directives to the pilot from air traffic controllers, updates on the weather and winds at destination, traffic alerts, and any changes that may require plan modifications. Because there is so much necessary information carried by limited senses, lapses may result (Nagel, 1988). Compounding the problem, the sensory information is received by finite cognitive resources. Further, when a pilot must make a decision even more load is placed on senses and cognitive resources as the pilot seeks additional information (Nagel, 1988).

Asserting that "the notion of skill is inseparable from that of error" (p. 18), Edwards (1988) recognized the omnipresence of error and discussed the need for its control. His approach to understanding error was through relating the type of error to the point in human information processing where a breakdown may occur. As an example, he suggested that at the point when information is received by the sense organs and converted into neural signals, information could be masked or may suffer distortion.

In addressing errors in SA that could negatively affect the decision-making process, Endsley (1995) described them from the perspective of her theory. She separated errors into those due to incomplete SA and those due to inaccurate SA and then related these categories to problems that could occur at the levels of perception, comprehension, and projection. For example, incomplete SA occurs at the level of perception if the situation itself fails to provide information that should be perceived (e.g., fog obscures the outside view, instruments are not functioning). It occurs at the level of understanding, if the person does not know the meaning of what he or she has perceived, and occurs at the level of projection, if the person has insufficient relevant knowledge to be able to project possible future consequences.

Because there are so many elements within the cockpit environment, the argument has been made that SA is not a total awareness of everything in the crew members' environment, but is an "appropriate" awareness of the situation (p. 146, Smith & Hancock, 1995). In the dynamic environment of flight, crew members must select the important elements in the situation from among those that have little or no relevance and must keep doing so as the situation changes. Some of the elements that they must attend to are the required cockpit tasks.

Defining a task as "a process performed to achieve a goal" (p. 308), Chou, Madhavan, and Funk (1996) analyzed errors in managing cockpit tasks by applying an error taxonomy to the data in accident reports and Aviation System Reporting System (ASRS) narratives. This taxonomy had three error categories: task initiation, task termination, and task prioritization. Their search through the ASRS narratives was guided by words like, "forget" and "memory lapse." Chou et al. (1996) concluded that task prioritization was the greatest problem for cockpit crews. They recommended that an aid be developed to monitor task state and status, to compute priorities, to remind pilots of tasks in progress, and to suggest attending to those that are not progressing well. It is noteworthy that many of these actions (monitoring, reminding, backing up other crew members on their tasks) are the same as those frequently mentioned by pilots as indicators of crew or team SA (Prince & Salas, 1997).

No matter what their differences, each of the schemes presented here relating error to SA (i.e., Nagel, 1988; Edwards, 1988; Endsley, 1995) agreed that these errors can be attributed to a number of causes at the level of the individual and his or her environment. Providing an example of the application of this approach, Chou et al. (1996) sought out errors in task management that accident investigators or crew members attributed to forgetting or memory lapses and suggested they could be avoided by providing a memory aid to the pilot. That is, the individual and the individual's immediate environment have been used to explain the cause of error as well as to explore methods of reducing that error.

## CRM Training and Error

In the 1970s, several accidents occurred that claimed hundreds of lives. These accidents raised questions and concerns over safety in the air transport community and created an urgent need to correct the problems that were the causes. Particularly disturbing was the discovery that human error had a central role in a high percentage of the accidents.

With the documentation of human error as the leading causal factor in accidents (Lauber, 1988), regulatory agencies and airline organizations looked for ways to reduce or eliminate this problem. At one time, it had been believed that human redundancy in the cockpit (i.e., additional crew members) would help reduce the incidence of error (Nagel, 1988; Foushee & Helmreich, 1988). Unfortunately, mere redundancy did not prove to be the solution, as was demonstrated by a rash of high profile accidents caused by human error that occurred in cockpits with multiple crew members (Kayton, 1993). In explaining the failure of redundancy to eliminate error, Helmreich and Foushee (1993) pointed out that traditionally, neither the selection process for pilot training nor the prevailing aviation culture was concerned with the need for pilots to think of themselves as part of a crew; rather, from the beginning of aviation, emphasis was placed on independence.

While exploring methods of error reduction for highly trained professional teams who also work in dynamic, demanding environments, Moray (1994) examined the value of human redundancy. He argued that even though redundancy in engineering has been successful, human redundancy is not predictable because of social interaction dynamics. Having a second person to check on the task performance of another can lead to the desired result of improved performance or it can result in an abdication of responsibility by both. Moray warned that the results of employing human workers for redundancy are impossible to predict unless the dynamics of the team are known.

Team dynamics are not determined solely by assigned roles or tasks within the team. Research on leadership in the cockpit (Ginnett, 1993) demonstrated that the cockpit environment, with its assigned positions and task requirements, is not sufficiently structured to ensure teamwork will occur when a crew begins to work together. Ginnett observed nine crews in air carrier operations with captains who were selected for their effectiveness or ineffectiveness as team leaders. He found that certain behaviors of captains are associated with the level of performance demonstrated by their crews and that these behaviors can be observed from the very beginning of crew interactions. For example, effective captains ensure a flow of information from other crew members to themselves by engaging the crew members in conversation.

Evidence from a controlled experiment implicated the entire crew in error commitment, rather than just one individual (Ruffell Smith, 1979). For this experiment, line-qualified crew members flew a scenario designed to be challenging and to require "ever-increasing information retrieval" (p. 5). It also included distractions designed to interfere with essential flight duties. Some of the errors committed by crews were: misreading charts, misunderstanding messages, neglecting speed, and failing to recognize that another crew member was overloaded.

Ruffell Smith found clear differences between crews in the number of errors committed and in the time taken to detect errors. In some cases, errors were made in the handling of one task because all the crew members were involved in another task. To illustrate this, Ruffell Smith cited an example of a crew preparing to land on a short runway after a diversion from their original flight plan. Crew members discussed the power needed, gross weight, and flap settings as they obtained information about the weather and prepared to dump fuel to reduce the aircraft weight. Ruffell Smith noted that because of the crew members' involvement in this problem, they stopped monitoring the aircraft parameters and, as a result, overboosted their engines. One of Ruffell Smith's conclusions was that continual monitoring by the entire crew is an important aspect of performance.

With the failure of simple crew member redundancy to reduce human error, but with evidence that some crew members' actions were effective in preventing and correcting errors (Ruffell Smith, 1979), developing training to encourage positive crew interactions seemed a promising direction to take. CRM training, according to Caro (1988), was to address "situational, socio-psychological, and other factors that influence aircrew performance" (p. 258). Unfortunately, when the first CRM programs were developed, there was little scientific information to offer on the process that defines teamwork despite years of small group research (Foushee & Helmreich, 1988). However, there was information about the effect of the individual characteristics of team members on team outcomes; for that reason, most of the early programs were designed to have an impact on the attitudes and styles of individual crew members (Foushee & Helmreich, 1988). The method used to improve the functioning of the crew was to change attitudes "through an increasing awareness of the importance of these (group function) factors" (Foushee & Helmreich, 1988, p. 223). Another technique was to make crew members aware of their style of interacting with others and encourage them to make changes (if needed) to more appropriate interaction styles (Hackman, 1993). Most programs also demonstrated the relation of team performance to human errors and accidents by including a discussion of accidents, focusing on errors, and discussing the possible remedies or preventative actions.

Training designed to have an effect on attitudes about group functioning was done with the expectation that changed attitudes would lead to improved interactions that resulted in better performance (or fewer errors). There was a fundamental problem with this; it was assumed that individuals with changed attitudes would know what to do to improve crew interactions. Further, it was expected that they would be able to perform the actions needed for effective performance as a team when they returned to the cockpit. Unfortunately, there was no evidence to support these expectations (Prince & Salas, 1993). Similarly, there was no evidence to support the value of training to change interpersonal styles for subsequent performance (Hackman, 1993).

## SA and CRM

Within air transport communities, pilots rarely fly alone. Although errors have been related most often to individual SA, they have important implications for the team. A tragic example of crew error from the operational community occurred in the early 1980s when an Air Florida plane crashed into a bridge over the Potomac. Many of those who listened to the cockpit voice recorder or who read the transcript of that recording believed that the first officer was aware of a problem with the plane that the captain had not recognized. The suggestion has been made that had the first officer told the captain of the problem the accident may have been averted. One important outcome of the research reported by Ruffell Smith (1979) was its demonstration of differences in the crews' abilities to avoid or correct errors related to awareness in the cockpit. It is apparent that using the awareness of all the crew members may reduce the incidence of error or ameliorate its results.

Until recently, the largest barrier to training SA in CRM has been the lack of knowledge about crew SA. An illustration of the state of research on team SA for cockpit crews, 15 years after the first CRM program began, is found in an annotated bibliography of the SA literature (Vidulich, Dominguez, Vogel, & McMillan, 1994). Out of 233 articles, chapters, and reports on SA in the bibliography, fewer than a dozen references focus on the SA relevant to the air carrier cockpit. Of these, only slightly more than half report data on research with crews. This comparative neglect of the subject, despite the common use of multiple crew members in airplane cockpits, has handicapped training developers.

In CRM programs developed after CRM skills were first identified, team skills are introduced as part of the training and often with each skill treated as a separate topic. One skill area that is included in many programs is SA. In initial training where CRM skills are introduced to crew members in the classroom, the skills are presented one by one and may or may not be integrated at the end of the course. In some organizations, one or two different CRM topics are "refreshed" each year during recurrent training. This separation of skills for training deemphasizes the interplay of the various types of actions and interactions that crew members use to maintain effective awareness. Skill areas are discussed in classes (usually through discussion of incidents and accidents). If the training continues in realistic flight scenarios (e.g., Line Oriented Flight Training [LOFT]), there is usually an effort made by instructors to debrief the crew members on their CRM behaviors; however, this process is uneven (Dismukes, Jobe, & McDonnell, 1997). Additionally, there is little real guidance to help the instructor select and evaluate any of the skill behaviors.

## NEW APPROACHES TO ERRORS, CRM, AND SA

The newer approaches to errors, CRM, and SA have placed the research into more complex situations and have taken a broader view. Error, originally studied as a problem with the individual and his or her environment, is now considered in light of the entire system or subsystems. CRM researchers have been responsible for identifying skills that define team process and are beginning to recognize the interactions of these skills to achieve effective crew performance. SA research includes the individual within the crew as well as crew interactions and their effect on SA.

### A Systems Approach to Error

There are other work environments (e.g., medicine and the nuclear power industry) besides the cockpit where workers are generally highly selected and trained, work in dynamic, complex surroundings with rapidly changing technology, and make important decisions. There is also a concern in these professions with error and its reduction (Rasmussen, 1996; Reason, 1994; Moray, 1994). Because of the similarities in the demands of the work environments, research findings in these settings have some relevance for aviation.

After studying error in the medical profession, Moray (1994) described two approaches to error. The first, the traditional approach, emphasized the individual's carelessness and inattention as the contribution to error. The second approach considered the systems within which individuals function an error source. Moray and two other prominent researchers (i.e., Rasmussen, 1996; Reason, 1994) questioned the value of assigning individual blame for errors because it leads to attempts to correct the errors through punishment, encouragement to act in a safer way, or retraining and rewriting procedures (or adding new ones) that directly address a specific error (Reason, 1994). According to Reason, "momentary inattention, distraction, pre-occupation, forgetting, and so on" that are the causes of errors are "transitory", "unintended", and "unpredictable" (p. xiii). Even if punishment, encouragement, or retraining have an effect, it is only temporary if organizational pressures on the job discourage change (Rasmussen, 1996). Equally, trying to control error through the introduction of procedures and rules is not often successful, according to Rasmussen, because it separates behavior from its social context. A rule or instruction is often designed for a particular task in isolation whereas, on the job, several tasks are active in a time-sharing mode. The dynamic context with its value systems forces the crew members to determine which of various possible procedures to use.

Moray (1994) has stated that error reduction must be approached through the system because the environment may push individuals into committing errors despite their experience and training. He defined a system as "any collection of components and the relations between them, whether the components are human or not, when the components have been brought together for a well defined goal or purpose" (p. 70). He described the structure of a hierarchical system as consisting of: the equipment, its ergonomics, individual behaviors, team behaviors, management, regulatory agencies, and society's pressures (p. 70). For the team component, Moray distinguished a team from a group (an informal collection of people) by the existence of a common team task and specified roles. He asserted that it is the breakdown of the team into a group that is a common cause of error.

Changes in any one subsystem may have far reaching and unpredictable results because of the interactivity of the subsystems (Moray, 1994). One familiar example in aviation is the introduction of cockpit automation (the equipment subsystem) to assist the pilot (individual subsystem). Some automated systems affect the team subsystem by changing the amount of communication among crew members (Costley, Johnson, & Lawson, 1989) and have an unintended effect on the individual subsystem by altering the amount of stress and fatigue reported by the crew members (Wiener, 1993). Although designed to aid crews, new equipment can create problems with awareness until crew members become accustomed to its use. Crew members at all levels of experience have stated that anything new in the cockpit demands some of their attention and lowers their awareness of other elements in the situation (Prince & Salas, 1997).

Although the organization's management has an effect on the crew members' interactions, it must be acknowledged that management, in turn, is affected by regulatory agencies and society's pressures. Looking at human error and its reduction from a different angle, Rasmussen (1996) made a case for considering the environment in which the system exists and how that may determine management policies that effect the design and functioning of the subsystems. He pointed out that court opinions in cases involving the most notorious nuclear power industry accidents cite, as the accident cause, pressure from a competitive environment that pushes normal organizational behavior toward accidents—not human error.

As airlines operate in an increasingly competitive environment (exemplified by drawing upper management from financial or managerial experts and not aviation experts [Bryant, 1997]), they invite the dangers described by Rasmussen. Commercial success often implies operating at the fringes of the usual accepted practice as focus moves to short-term financial criteria. One solution, as offered by Rasmussen (1996), is to look for system problems that are likely to put people at risk for error and train them to function in response to those general problem areas. The influence of the operating environment on management's policy and its effect on crew members can be illustrated by on-time departures. In the air carrier industry there is pressure on management for on-time service that is created by the expectations of the traveling public, regulatory agencies, and the costs of running a company (e.g., crew rest considerations, overtime). This can lead to crew members rushing preflight activities in attempts to meet this goal. Yet, these pre-flight preparations have been found to be valuable to team SA (Prince & Salas, 1997) because they set up the team and its expectations for the flight. Additionally, rushed preparations may cause crew members to skip a necessary item in their hurry to complete the procedures and lead to a potential for error. A single checklist item that is missed, such as the position of the flaps, can lead to disaster. Although the crew members may be seen to commit an error, it may be done in an atmosphere where the objective of on-time departure is foremost in their attention. That is, attention implies withdrawal from some things in order to deal effectively with others. Rasmussen (1996) suggested facing this organizational reality by designing training that makes boundaries of acceptable performance explicit and helps individuals develop the coping skills they need at those boundaries instead of fighting specific deviations from a prescribed path.

## Advances in CRM Training

In the late 1980s, a major change in the emphasis in CRM training occurred. Programs began training about team-process skills rather than making crew members aware of the need for teamwork. This resulted from increased knowledge of the process of team functioning (Hackman, 1993; Helmreich & Foushee, 1993; Prince & Salas, 1993). The knowledge was an outcome of research that included a variety of methods (e.g., critical incident interviews, scenario analysis, controlled experiments). Interviews with crew members elicited behaviors observed in experiences with effective and ineffective teamwork in the cockpit. These behaviors were used to identify important skill areas for team process (Prince, Chidester, Bowers, & Cannon-Bowers, 1992). Using another method of inquiry, researchers observed videotapes of crews flying difficult scenarios. They separated crews into those that made few errors and those with a high number of errors and analyzed communications to find behaviors or patterns of behaviors that discriminated between the groups. Some of the differences found were in the way crew members communicated with one another (e.g., commands and replies in Foushee & Manos, 1981; and patterns of communications in Kanki, Lozito, & Foushee, 1989) and in their approach to making decisions (e.g., considering conditions at the destination airport in Orasanu, 1990) that were related to the number of errors made by the crew members. Another method used to investigate CRM skills involved conducting controlled experiments, particularly those used to evaluate training methods and to develop measurement instruments (Prince & Salas, 1999).

Experiments conducted to determine optimal training methods for crew interaction skills showed that training that included practice and feedback was superior for transfer over other methods (e.g., lecture, discussion, modeling; Prince & Salas, 1999). Research also showed that the practice situation for a skill required crew members to make the kinds of decisions they must make in the cockpit and interact with one another in a manner similar to that required in the cockpit. This was referred to as cognitive fidelity (Brannick, Prince, Salas, & Stout, 1998). That is, placing crew members in "game-playing" situations, even those that require teamwork, planning, and decision making, is not as effective for training as placing them in situations that are similar to those they face in the cockpit (Brannick et al., 1998). Most encouraging was the evaluation of training programs that used these methods. These evaluations documented increases in the use of CRM skills as a consequence of training (Salas, Fowlkes, Stout, Milanovich, & Prince, 1999).

While this research was proceeding, some training developers, perhaps as a reaction to the "soft" training that was part of the initial CRM, began to emphasize specific behaviors. This was done partially by changing procedures and checklists to mandate certain crew actions at precise times. A move toward this exact identification of actions can be seen in some checklists used in Line Oriented Simulation (LOS) scenarios. Rather than watching a crew for their demonstrations of skill behavior throughout the scenario, the observer is given clear actions within distinct events to look for and document. There is already some concern that focusing CRM training on skills and specific behaviors may be causing a loss in the original objective of reducing human error (Helmreich, 1996). The expression of this concern is joined by a suggestion to change the training to clearly focus on errors and how they may be overcome (Helmreich, 1996).

### SA in Crews: Research Overview

There are several explanations for the lack of research on SA as it is evidenced in a crew. These include: access to crews, complexity in both the task and the environment, and complexity of the construct. Research in individual SA has shown evidence that experience with the specific task in the task environment is important, both to awareness (Waag & Bell, 1994) and to situation assessment (Klein, 1989). This means that to study crew SA, one must study people in their work environment. However, access to crew members as they work on their jobs (particularly when working space is limited as it is in a cockpit) is difficult to obtain.

Because of the dynamic nature of the cockpit, the multitude of tasks and the ability of the situation to present the crew with unusual demands (such as the simultaneous completion of tasks that are normally done in sequence), some of the studies of awareness in the air carrier environment have been focused on a single aspect of SA (e.g., mode awareness by Sarter & Woods, 1995; situation assessment as it applies to specific problem events by Orasanu, 1993). Each aspect is vital to SA, but is only a portion of what is necessary for "the continuous extraction of environmental information, integration of this information with previous knowledge to form a coherent mental picture and the use of that picture in directing further perception" (p. 11, Dominguez, 1994).

Another difficulty facing research efforts on team SA is the complexity of the topic as seen in the attempts to define team SA. Two definitions of team SA demonstrate the divergence of approaches. Endsley (1995) defined it as "the degree to which every team member possesses the situation awareness required for his or her responsibilities" (p. 39). Schwartz

(1990), however, described the team SA as being moderated by the pilot in command who receives information from the entire crew. There is some question as to whether team SA refers to: the collective SA of all crew members, only the similar state of awareness of the crew members (i.e., when crew members are aware of the same elements in the situation), the awareness of the crew members as moderated by that of the primary decision maker, or some other possibility. This dilemma occurs under the following circumstances: two crew members may need to have awareness of some of the same situation elements at the same time, each crew member may be involved in separate tasks and may need to be aware of different aspects of the environment, or they may need to be cognizant of the awareness level of the other crew member. It is not uncommon for all of these conditions to exist at the same time.

Even though this concept is not understood well, training has been adopted because the need to improve SA in the team context is so pressing. Since 1990, researchers have pursued a line of research that has the goals of expanding knowledge for training content, providing feedback, and evaluating crew members by the SA evidenced in the crew.

Research began with a literature review (Salas et al., 1995). It was determined that, at its simplest level, team SA is a construct that includes the individual SA of each team member and team processes. Two different sets of interviews were conducted and the data from these provided behaviors associated with SA and effective team performance. The interviews also helped identify other skill areas essential for building and maintaining SA within a team. The interviews allowed documentation of the differences between aviators and their understanding of SA that is related to experience (Prince & Salas, 1997). After collecting the interview data, ASRS reports were analyzed for SA related incidents (Jentsch, Barnett, & Bowers, 1997). These yielded information on circumstances that frequently were associated with a problem with SA in a multiplace cockpit.

Finally, a series of controlled experiments were conducted. The experiments' goals included: developing a measurement tool for crew SA (Prince, Prince, Brannick, & Salas, 1997); testing the tool with a second one (Prince, Salas, & Stout, 1995); exploring the relation between flight hours, situation difficulty, and the SA for the crews of two pilots (Prince et al., 1996); determining the relation between SA as measured by crew actions and subsequent crew decision making (Jentsch, Bowers, Settin-Wolters, & Salas, 1995); and training team SA behaviors (Brannick et al., 1998).

The data gained from all of this research yielded specific information about training and SA to those who operate in a crew. This can help in selecting training content, devising distinct training opportunities, and evaluating training elements.

## GUIDANCE FOR SA TRAINING IN CREWS

When the literature and research on errors, CRM, and SA is integrated, it provides a direction for the development and implementation of training for crew SA. This section discusses some of the general guidance offered and examples of specific training guidelines that can be extracted from this massed information.

### The Contribution From the Error Literature

Three recommendations for training are evident in the recent error research. The first of these is to concentrate effort at the subsystems above the individual in order to reduce error. From one body of work, it is evident that teams need to be built and maintained as teams with attention paid to providing structure, but allowing flexibility (Moray, 1994). This helps ensure that the team will not break down into a group where the only advantage of multiple crew members is the expectation of redundancy, a questionable method of error reduction.

The second recommendation comes from the work of Rasmussen (1996) who said that it is necessary to recognize the realities of the system (including pressures engendered by a competitive environment) in which crew members work and to determine the coping skills that crew members need to operate effectively in these conditions. Thus, training developers need to know the situations created by the operating environment so crew members can be made aware of the value of developing the skills needed.

The third direction for training supplied by the error research is to give teams the skills to work in realistic, difficult situations. This supports the first two training suggestions. Teams need to recognize and plan for difficult situations as well as learn the knowledge, skills, and attitudes that will help them operate effectively when the plan is not sufficient. These are referred to as coping skills (Rasmussen, 1996).

### The Contribution of CRM Training Research

CRM training research has made three major contributions to the training of crew SA. The first was to identify those interaction skills that are partially responsible for team processes and outcomes. The behaviors and actions required to maintain crew SA as well as those related to planning, making decisions, communicating, and leading or managing the crew are included. This adds to the training recommendations from the error research by specifying the actions of crew members that build and maintain a team and it defines those skills that may be considered coping skills.

The second contribution is the exploration and development of instructional strategies for training team coordination, including crew SA (see Salas & Cannon-Bowers, 1997). Research has shown that the transfer of training for team process skills is most likely to occur if the training is provided through information, demonstration, practice, and feedback (Prince & Salas, 1999). This suggests that a training strategy include: classroom training, selected reading materials or computer-based training for information and demonstration, practice of the skills in situations relevant to the cockpit (e.g., realistic simulations), and instructor feedback. It also implies a consistency in the training materials through each phase of the training. Scenario development guidance for this training suggests testing the scenarios to ensure that they call for the skill or skills being trained. It also requires giving crew members specific feedback on their use of the skills as part of the training (Brannick et al., 1998). CRM research has also shown that the skill practice must be relevant for the cockpit; that is, it must require crew members to interact in a way similar to the way they do on the job.

The third contribution of this research is the development of measurement tools for the team process skills (see Brannick et al., 1997). These tools allow researchers to verify the value of the skills, help trainers diagnose crew members' skill needs, and make possible the evaluation of the training.

## Contributions of Crew SA Research

The work that has been done on SA in crew environments has both refined the knowledge of the concept and demonstrated the importance of the specific behaviors that define it. It has resulted in a number of explicit recommendations for training and has carried forward the work accomplished in CRM and error reduction. For example, in skill development, error research recommends the development of coping skills (Rasmussen, 1996), but is not explicit about what they might be. The CRM research has identified the team process skill areas, including the specific actions of crew members that are directly related to SA. The SA research has pinpointed those team process skill areas (i.e., preparation, communication, leadership, and adaptability) that are most supportive of the maintenance of SA in a crew situation.

One experiment has demonstrated the relation between using the actions associated with SA in the routine segments of flight and subsequent decision making. Building on the measurement methods for CRM, the research produced a measurement instrument (observation of behaviors) that is correlated with a second method (a probe method) useful for feedback and evaluation.

Research contributed to information about the natural development of SA and uncovered specific differences in the crew members' understanding of awareness and the elements they use to maintain it based on their flight experience. Researchers also demonstrated that experienced pilots use more behaviors related to awareness in a difficult scenario than do inexperienced pilots. Finally, the training transfer of behaviors associated with SA has been documented (Brannick et al., 1998).

### Guidelines for Training

The research results discussed in this chapter were used to develop guidelines for team SA training that include examples of its relevance to operations and training tips (Prince, 1998). The guidelines themselves are presented in the Appendix. All of the guidelines are based on research and most are based on results of research conducted on team SA. For example, all of the guidelines for scenario-based training came from research done with both low-fidelity table top trainers and high-fidelity simulators. Five of the guidelines are presented to illustrate the connections between the recommendations from the research on errors, CRM, and crew SA.

1. **Emphasize the active role that crew members need to take to maintain SA.** Active is the key word is this guideline. Error research has shown that the crew must function as a team. It has shown that there are actions crew members take that make a difference in occurrence of errors related to SA. Simply being placed in the cockpit with others does not ensure that the crew will not commit errors; instead, they must take the actions that have been identified with effective teamwork. CRM research identified many of the necessary actions that are related to awareness. Finally, SA research demonstrated that the crew members who are actively involved in the flight (as shown by their knowledge of multiple flight factors) perform more behaviors associated with crew awareness.

2. **The best way to train for effective performance is to build up crew members' teamwork skills.** Research showed that teamwork skills (including crew SA) are correlated with performance. For the competitive environment, Rasmussen (1996) recommended building coping skills so teams could respond to the realistic demands of their work environment. Team process skills identified in CRM research can be considered coping skills because they help crew members with their task performance, particularly under difficult conditions. Research revealed that the development and maintenance of a high level of SA in crews is assisted by other team process skills (e.g., communication, preparation and planning, leadership, and adaptability). This testifies to the need for the development of team process skills.

3. **Include situations that are a threat to crew SA when developing scenarios for training.** According to error research, training should help crews develop coping skills for the difficult conditions they must face; these situations include threats to SA. CRM research showed that training scenarios cannot be accepted at face value for their ability to train skills, but must be tested to be sure that the skills to be trained are elicited by the scenario. SA research showed that a scenario that does not challenge a crew (not sufficiently threatening) will not call for the level of SA required in a more challenging scenario.

4. **Present the information that crew SA must be maintained throughout the flight.** Moray (1994) noted that error is likely to occur when teams break down and become simply groups of people. Maintaining team process skills helps the crew remain a team. CRM research demonstrated that SA actions during the routine segments of flight are related to the decision-making performance of crew members in a subsequent event. SA research showed that overall performance ratings were correlated with ratings on SA actions taken during the flight.

5. **Develop a debrief guide for instructors on what they need to look for besides did it or didn't do it.** CRM training research showed that specific skill (e.g., behavioral) feedback is important to skill training. In order for crew members to develop skills, whether coping, team process, or team SA, they must receive specific debriefs on their performance.

## CONCLUSIONS

We have traced the research in three areas important to aviation (i.e., errors, CRM, and crew SA) from the early attempts to understand these areas by focusing on the individual to the more recent interest in the areas' application to the operational environment. The latter necessitates a consideration of the teams in which individuals work. We showed the complexity of the studied constructs as they occur on the line (e.g., interactivity of subsystems, dynamics within the team, interaction of team processes, and the importance to the team of the individual crew member's on-going flight knowledge).

We also showed that the research results are rich sources for developing more effective training for crew members. According to the combined research, the resulting training should encourage crew members to function as a team. This is done by demonstrating the team's importance and relevance to handling their tasks and by developing the skills that are necessary for interactions in their work environment.

Research in all three areas sought to improve performance in existing systems and all three have error reduction as a goal. These areas differ in the scope of their subjects. Error research has been conducted in a variety of settings and across the entire system in which performance occurs. CRM has been concentrated primarily in aviation and has been limited to the individual and team levels. SA for crews has been studied almost exclusively as a part of CRM. The final conclusions, however, demonstrate an agreement as to the elements important to an effective training strategy.

The future for crew SA research and practice looks bright. Although progress has been made in uncovering the important behavioral elements of SA, much work remains to be done. The challenges of relating (and measuring) cognition to crew performance has proven to be formidable, but research must continue in order to understand what crews think, feel, and do.

## APPENDIX

### Guidelines for Training SA

PRETRAINING CONSIDERATIONS

1. Recognize that crew members with different levels of experience and positions (e.g., 1,500 hours and 15,000 hours; captains and second officers) are likely to define and conceptualize SA differently based on their experience. Consider conducting separate classes; making sure instructors are prepared with this information; and examining already developed training materials for relevance.

2. Design the course to emphasize the active role that crew members must take to gain and maintain crew SA.

3. Design all the course elements (information, scenario events, debriefing guides) to reflect the same training philosophy and the same skills.

TRAINING CONTENT

1. Inform crew members that 40% of the reported incidents in the ASRS data base occurred when only one crew member had a problem with SA.

2. Demonstrate to crew members that crew SA requires a high level of flight knowledge about the specific flight (familiarity with the details of the flight) among the crew members and this requires involvement of all crew members.

3. Emphasize the active role that crew members need to take to maintain SA (situation assessment).

4. Refresh crew members' knowledge with information about CRM-related actions that help them develop and maintain crew awareness. These

are in the categories of leadership, communications, preparation and planning, and adaptability. (Provide specific examples of actions, discuss the actions and how they may affect SA, and emphasize the importance of *verbalizing actions/intended actions*).

5. Expand on the CRM-related actions listed above with the following information. (Additional information on planning, preparation, and adaptability is given here particularly for the training course that is designed for crew members with low experience levels, such as new first officers in a regional air line).

6. Introduce the specific actions that can be observed and that indicate the level of a crew's SA. (e.g., discuss and demonstrate the importance of identifying problems and potential problems to one another to help raise the SA of the entire crew, show alertness to the present flight status, e.g., we've got 30 miles before we need to start down, with this temperature, we'll need to put on the wing antiice after take-off), and show alertness to the task performance of self and others.)

7. Present the information that crew SA must be maintained throughout the flight. SA in routine phases is related to decision making in nonrouting phases.

8. Introduce crew members to the idea that the main purpose of building a team in the cockpit is to reduce the consequences of human error.

9. Emphasize that good team SA helps reduce human error caused accidents.

10. Give crew members examples of basic problems in SA and how those problems can be connected to classes of errors.

11. Emphasize to crew members that when pointing out a loss of SA to another crew member, they need to consider the other person and to remember that what appears as a loss of SA in another may be their own confusion.

12. Inform crew members of the need to be alert to certain situations that have been found to be particular threats to SA (e.g., high workload, a breakdown in communication or coordination, use of improper procedures, maintenance problem or equipment malfunction, fatigue, unusual weather conditions, low experience level of others, or any combination of these).

## SCENARIO-BASED TRAINING

1. Consider the difficulty level of the scenario when building a scenario for training or evaluating SA. Easy scenarios may not be demanding enough to allow enough discrimination among crews to diagnose problems with SA.

2. Use a scenario on a low-fidelity training device for training specific SA actions that you want to see a crew demonstrate in a LOS scenario.

3. Use a simple table-top training device to have crew members fly a scenario. The scenario can be stopped to ask questions of the crew members and to direct their attention to important elements for awareness.

4. Introduce scenarios in the classroom setting. Form crews with the class members and have several crews fly different short scenarios while class members watch and take notes for debriefing the crews.

5. Include situations that are a threat to crew SA when developing scenarios for training. These were listed in the section, Guidelines for Training Program Content.

6. Develop a debrief form that can be used by both class members and instructors for observing crews in the scenarios used for training.

7. Develop a debrief guide to refresh instructors on the scenario's purpose and what they need to look for in the scenario beyond, "did it" or "didn't do it."

8. Select examples to debrief that are explicit and demonstrate specific concepts.

9. Debrief exercises to make connections for crew members on major concepts. Don't assume they understand, make sure they do.

10. Use one training method to convey both individual and team situation assessment/awareness concepts.

## EVALUATION

1. Train instructors or crew members or both to observe crew member actions. These include those actions helpful for SA and those related to errors in SA.

2. Consider the experience level of those being trained when making up a class for observation training.

3. When developing a videotape for classroom demonstration or for observation training, include both good and poor actions on the part of the crew members.

4. Include reenforcement of some of the actions of interest on the training videotape (i.e., some consequence for the action is shown) when developing a videotape for classroom demonstration or for observation training.

5. When developing a videotape for classroom demonstration or for observation training, develop a specific list of behaviors to look for in the videotape.

# REFERENCES

Adams, M. J., Tenney, Y. J., & Pew, R. W. (1995). Situation awareness and the cognitive management of complex systems. *Human Factors, 37*(1), 85–104.

Brannick, M. T., Prince, C., Salas, E., & Stout, R. J. (1998). *Development and evaluation of a team training tool.* Unpublished manuscript.

Bryant, A. (1997, January). Pilots just want a little respect. *The New York Times.*

Caro, P. (1988). Flight training and simulation. In E. L. Weiner & D. C. Nagel (Eds.), *Human factors in aviation* (pp. 229–262). San Diego, CA: Academic Press.

Chou, C., Madhavan, D., & Funk, K. (1996). Studies of cockpit task management. *International Journal of Aviation Psychology, 6,* 307–320.

Costley, J., Johnson, D., & Lawson, D. (1989). A comparison of cockpit communication B737–B757. In *Proceedings of the Fifth International Symposium on Aviation Psychology* (Vol. 1, pp. 413–418) Columbus: The Ohio State University

Dismukes, R. K., Jobe, K. K., & McDonnell, L. K. (1997). *LOFT debriefings: An analysis of instructor techniques and crew participation* (NASA Technical Memorandum 110442). San Jose, CA: NASA–Ames.

Dominguez, C. (1994). Can SA be defined? In M. Vidulich, C. Dominguez, E. Vogel, & G. McMillan (Eds.), *Situation awareness papers and annotated bibliography* (Report AL/CF-TR-1994-0085) Wright-Patterson AFB, OH: Armstrong Laboratory.

Edwards, E. (1988). Introductory overview. In E. L. Wiener & D. Nagel (Eds.), *Human factors in aviation* (pp. 3–26). San Diego, CA: Academic Press.

Endsley, M. R. (1995). Towards a theory of situation awareness. *Human Factors, 37*(1), 32–64.

Foushee, H. C., & Helmreich, R. L. (1988). Group interaction and flight crew performance. In E. L. Wiener & D. C. Nagel (Eds.), *Human factors in aviation* (pp. 189–227). San Diego, CA: Academic Press.

Foushee, H. C., & Manos, K. L. (1981). Information transfer within the cockpit: Problems in intracockpit communications. In C. E. Billings & E. S. Cheaney (Eds.), *Information transfer problems in the aviation system* (NASA Report No. TP-1875). Moffett Field, CA: NASA–Ames Research Center.

Ginnett, R. C. (1993). Crews as groups: Their formation and leadership. In E. L. Wiener, B. Kanki, & R. L. Helmreich (Eds.), *Cockpit resource management* (pp. 71–98). San Diego, CA: Academic Press.

Hackman, J. R. (1993). Teams, leaders, and organizations: New directions for crew-oriented flight training. In E. L. Wiener, B. G. Kanki, & R. L. Helmreich (Eds.), *Cockpit resource management* (pp. 47–69). San Diego, CA: Academic Press.

Helmreich, R. (1996, October). *The evolution of crew resource management.* Paper presented at the IATA Human Factors Seminar, Warsaw, Poland.

Helmreich, R. L., & Foushee, H. C. (1993). Why crew resource management? Empirical and theoretical bases of human factors training in aviation. In E. L. Wiener, B. G. Kanki, & R. L. Helmreich (Eds.), *Cockpit resource management* (pp. 3–45). San Diego, CA: Academic Press.

Jentsch, F., Barnett, J., & Bowers, C. A. (1997). *Loss of aircrew situation awareness across validation.* Poster presented at the 41st Annual Meeting of the Human Factors and Ergonomics Society, Albuquerque, NM.

Jentsch, F., Bowers, C. A., Settin Wolters, S., & Salas, E. (1995). Crew coordination behaviors as predictors of problem detection in decision making. In *Proceedings of the 39th Meeting of the Human Factors and Ergonomics Society. Human Factors and Ergonomics, 2,* 1350–1353. Santa Monica, CA.

Kanki, B., Lozito, S., & Foushee, H. C. (1989). Communication indices of crew coordination. *Aviation, Space, and Environmental Medicine*, 56–60.

Kanki, B., & Palmer, M. T. (1993). *Communication and crew resource management*. In E. Wiener, B. Kanki, & R. Helmreich (Eds.), Cockpit resource management (pp. 99–136). San Diego, CA: Academic Press.

Kayton, P. J. (1993). The accident investigator's perspective. In E. L. Wiener, B. Kanki, & R. L. Helmreich (Eds.), *Cockpit resource management* (pp. 283–314). San Diego, CA: Academic Press.

Klein, G. (1989). Recognition-primed decisions. In W. B. Rouse (Ed.), *Advances in man-machine systems research* (Vol. 5, pp. 47–92). Greenwich, CT: JAI.

Lauber, J. (1988). *Airline safety in a transitional era*. Paper presented at the Annual Airline Operational Forum Of The Air Transportation Association Of America, Williamsburg, VA.

Moray, N. (1994). Error reduction as a system problem. In M. S. Bogner (Ed.), *Human error in medicine*. Hillsdale, NJ: Lawrence Erlbaum Associates.

Nagel, D. C. (1988). Human error in aviation operations. In E. L. Wiener & D. C. Nagel (Eds.), *Human factors in aviation*. New York: Academic Press.

Norman, D. (1981). Categorization of action slips. *Psychological Review, 88*, 1–55

Orasanu, J. (1990). *Shared mental models and crew decision making* (CSI Report 46). Princeton, NJ: Princeton University, Cognitive Science Laboratory.

Orasanu, J. (1993). Decision making in the cockpit. In E. L. Wiener, B. Kanki, & R. L. Helmreich (Eds.), *Cockpit resource management* (pp. 137–172). San Diego, CA: Academic Press.

Prince, C. (1998). *Guidelines for situation awareness training*. Orlando, FL: NAWCTSD/UCF/FAA Partnership for Aviation Training.

Prince, C., Chidester, T. R., Cannon-Bowers, J. A., & Bowers, C. A. (1992). Aircrew coordination: Achieving teamwork in the cockpit. In R. W. Swezey & E. Salas (Eds.), *Teams: Their training and performance* (pp. 329–353). Norwood, NJ: Ablex.

Prince, A., Prince, C., Brannick, M., & Salas, E. (1997). The measurement of team process behaviors in the cockpit: Lessons learned. In M. T. Brannick, E. Salas, & C. Prince (Eds.), *Team performance assessment and measurement: Theory, methods, and applications* (pp. 289–310). Mahwah, NJ: Lawrence Erlbaum Associates.

Prince, C., & Salas, E. (1993). Training and research for teamwork in the military aircrew. In E. L. Wiener, B. G. Kanki, & R. L. Helmreich (Eds.), *Cockpit resource management* (pp. 337–366). San Diego, CA: Academic Press.

Prince, C. & Salas, E. (1997). Situation assessment for routine flight decision making. *International Journal of Cognitive Ergonomics, 1*, 315–324.

Prince, C., & Salas, E. (1999). Team processes and their training in aviation. In D. Garland, J. Wise, & D. Hopkin (Eds.), *Aviation Human Factors*. Mahwah, NJ: Lawrence Erlbaum Associates.

Rasmussen, J. (1996, August). *Risk management in a dynamic society: A modeling problem*. Key note address: Conference on Human Interaction with Complex Systems, Dayton, Ohio.

Reason, J. (1994). Foreword. In M. S. Bogner (Ed.), *Human error in medicine*. Hillsdale, NJ: Lawrence Erlbaum Associates.

Ruffell Smith, H. P. (1979). *A simulator study of the interaction of pilot workload with errors, vigilance, and decisions* (NASA TM–78482). Moffett Field, CA: NASA–Ames Research Center.

Salas, E., Cannon-Bowers, J. A., & Blickensderfer, E. L. (1997). Enhancing reciprocity between training theory and practice: Principles, guidelines, and specifications. In J. K. Ford, S. W. J. Kozlowski, K. Kraiger, E. Salas, M. S. Teachout (Eds.), *Improving training effectiveness in work organizations* (pp. 291–322). Mahwah, NJ: Lawrence Erlbaum Associates.

Salas, E., Fowlkes, J. E., Stout, R. J., Milanovich, D., & Prince, C. (1999). Does CRM training enhance teamwork skills in the cockpit?: Two evaluation studies. *Human Factors, 41*(2), 326–343.

Salas, E., Prince, C., Baker, D. P., & Shrestha, L. (1995). Situation awareness in team performance: Implications for measurement and training. *Human Factors, 37*(1), 123–136.

Sarter, N., & Woods, D. (1995). How in the world did I ever get into that mode? Mode error and awareness in supervisory control. *Human Factors, 37*(1), 5–19.

Schwartz, D. (1990). *Training for situation awareness.* Houston, TX: Flight Safety International.

Senders, J. W. (1994). Medical devices, medical errors, and medical accidents. In M. S. Bogner (Ed.), *Human error in medicine.* Hillsdale, N.J.: Lawrence Erlbaum Associates.

Smith, K., & Hancock, P. A. (1995). Situation awareness is adaptive, externally directed consciousness. *Human Factors, 37*(1), 137–148.

Vidulich, M., Dominguez, C., Vogel, E., & McMillan, G. (1994). *Situation awareness papers and annotated bibliography* (Report AL/CF–TR–1994–0085). Wright Patterson AFB, OH: Armstrong Laboratory.

Waag, W., & Bell, H. (1994, June). *A study of situation assessment and decision making in skilled fighter pilots.* Paper presented at the 2nd Conference on Naturalistic Decision Making, Dayton, OH.

Wiener, E. L. (1993). Crew coordination and training in the advanced-technology cockpit. In E. L. Wiener, B. Kanki, & R. L. Helmreich (Eds.), *Cockpit resource management* (pp. 199–230). San Diego, CA: Academic Press.

# Chapter 16

# Training for Situation Awareness in Individuals and Teams

Mica R. Endsley
*SA Technologies, Inc.*

Michelle M. Robertson
*Liberty Mutual Research Center*

## NEED FOR SA TRAINING IN AVIATION

In the aviation domain, maintaining a high level of *situation awareness* (SA) is one of the most critical and challenging features of a pilot's job. Problems with SA were found to be the leading causal factor in a review of military aviation mishaps (Hartel, Smith, & Prince, 1991). In a study of accidents among major airlines, 88% of those involving human error were attributed to problems with SA as opposed to problems with decision making or flight skills (Endsley, 1995b). Although similar studies have not been performed for general aviation accidents, SA is reported to be a considerable challenge in this population as well, particularly because general aviation pilots are frequently less experienced and less current than operators for major airlines (Hunter, 1995).

Due to the important role that SA plays in the pilot decision-making process and its substantial role in aviation accidents and incidents, the development and validation of methods for training to improve SA in aircraft pilots is a subject that is beginning to receive focus in the operational and human factors communities. SA training can be focused on improving individual SA or on improving SA at the team level. Each approach should be considered as complimentary and potentially useful as an addition to activities directed at improving SA through system design. As a practical matter, pilots will always need to learn to develop the best SA possible with whatever system they are flying.

## SA CHALLENGES AND SKILLS

One way of identifying methods for improving SA is to examine the ways SA errors occur. A second method is to identify the ways in which pilots successfully develop and maintain SA as compared to pilots who do a poorer job at these tasks. A number of studies have been performed pertinent to these issues.

### Errors in SA

An analysis of SA errors in aviation was conducted (Jones & Endsley, 1996) using reports from NASA's Aviation Safety Reporting System (ASRS) using an SA error taxonomy based on a model of SA (Endsley, 1995c), as shown in Table 16.1. Gibson, Orasanu, Villeda, and Nygren (1997) also performed a study of SA errors based on ASRS reports. They found problems with workload or distraction (86%), communications or coordination (74%), improper procedures (54%), time pressure (45%), equipment problems (43%), weather (32%), unfamiliarity (31%), fatigue (18%), night conditions (12%), emotion (7%), and other factors (37%). The consequences of SA loss included altitude deviations (26%), violations of FARs (25%), heading deviations (23%), traffic conflicts (21%), and non-adherence to published procedures (19%). Dangerous situations were found to result from 61% of the cases. Clearly, the loss of SA must be taken seriously in aviation.

### Comparisons of Pilots

Several other researchers have investigated the differences in SA between pilots who perform well and pilots who do not. Prince and Salas (1998) studied the situation assessment behaviors of General Aviation (GA) pilots (*M* experience level = 720 hours), airline pilots (*M* experience level = 6,036 hours), and commercial airline check airmen (*M* experience level = 12,370 hours). They found several key differences with experience level:

1. Increasing levels of preflight preparation—GA pilots talked about personal preparation before the flight and line pilots emphasized knowing the equipment, its limits, and the flight briefing. Check airmen focused on planning and preparation specific to the flight and gathered as much information as possible about the conditions and flight elements (e.g., weather, ATC, airport status) in order to prepare in advance.

2. More focus on understanding and projection—GA pilots described themselves as passive recipients of information with an emphasis on information in the immediate environment (Level 1 SA). Line pilots dealt more

TABLE 16.1
Causal Factors Related to Errors in Situation Awareness

---

Loss of Level 1 SA—Failure to correctly perceive the situation (76.3%)
* *Information not available (11.6%)*
  * system & design failures
  * failure of communication
  * failure of crew to perform needed tasks
* *Information difficult to detect (11.6%)*
  * poor runway markings
  * poor lighting
  * noise in the cockpit
* *Information not observed (37.2%)*
  * omission from scan
  * attentional narrowing
  * task related distractions
  * other distractions
  * workload
* *Misperception of information (8.7%)*
  * prior expectations
* *Memory error (11.1%)*
  * disruptions in routine
  * high workload
  * distractions
Loss of Level 2 SA—Failure to correctly comprehend the situation (20.3%)
* *Lack of/incomplete mental model (3.5%)*
  * automated systems
  * unfamiliar airspace
* *Incorrect mental model (6.4%)*
  * mismatching information to expectations of model or model of usual system
* *Over-reliance on defaults values in the mental model (4.7%)*
  * general expectations of system behavior
Loss of Level 3 SA—Failure to correctly project situation (3.4%)
* *Lack of/incomplete mental model (0.4%)*
* *Over-projection of current trends (1.1%)*
* *Other (1.9%)*

---

*Note.* From Jones, D. G. and Endsley, M. R. Sources of situation awareness errors in aviation. *Aviation, Space, and Environmental Medicine, 67*(6), 507–512.

at the level of comprehension (Level 2 SA) and emphasized their active role in seeking out information. Check airmen were more likely to deal with Level 3 SA, seeking to be proactive. They dealt with large numbers of details and the complex relations between factors in this process.

After conducting critical incident reviews with the pilots, Prince and Salas (1998) identified four major actions that are important for team SA in commercial pilots: (a) identifying problems or potential problems, (b) demonstrating knowledge of the actions of others, (c) keeping up with flight details, and (d) verbalizing actions and intentions. Prince, Salas, and Stout

(1995) found that those aircrews who performed better on an objective measure of SA demonstrated more actions in these areas. They seemed to solve problems faster and recognized problem situations developing.

Orasanu and Fischer (1997) studied the characteristics of commercial aircrews in making various types of decisions through an analysis of ASRS data and observations from simulator studies. They found that in making go/no-go decisions about an approach, the better performing crews attended more to cues signaling deteriorating weather and sought out weather updates allowing them to plan for a missed approach in advance. When they studied a choice-type task that involved picking an alternate airport, however, the better performing crews took longer. These crews were much more attuned to the constraints imposed by a hydraulic failure and reviewed other alternate airports in light of the constraints. They gathered more information allowing them to make a better decision, whereas poorer performing crews went right to evaluating options. Analysis of a hydraulic failure, which represented a scheduling-type task, showed the better performing crews took active steps to manage what would become a high workload task. They planned in advance for actions that would occur in the high workload periods and thus were more effective in these situations.

From this research, Orasanu and Fischer (1997) focused on a two-step decision model: situation assessment and action selection. Time availability, risk level, and problem definition are indicated as critical components of the situation assessment phase. Situation ambiguity and the availability of responses were hypothesized to be critical factors dictating the difficulty of the decision. When cognitive demands are greater, the higher performing crews managed their effort by performing actions that would buy them extra time (e.g., holding) and by shifting responsibilities among the crew. Good situation assessment, contingency planning, and task management were highlighted as critical behaviors associated with success. Less effective pilots appeared to apply the same strategies in all cases rather than matching their strategy to the situation.

In examining accident reports, Orasanu, Dismukes, and Fischer (1993) also reported that pilots who had accidents tended to interpret cues inappropriately and often underestimated the risk associated with a problem and overestimated their ability to handle dangerous situations. Wiggens, Connan, and Morris (1995) found that general aviation pilots who performed poorly when deciding to continue into inclement weather were poorly gauged in terms of matching their skill level to the situation. "In the absence of extensive task-related experience, pilots are more inclined to rely on their self-perceived risk-taking behaviour [sic] than their self-perceived ability to resolve various decisions" (p. 848). The more experienced pilots demonstrated behaviors that were much more related to perceptions of their own ability. It would appear that inexperienced pilots may be deficient in their ability to properly assess risk and capabilities (Level 2 SA) from the situational cues at hand.

In a study of individual differences in SA abilities, Endsley and Bolstad (1994) found that military pilots with better SA were better at attention sharing, pattern matching, spatial abilities, and perceptual speed. O'Hare (1997) found evidence that elite pilots (defined as consistently superior in gliding competitions) performed better on a divided attention task purported to measure SA. Gugerty and Tirre (1997) found evidence that people with better SA performed better on measures of working memory, visual processing, temporal processing, and time-sharing ability. Although some of these characteristics may not be trainable, at least attention-sharing has shown some indication that it can be improved through training (Damos & Wickens, 1980). Reducing loads on working memory may also help.

## TARGET AREAS FOR IMPROVING INDIVIDUAL PILOT SA

From these various studies several key factors can be identified that indicate where individual pilot SA can be improved.

### Task Management

Interruptions, task-related distractions, other nontask-related distractions, and overall workload pose a high threat to SA. Good task management strategies appear critical for dealing with these problems. Schutte and Trujillo (1996) found that the best performing crews in nonnormal situations were those whose task management strategies were based on the perceived severity of the tasks and situations. Those who used an event interrupt strategy (dealing with each interruption as it came up) and those who used a procedural-based strategy performed worse. The ability to accurately assess the importance and severity of events and tasks is an important component of Level 2 SA. This understanding also allows pilots to actively manage their task and information flow so as not to end up in situations where they are overloaded and miss critical information.

### Development of Comprehension (Level 2 SA)

In addition to problems with properly assessing the importance or severity of tasks and events, pilots will also perform poorly if they are unable to properly gauge the temporal aspects of the situation, the risk levels involved, and both personal and system capabilities for dealing with situations. Simmel and Shelton (1987), analyzed accident reports and noted that the ability to accurately determine the consequences of nonroutine events appeared to be the problem for the pilots. Each of these factors (timing, risk, capabilities, consequences, and severity) are major components of Level 2 SA for pilots (Endsley, Farley, Jones, Midkiff, & Hansman, 1998). The research suggests that more inexperienced pilots are less able to make these important assessments and remain more focused at Level 1 SA.

### Projection (Level 3 SA) and Planning

Amalberti and Deblon (1992) found that a significant portion of experienced pilots' time was spent anticipating possible future occurrences. This gave them the knowledge (and time) necessary to decide on the most favorable course of action to meet their objectives. Experienced pilots also appeared to spend significant time in preflight planning and data gathering and engaged in active contingency planning in flight. Each of these actions served to reduce workload during critical events. Using projection skills (Level 3 SA), these pilots actively sought out important information in advance of a known immediate need for it and planned for various contingencies. Not all planning is equally effective, however. Taylor, Endsley, and Henderson (1996) found that teams who viewed only one plan were particularly susceptible to Level 2 SA errors: failure to recognize cues that things were not going according to plan. Active planning for various contingencies and not just the expected is critical.

### Information-Seeking and Self-Checking Activities

Pilots with high levels of SA actively seek out critical information. As a result, they are quicker to notice trends and react to events. Furthermore, it has been noted that these pilots are good at checking the validity of their own situation assessments, either with more information or with others (Taylor et al., 1996). This strategy was found to be effective in dealing with false expectations and incorrect mental models. Other researchers also suggest a "Devil's Advocate" strategy where people are encouraged to challenge their interpretations of situations (Klein, 1995; Orasanu, 1995).

## TRAINING TO IMPROVE THE SA OF INDIVIDUALS

Relatively few programs have attempted to specifically train SA to date. Most work has been directed toward design and automation issues and the more fundamental research issues concerning factors underlying SA, although there are a few exceptions. Many major airlines, for example, have recently introduced short training courses on SA that mainly serve to acquaint pilots with what SA is and some of the factors that can impact it. Although these type of SA courses may be helpful and are certainly a good first step for airlines to undertake, no validation work has been done as to their effectiveness.

Effective improvement of SA through training approaches will most likely be achieved by improving the skills and knowledge bases that are critical for achieving good SA in flight. Some basic approaches to training individual SA, outlined in Endsley (1989), include higher-order cognitive skills training, intensive preflight briefings, the use of structured feedback, and SA-oriented training programs.

## Higher-Order Cognitive Skills Training

Training programs devoted specifically to teaching higher-order cognitive skills related to SA include those for teaching attention-sharing, task management, contingency planning, information seeking and filtering, self-checking, and other metaskills identified through research to be important for SA. Most of these skills have been identified as being important for achieving and maintaining good SA in flight. Training that is directed at these constituent skills associated with SA should therefore be explored as methods of helping pilots perfect their capabilities in these areas.

## Intensive Pre-Flight Briefings

As the acquisition and interpretation of information is highly influenced by expectations, preflight briefing is critical. Prince and Salas (1998) found that the issues focused on in the preflight preparation varied considerably for pilots of different experience levels. The idea behind intensive preflight briefings is to use multimedia tools to help pilots develop a clear picture of their flight: where the hazard areas are, where the weather is, and what a new airport's approach pattern looks like. By being able to "fly-through" the flight in advance, pilots develop a better mental picture of the environment. This is extremely helpful when flying into a new airport; for instance, being able to picture where to look for needed cues and runway configurations. More importantly, such a tool may be incorporated within a contingency planning assistant and prompt pilots to look for potential hazards, such as deteriorating weather conditions, heavy air traffic, or in-flight mechanical failures and develop appropriate contingency plans.

## SA-oriented Training Programs

Current training programs can be greatly enhanced by incorporating training that specifically focuses on the development of pilot SA. Current initial training for pilots focuses primarily on the basic skills of flying the aircraft. At some point after the psychomotor skills and basics of flight have been mastered, a training regime focused on SA and decision making in flight would be of the most benefit. This type of training should focus on developing the schemata and mental models that allow experienced pilots to have a much better understanding of the importance, consequences, timing, risk levels, and capabilities associated with different events and options. This type of training would specifically focus on creating Levels 2 and 3 SA from the basic information available in flight.

It also would help to train pilots in the critical cues that signify prototypical classes of situations. More experienced pilots know where to look for cues and understand the significance of these cues when they occur. Kass, Herschler, and Companion (1990) showed success with training subjects to recognize critical cues in a simulated battlefield environment. Other research, however, has shown that the cues attended to in early training can largely affect what is attended to later on (Doane, Alderton, Sohn, & Pellegrino, 1996). It is therefore important that a broad range of representative sets of situational cues be included when training pilots in this way.

## Structured Feedback

Feedback is critical to the learning process. In order to improve SA, pilots need to receive feedback on the quality of their SA. For example, inexperienced pilots may fail to appreciate the severity of deteriorating weather because unless they have an accident, or at least a bad scare, they may have come through similar weather in the past just by luck. Unfortunately, this result also reinforces poor assessments. Due to the probabilistic link between SA and outcome, it is difficult for pilots to develop a good gauge of their own SA in normal flight. Feedback on SA can be used to help train SA, however, through the Situation Awareness Global Assessment Technique (SAGAT), a measure of SA developed for design evaluation (Endsley, 1995a). SAGAT uses freezes in the flight simulation to query pilots about important aspects of their SA. The accuracy of the pilot's perceptions are then compared to the real situation to provide an objective and direct measure of pilot SA. This technique can be adapted for a training application by providing feedback to the pilot on how accurate he or she was on the responses given (e.g., you thought you were here when actually you are there; you have traffic at one o'clock, but were unaware of it; and you are actually very close to stall speed). This type of technique can be integrated with the higher-order cognitive skills training and SA-oriented training programs to help pilots fine-tune their information acquisition strategies and schemata.

## Validation of Training Techniques

Each of these approaches shows promise for improving the pilots' ability to develop and maintain SA during flight. Most importantly, these techniques, or others developed for training SA, should be carefully tested and evaluated. Control groups should be employed to determine whether the SA training concepts discussed result in significant changes in the pilots' ability to acquire and maintain SA in challenging aviation environments. SA measurement techniques discussed in this book should be as applicable to such an endeavor as they are to the evaluation of changes in system design. The measurement of SA in evaluating training techniques should provide greater sensitivity in the analysis of the effectiveness of the techniques (as sensitive performance measures may be difficult to obtain) and allow a determination of whether the techniques examined actually effected SA, or potentially effected pilot performance through some other mechanism. This is an important issue that tends to be neglected in SA research. The degree to which trained skills transfer to improved performance on the flight deck should be measured.

## TRAINING TO IMPROVE THE SA OF TEAMS

In addition to efforts directed at improving the SA of individuals, there is considerable interest in improving the SA of teams. This approach has the advantage of enhancing recovery from SA losses. Taking the view that SA problems may be inevitable, good team processes may help pilots recognize SA problems and take steps to deal with them. Prince and Salas (chap. 14, this volume) deal extensively with the issue of training team SA in pilots. Many their strategies are focused on the communications and coordination of the aircrew and feature Crew Resource Management (CRM) principles.

In previous work (Endsley & Robertson, in press), we conducted a three year study to develop, implement, and evaluate a program for enhancing team SA in aircraft maintenance technicians.

## Team SA Training Course Development

The Team SA Training Course was developed based on an analysis of SA requirements and problems in aviation maintenance teams (Endsley & Robertson, 1996a; Endsley & Robertson, 1996b). The analysis investigated SA across multiple teams involved in aircraft maintenance. It identified several teams within the aviation maintenance setting, each of which involved leads and supervisors as well as line personnel: aircraft maintenance technicians (AMT), stores, maintenance control, maintenance operations control, aircraft-on-ground, inspection, and planning. The analysis produced a delineation of SA requirements for each of these groups and an understanding of the way in which each group interacted with the others to achieve SA pertinent to their specific goals. SA appears to be crucial to the ability of each group to perform tasks (each task is interdependent on others being performed by other team members), their ability to make correct assessments (e.g., whether a detected problem should be fixed now or later [placarded]), and their ability to correctly project into the future to make good decisions (e.g., time required to perform task and availability of parts). As a part of the analysis, certain shortcomings, both in the technologies employed and in the organizational and personnel system, were identified that may compromise team SA in the maintenance environment.

From the analysis, five major areas for improving SA in aviation maintenance were identified:

1. It found that there were significant differences in the perceptions and understanding of situations between teams that were related to differences in the mental models held in these different teams. The same information would be interpreted quite differently by different teams leading to significant misunderstandings and system inefficiencies.

2. It noted problems with not verbalizing the information that went into a given decision (the rationale and supporting situation information). Only the decision would be communicated between teams. This contributed to suboptimal decisions in many cases as good solutions often required the pooling of information across multiple teams.

3. A problem with lack of feedback in the system also was present. The results of a given decision would not be shared back across teams to the team initiating an action. This contributed to the inability of people to develop robust mental models.

4. The importance of teamwork and the need to use shift meetings to establish both shared goals and a shared understanding of the situation was noted. The conduct of shift meetings for accomplishing these objectives was found to be highly variable in this environment.

5. Several problems that can reduce individual SA were also noted in this domain, including task-related and other distractions, negative effects of noise and poor lighting, vigilance, and memory issues.

The Team SA Training Course was developed to address these five SA training goals and objectives. In addition, the course also provided a review of Maintenance Resource Management (MRM) principles that are considered to be prior knowledge requirements for the trainees. The Team SA Training Course was designed to be presented as an 8-hour classroom delivery course. The course was designed to be presented to personnel from across all maintenance operations departments (also called technical operations in some airlines). The course is best taught to a class composed of a mixed cross section from different maintenance operations organizations (e.g., stores, AMTs, inspectors, and maintenance operations control). This is because the course focuses on helping to reduce the gaps and miscommunications that can occur between these different groups and it is anticipated that much of the course's benefit comes from the interaction that occurs when trainees share different viewpoints and information going through the exercises.

The instructional strategy used for the course features adult inquiry and discovery learning. This allows a high level of interaction and participation amongst the trainees creating an experiential learning process. The Team SA Training Course strongly encourages participation in problem solving, discussion groups, and responding to open ended questions, thus promoting the acquisition and processing of information.

### Evaluation

For its initial evaluation, the Team SA Training Course was delivered by a major airline at four of its large maintenance bases. Most of the maintenance organization personnel in the airline had already received MRM training, a precursor to the Team SA Training Course. The course was delivered over a 2 day period by this airline. (It was expanded from the original 8-hour course design by this airline to allow for more group exercises, interaction, and case studies.)

Seventy-two people from nine different maintenance locations attended the training sessions where the evaluation occurred. Participation in the course was voluntary and participation in the course evaluation was also voluntary and confidential. Participants were present from a full cross section of shifts. The majority of the participants were male (86%). The participants came from a wide range of technical operations departments and job titles. The most frequent job title was that of line mechanic (AMT), followed by leads and supervisors. A good cross section of other organizations within the Technical Operations Group were also represented, including inspection, planning, and documentation support personnel. Attendees were very experienced at their jobs ($M = 10.41$ years) and within the organization ($M = 12.16$ years).

The Team SA training evaluation process consisted of three levels: value and usefulness of the training, pre- and posttraining measures, and changes in behavior on the job. A questionnaire was administered immediately following the course to get participants' subjective opinions on the value and usefulness of the course. Additionally, the amount of learning in attitudes and behaviors related to SA was also measured. An evaluation form was provided immediately prior to the training to assess the knowledge and behaviors of the trainees related to SA. It was administered again immediately following the course to measure changes in attitudes and self-reported intentions to change behavior as a result of the training. The form was administered again 1 month later to assess changes in behavior on the job.

*Value and Usefulness.* The posttraining course evaluation was used to measure the level of usefulness and perceived value of the course. Course participants scored each subsection of the course on a 5-point scale that ranged from 1 (*waste of time*) to 5 (*extremely useful*). On average, they rated each of the topics as *very useful* (*M* scores between 3.5 and 4.7). In addition to rating topics in the course, participants also answered several questions related to the course as a whole, shown in Fig. 16.1. The mean rating for the course overall was 4.3, corresponding to better than *very useful*. A whopping 89% of the participants viewed the course as either *very useful* or *extremely useful* and represented a high level of enthusiasm for the course. There were no low ratings of the course as a whole. Over 94% of the participants felt the course was either *very useful* or *extremely useful* for increasing aviation safety and teamwork effectiveness (*M* rating of 4.4). Over 89% felt the course would be either very or extremely useful to others (*M* rating of 4.3). When asked to what degree the course would affect their behavior on the job, 83% felt they would make a "moderate change" or a "large change" as shown in Fig. 16.2.

*Changes in Behavior and Attitudes.* The mean change in the posttest compared to the pretest on each behavior described in the pre- and post-training self-reported SA behavioral measure form was also assessed. A factor analysis on the questionnaire revealed a moderate degree of homogeneity. That is, responses on the items were somewhat interrelated, however, no large groupings of related factors were revealed to explain a large portion of the variance. (Only one factor accounted for more than 10% of the variance, with most accounting for less than 5%.) The questionnaire was therefore treated as a set of independent items. Changes on each item were compared for each subject using a paired-comparison analysis (pretest to post-test).

The Wilcoxon nonparametric statistical analysis revealed that attitudes and self-reported behaviors changed significantly on 7 of the 33 items (*p*

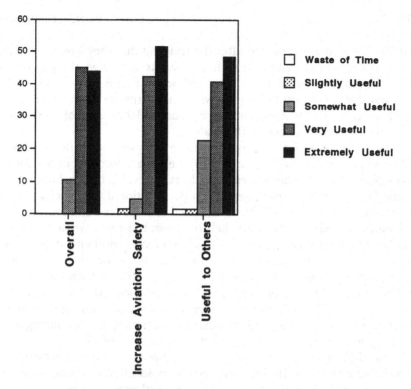

FIG. 16.1. Team training course evaluation. *Note.* From "Situation awareness in aircraft maintenance teams," by M. R. Endsley and M. M. Robertson, in press, *International Journal of Industrial Ergonomics.* Reprinted with permission.

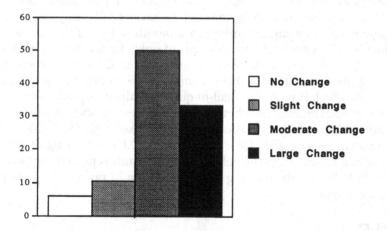

FIG. 16.2. Perceived effect of course on behavior. *Note.* From "Situation awareness in aircraft maintenance teams," by M. R. Endsley and M. M. Robertson, in press, *International Journal of Industrial Ergonomics.* Copyright Endsley and Robertson. Reprinted with permission.

< 0.05). Participants reported after the training that they would be more likely to keep others up-to-date with their status as they go along in doing their jobs (an increase of 15%). They also were slightly more likely to report that they would try to keep up with what activities others were working on over the course of the shift (an increase of 10%). Both of these items relate to improved SA across the team.

Participants reported they would be more likely to try to understand others' viewpoints when engaged in a disagreement with other departments (an increase of 15%). This relates to an effort to develop better shared mental models regarding other departments. Additionally, participants reported changes in several behaviors related to improved communications and teamwork. They were more likely to report improved written communication when sending an aircraft with a problem to another station (an increase of 21%). Participants were more likely to report that they would make sure to pass on information about an aircraft and work status to the next station (an increase of 13%). They were also more likely to report that they make sure all problems and activities are discussed during shift meetings (an increase of 11%) and encourage others to speak up during shift meetings to voice concerns or problems (an increase of 12%).

These differences between the pretest and post-test measures on SA related behaviors and attitudes indicate that in addition to participants responding positively to the course, they reported actual changes in behaviors they would make on the job as a result of the course, thus improving SA on the job both between and within maintenance teams.

*Changes on the Job.* In order to assess whether participants actually made the intended changes in their job behaviors following the course, the same form was again administered 1 month following the course. At the time of this analysis, the participants of only one course had been on the job for a full month after the training session. Of these participants (17), six responses were available for this analysis (representing a return rate of 35% which is typical of mail-in questionnaires). A paired comparison of responses on each item between the post-test questionnaire and the 1-month questionnaire was made using the Wilcoxon test. This analysis revealed no changes on any of the test items at a 0.05 level of significance. Therefore, it appears that the behaviors participants reported they would engage in following the training were carried out in practice, at least for this small sample.

### Summary

Overall the SA Team Training Course was highly successful. The course content associated with all of the major training objectives was rated very highly with the vast majority of participants rating each area as *very useful* or *extremely useful*. The course was viewed as between *very useful* and *extremely useful* overall for increasing aviation safety and in terms of usefulness to others. The majority of participants felt that the course would result in changes in their behaviors on the job. The results of the follow-up questionnaire, administered 1 month after the training course, supported these intentions and showed that participants were making the changes they had intended to make following the training.

The evaluation represents only an initial evaluation of the Team SA Training Course in its prototype implementation phase based on the responses of an initial group of course participants. To further validate these findings, additional evaluation is needed with succeeding groups of trainees in the course. More follow-up research is also needed to validate the results of the on-the-job behavior changes and the degree to which the training impacts critical maintenance performance measures at the airline. These activities are in progress. Despite these limitations, this example shows a process by which team training courses can be developed for different domains. The preliminary results at least indicate that there may be considerable value in pursuing courses such as these.

## CONCLUSIONS

This chapter attempts to build on existing research to scientifically explore potential techniques for training SA. As good SA is at the heart of good decision making and performance in aviation, training focused on improving SA should be highly effective at reducing the level of accidents and incidents in aviation. Research designed to explore this possibility has only recently been undertaken. Much work is still needed to determine whether programs built from these research findings will bring fruition. It is possible that the development of SA in a given domain requires skills and knowledge bases only acquired through trial and error learning. It is also highly possible, however, that training programs directed at enhancing the most fundamental cognitive skills and at boot-strapping the acquisition of needed knowledge bases can improve on the hit-or-miss processes currently in play.

Although most efforts at training SA have been directed at the pilot community, SA training may be a viable approach for other communities as well. Even though it is likely that the issues confronting personnel may be rather different from those found in aviation, some of the approaches outlined here for improving SA at the individual or team level may also be applicable to other domains. Considerable research is needed, however, to determine the sources of SA differences between high and low SA individuals in these other domains and the skills and strategies that may be appropriate for maintaining high SA. Within aviation, we are only just beginning to explore the ways in which individuals and teams develop and maintain SA within the challenging operational environment. Far more research is needed in this area.

# REFERENCES

Amalberti, R., & Deblon, F. (1992). Cognitive modeling of fighter aircraft process control: a step towards an intelligent on-board assistance system. *International Journal of Man-machine Systems, 36,* 639–671.

Damos, D., & Wickens, C. D. (1980). The acquisition and transfer of time-sharing skills. *Acta Psychologica, 6,* 569–577.

Doane, S. M., Alderton, D. L., Sohn, Y. W., & Pellegrino, J. W. (1996). Acquisition and transfer of skilled performance: Are visual discrimination skills stimulus specific? *Journal of Experimental Psychology, 22*(5), 1218–1248.

Endsley, M. R. (1989). Pilot situation awareness: The challenge for the training community. In *Proceedings of the Interservice/Industry Training Systems Conference (I/ITSC)* (pp. 111–117). Ft. Worth, TX: American Defense Preparedness Association.

Endsley, M. R. (1995a). Measurement of situation awareness in dynamic systems. *Human Factors, 37*(1), 65–84.

Endsley, M. R. (1995b). A taxonomy of situation awareness errors. In R. Fuller, N. Johnston, & N. McDonald (Eds.). *Human Factors in Aviation Operations* (pp. 287–292). Aldershot, England: Avebury Aviation, Ashgate Publishing, Ltd.

Endsley, M. R. (1995c). Toward a theory of situation awareness in dynamic systems. *Human Factors, 37*(1), 32–64.

Endsley, M. R., & Bolstad, C. A. (1994). Individual differences in pilot situation awareness. *International Journal of Aviation Psychology, 4*(3), 241–264.

Endsley, M. R., Farley, T. C., Jones, W. M., Midkiff, A. H., & Hansman, R. J. (1998). *Situation awareness information requirements for commercial airline pilots* (ICAT-98-1). Cambridge: Massachusetts Institute of Technology International Center for Air Transportation.

Endsley, M. R., & Robertson, M. M. (1996a). *Team situation awareness in aircraft maintenance.* Lubbock: Texas Tech University.

Endsley, M. R., & Robertson, M. M. (1996b). Team situation awareness in aviation maintenance. In *Proceedings of the 40th Annual Meeting of the Human Factors and Ergonomics Society* (Vol. 2, pp. 1077–1081). Santa Monica, CA: Human Factors & Ergonomics Society.

Endsley, M. R., & Robertson, M. M. (in press). Situation awareness in aircraft maintenance teams. *International Journal of Industrial Ergonomics.*

Gibson, J., Orasanu, J., Villeda, E., & Nygren, T. E. (1997). Loss of situation awareness: Causes and consequences. In R. S. Jensen & R. L. A. (Eds.), *Proceedings of the Eighth International Symposium on Aviation Psychology* (Vol. 2, pp. 1417–1421). Columbus: The Ohio State University.

Gugerty, L., & Tirre, W. (1997). Situation awareness: a validation study and investigation of individual differences. In *Proceedings of the Human Factors and Ergonomics Society 40th Annual Meeting* (Vol. 1, pp. 564–568). Santa Monica, CA: Human Factors & Ergonomics Society.

Hartel, C. E., Smith, K., & Prince, C. (1991, April). *Defining aircrew coordination: Searching mishaps for meaning.* Paper presented at the Sixth International Symposium on Aviation Psychology, Columbus, OH.

Hunter, D. R. (1995). The airman research questionnaire: characteristics of the American pilot population. In R. S. Jensen & L. A. Rakovan (Eds.), *Proceedings of the Eighth International Symposium on Aviation Psychology* (Vol. 2, pp. 795–800). Columbus: Ohio State University.

Jones, D. G., & Endsley, M. R. (1996). Sources of situation awareness errors in aviation. *Aviation, Space, and Environmental Medicine, 67*(6), 507–512.

Kass, S. J., Herschler, D. A., & Companion, M. A. (1990). Are they shooting at me?: An approach to training situation awareness. In *Proceedings of the Human Factors Society 34th Annual Meeting* (Vol. 2, pp. 1352–1356). Santa Monica, CA: Human Factors Society.

Klein, G. A. (1995). *A user's guide to naturalistic decision making* (DASWO-1-94-M-9906). Fairborn, OH: Klein Associates.

O'Hare, D. (1997). Cognitive ability determinants of elite pilot performance. *Human Factors, 39*(4), 540–552.

Orasanu, J. (1995). Situation awareness: Its role in flight crew decision making. In R. S. Jensen & L. A. Rakovan (Eds.), *Proceedings of the Eighth International Symposium on Aviation Psychology* (Vol. 2, pp. 734–739). Columbus: Ohio State University.

Orasanu, J., Dismukes, R. K., & Fischer, U. (1993). Decision errors in the cockpit. In *Proceedings of the Human Factors and Ergonomics Society 37th Annual Meeting* (Vol. 1, pp. 363–367). Santa Monica, CA: Human Factors & Ergonomics Society.

Orasanu, J., & Fischer, U. (1997). Finding decisions in natural environments: the view from the cockpit. In C. E. Zsambok & G. Klein (Eds.), *Naturalistic Decision Making* (pp. 343–357). Mahwah, NJ: Lawrence Erlbaum Associates.

Prince, C., & Salas, E. (1998). Situation assessment for routine flight and decision making. *International Journal of Cognitive Ergonomics, 1*(4), 315–324.

Prince, C., Salas, E., & Stout, R. (1995). Situation awareness: team measures, training methods. In D. J. Garland & M. R. Endsley (Eds.), *Experimental Analysis and Measurement of Situation Awareness*. Daytona Beach, FL: Embry-Riddle University Press.

Schutte, P. C., & Trujillo, A. C. (1996). Flight crew task management in non-normal situations. In *Proceedings of the Human Factors and Ergonomics Society 40th Annual Meeting* (pp. 244–248). Santa Monica, CA: Human Factors & Ergonomics Society.

Simmel, E. C., & Shelton, R. (1987). A assessment of nonroutine situations by pilots: A two-part process. *Aviation, Space and Environmental Medicine, 58*, 1119–21.

Taylor, R. M., Endsley, M. R., & Henderson, S. (1996). Situational awareness workshop report. In B. J. Hayward & A. R. Lowe (Eds.), *Applied aviation psychology: Achievement, change and challenge* (pp. 447–454). Aldershot, England: Ashgate Publishing Ltd.

Wiggens, M., Connan, N., & Morris, C. (1995). Self-perceptions of weather-related decision-making ability amongst pilots. In R. S. Jensen & L. A. Rakovan (Eds.), *Proceedings of the Eighth International Symposium on Aviation Psychology* (Vol. 2, pp. 845–850). Columbus: Ohio State University.

# Author Index

# Subject Index

Printed in the United States
by Baker & Taylor Publisher Services